Floral Diagrams

Floral morphology remains the cornerstone for plant identification and studies of plant evolution.

This book gives a global overview of the floral diversity of the angiosperms through the use of detailed floral diagrams. These schematic diagrams replace long descriptions or complicated drawings as a tool for understanding floral structure and evolution. They show important features of flowers, such as their relative position in an inflorescence, the positions of the different floral organs, their fusion, symmetry and structural details. In addition, the book contains a wealth of information as a comprehensive synthesis of the diversity of families. The relevance of the diagrams is discussed, and pertinent evolutionary pathways are illustrated. The range of plant species represented reflects the most recent classification of flowering plants based mainly on molecular data, which is expected to remain stable in the future. This book will be invaluable for researchers and students working on plant structure, development and systematics, as well as being an important resource for plant ecologists, evolutionary botanists and horticulturists.

LOUIS P. RONSE DE CRAENE is Director of the MSc course on the Biodiversity and Taxonomy of Plants at the Royal Botanic Garden Edinburgh. He is a world-leading expert in floral morphology, with emphasis on floral development. His research interests centre on the structural complexity and evolution of flowers and encompass a broad range of angiosperm families.

D1618933

Floral Diagrams

*An Aid to Understanding
Flower Morphology and
Evolution*

LOUIS P. RONSE DE CRAENE
Royal Botanic Garden Edinburgh

CAMBRIDGE UNIVERSITY PRESS
Cambridge, New York, Melbourne, Madrid, Cape Town, Singapore,
São Paulo, Delhi, Dubai, Tokyo

Cambridge University Press
The Edinburgh Building, Cambridge CB2 8RU, UK

Published in the United States of America by Cambridge University Press, New York

www.cambridge.org
Information on this title: www.cambridge.org/9780521493468

© L. P. Ronse De Craene 2010

First published 2010

Printed in the United Kingdom at the University Press, Cambridge

A catalogue record for this publication is available from the British Library

ISBN 978-0-521-49346-8 Hardback
ISBN 978-0-521-72945-1 Paperback

To Catherine, Camille and Alexandre, with love

Contents

Foreword

This very welcome addition to the literature on the structure and evolution of flowers provides a valuable and practical new perspective on a classical botanical theme. It focuses on the relationships between flower structure and the evolutionary diversification of plants as reflected in the latest system of classification.

Floral diagrams provide one of the best examples of the idea that a picture is worth a thousand words. They provide a stylised system for describing and communicating the arrangement of floral organs with great simplicity, regardless of the structural complexity of the particular flower. It is therefore no surprise that floral diagrams have stood the test of time and remain as effective today as when they first began to be used. The German botanist August W. Eichler is generally credited with their introduction in the late nineteenth century and they were rapidly adopted, soon becoming a familiar feature of numerous botanical textbooks. Eichler was also a pioneer in the field of classification of flowering plants and one of the first botanists to base a system of classification upon evolutionary principles. Whilst the utility of floral diagrams has remained unchanged since their invention, we now use very different methods to establish the evolutionary relationships between different groups. For most of the twentieth century plant classification relied on the comparison of morphological characters and numerous different schemes competed for attention. The advent of classifications based upon the analysis of DNA sequence data rather than on traditional morphological characters resulted in revolutionary advances. An international collaboration by the *Angiosperm Phylogeny Group* has provided a new and much more stable framework. Although we might expect some minor changes to take place as new gene sequences begin to be integrated into the analysis, the classification we have today is unlikely to change much in the future.

Louis Ronse de Craene, a research botanist who has made a significant contribution to the study of evolutionary development in flowers, has therefore written a very timely book. Here he reconnects the tradition of floral diagrams as a concise shorthand notation with the latest understanding of plant evolutionary relationships. The book will provide an invaluable tool for anyone who wishes to understand the form of flowers and use them, for example, to identify unfamiliar plants.

Stephen Blackmore
Regius Keeper
Royal Botanic Garden Edinburgh

Preface

Flowers are extremely attractive to us as a source of inspiration and happiness. It is no wonder that various technical textbooks in plant science tend to enhance their front page with some glamorous illustrations of flowers. Despite this wide interest, our knowledge about the diversity of floral structures is still limited and relies mostly on research carried out in the nineteenth century.

A floral diagram is a basically schematic cross-sectional drawing of a flower. However, floral diagrams are more than just a two-dimensional representation. There are more than 250 000 species of angiosperms and their flowers vary in many ways. The arrangement of flowers in inflorescences, the number, position, identity and shape of floral organs and the symmetry of the flower as a whole are rarely identical between different families, genera or even species. Floral diagrams are a rich source of data for identification purposes and for understanding structures, but can also be used to express a hypothesis of evolution. The information contained in floral diagrams is potentially immense and replaces complex descriptions.

Students often struggle with the identification of flowers, mainly because they fail to look at the structures hidden in the bud. However, the spatial arrangements of organs in the flower, as well as the number of whorls, are an essential indication of systematic relationships. This information is particularly important for identifying plants in the field and tells us much about the key characters of a specific group of plants. The educational merits of floral diagrams in the classroom are obvious. Used together with floral formulae, they convey information in a rigorous and clear way. They are important for systematists or evolutionary botanists in providing information for databases dealing with morphological data (e.g. Morphbank: http://www.morphbank.net/) or for clarifying phylogenetic questions. In paleobotanical research floral diagrams are a useful resource in reconstructing the shape of fossilized flowers.

Researchers in evolutionary developmental genetics will find appropriate questions about the nature of floral organs to investigate. Finally, horticulturists or amateur botanists will find this book valuable to understand general patterns of flower construction and floral diversity.

It is more than 130 years since August Wilhelm Eichler (1839–1887), then professor of botany in Kiel, produced a book in two parts, in 1875 and 1878. This book, entitled *Blüthendiagramme Construirt und Erlaütert*, is a major reference work, concentrating the information about flowers known at that time. As such, it represents a treasure trove, detailed and often accurate, and even today extremely valuable as a source of data. Eichler's work was an inspiration for later generations of morphologists, such as Arthur W. Church (1865–1937) and Agnes Arber (1879–1960). A particularly fine example of a book using floral diagrams is *Types of Floral Mechanism* published by Church (1908) and intended as a series, but limited to a single volume by lack of interest and funds (Mabberley, 2000). Since Eichler's book was published, much progress has been made in documenting flower morphology, especially during the last decades of the twentieth century, when there was a renewed interest in floral morphology coupled with the use of the scanning electron microscope.

However, information about flowers is scattered in scientific papers that are not readily accessible to the general public, providing little scope for a broad overview of the flowering plants. Alternatively it dates from important work carried out in the nineteenth century that is in danger of being forgotten. Floral diagrams were used sparingly in different textbooks as illustrative material (e.g. Baillon, 1868–1894; Engler and Prantl, 1884–1909), but never to the extent of Eichler's book. More recent examples are Melchior (1964), Sattler (1973), Graf (1975) and Stützel (2006). The most recent major textbooks on angiosperm phylogeny (e.g. Judd *et al.*, 2002; Soltis *et al.*, 2005; Simpson, 2006) lack any floral diagrams. Spichiger *et al.* (2002) did include diagrams for major families, but the diagrams are oversimplified and riddled with mistakes.

The system of classification used by Eichler is outdated, as it is based on the Englerian concept that simple, unisexual catkin-like flowers are ancestral and that more elaborate bisexual flowers are derived. Recent changes in the phylogeny of flowering plants based mainly on molecular evidence have created the need for a new book on floral diagrams. It is essential to link obvious morphological characters to any molecular phylogeny. I hope this book fulfils this purpose.

Acknowledgments

The idea for this book has matured over several years. Discovering the variation in flowers has always been a major passion for me, and there is no better way to present this variation than by floral diagrams. Besides the possibility of describing flowers and their structures by photographs or drawings, floral diagrams bring another dimension that I enjoyed exploring. The book allowed me to compile more than twenty years of research on flowers.

My special thanks go to Paula Rudall, Livia Wanntorp, Erik Smets and David Harris for critically reviewing an early draft of the book and for their helpful suggestions. Several colleagues were inspirational and helpful through previous contacts, such as Peter Endress, Greg Kenicer, Peter Linder, Gerhard Prenner, Rolf Sattler, Dennis Stevenson, Wolfgang Stuppy, the late Cyrille Vandenberghen and Philip Smith, among others. Without the vast living collections at the Royal Botanic Garden Edinburgh, the extent of my book would be meagre indeed. I thank the horticultural staff, especially Fiona Inches, for their help in identifying specimens. Aleck Yang, Colin Belton, David Harris, Euridice Honorio, Paulina Hechenleitner, Peter Linder, Tony Miller, Mark Newman, Rolf Rutishauser, Tory Tokuoka and Peter Wilkie contributed to obtaining material for study. Frieda Christie and the technical staff of the Royal Botanic Garden Edinburgh have been very helpful in assistance with preparing and observing some specimens used in this study. The curators of the pickled collections of the Swedish Museum of Natural History, the National Herbarium of the Netherlands in Wageningen and the Royal Botanic Garden Kew were helpful in allowing me to visit the collections and use the material. Finally, I thank my daughter Camille for her creative help in designing parts of the book. The writing of this book would not have been possible without the constant support and patience of my dear wife Catherine. Finally, I owe much to the generosity of my parents who allowed me to pursue my career in botany.

PART I   INTRODUCTION TO FLORAL DIAGRAMS

1

Introduction to flower morphology

1.1 Definition of flowers

Flowers were defined in many different ways in the past and there is no general agreement about how a flower should be defined. In the past there was emphasis on two main contrasting hypotheses (reviewed in Bateman, Hilton and Rudall, 2006). The pseudanthial hypothesis accepts that flowers evolved from a branched, multiaxial structure, i.e. a condensed compound inflorescence (e.g. Eichler, 1875; Eames, 1961). This means that a flower is an assemblage of separately functioning entities that became grouped together. The euanthium hypothesis stated that the flower evolved from a simple uniaxial (euanthial) structure, i.e. a condensed sporophyll-bearing axis with proximal microsporophylls and distal megasporophylls (e.g. Arber and Parkin, 1907). Floral organs all have attributes of leaves, and leaf-like elements, such as stipules, leaf bases, petioles and blades, occasionally appear in flowers (e.g. Guédès, 1979). More recently, phylogenetic studies have supported the theory that flowers evolved once and that all flowers are thus homologous ('anthophyte hypothesis' reviewed in Bateman, Hilton and Rudall, 2006). The theory is supported by evolutionary developmental evidence that the same genes are acting on the flower and vegetative shoot, and that the flower is best interpreted as a short shoot with specialized leaves (Glover, 2007).

More specifically, a flower can be defined a determinate structure with a generally defined number of organs; it bears both staminate and pistillate parts and organs are set in four series: sepals, petals, stamens and carpels. However, several angiosperm flowers lack these defining features, and differentiation of a perianth or limits of flowers and inflorescences are often unclear. The definition of flowers implicitly refers to angiosperms, but should also include gymnosperms. The gymnosperm cone could also be described as a flower, though the organization of the cone is generally unisexual. A defining character for angiosperms is enclosure of ovules by carpels (angiospermy), separating the flowering plants from their closest relatives (gymnosperms). Flowers are usually grouped into inflorescences that may be simple to highly complex, and are – at least in bud – often subtended by a specialized leaf, the bract, together with one or more (generally two) smaller leaves (bracteoles) placed in a lateral position. In monocots, there is another single adaxial bract instead of the two bracteoles. Bracts and bracteoles are absent in some species, and this can influence the position of organs in the flower. In some groups of plants the limits between bracts and floral parts are unclear. The inflated axis, called a receptacle, bears floral organs in spirals or in whorls (or a mixture of both). The outer floral organs are sterile leaves called a perianth. When undifferentiated they are described as tepals. More often, there is differentiation into an outer whorl (calyx or sepals) and an inner whorl (corolla or petals). The androecium, or the totality of stamens bearing microsporangia, can be organized in a single whorl or into several whorls, with a specific position relative to the petals. The gynoecium consists of carpels bearing ovules (megasporangia). Carpels can be free, or more often fused into an entity enclosing the ovules or seeds. Besides carpels and stamens, some flowers have sterile structures (staminodes or carpellodes) or other emergences of the receptacle. These are often developed as nectaries and can be conspicuous.

Another characteristic besides leaves that flowers share with the vegetative parts of the plants is the phyllotaxis or order of initiation of floral organs (see p. 28). The transition in phyllotaxis from the vegetative shoot to the flower can be gradual or abrupt and is mediated by bracts and sepals. The calyx usually continues the same spiral sequence as vegetative leaves. There is generally a disruption in the initiation sequence between sepals and petals, leading to an alternation of whorls. The result of the phyllotactic sequence is a relatively stable position of floral parts relative to each other, which is fundamentally important in understanding and interpreting the structure of the flower.

Complex versus reduced flowers

The development of flowers is correlated with mode of pollination. There is a marked difference between flowers with a biotic (animal) pollination syndrome and those with an abiotic syndrome (wind or water). Animal pollination is accompanied by a series of adaptations to attract specific pollinators and to protect the floral parts from damage. Differentiation of protective sepals and carpels, nectaries, showy petals and stamens is part of the arsenal leading to effective fertilization. Depending on the pollinating animal, different strategies were developed to increase the success of pollination, occasionally leading to complex flowers or inflorescences. Wind pollination is accompanied by a syndrome of characters, such as small, unisexual flowers, loss of petals or reduction of the perianth, long styles and filaments, production of a large amount of pollen, lack of viscin in pollen, and reduction in the number of ovules (e.g. Linder, 1998; Friedman and Barrett, 2008).

Reversals in the pollination syndrome occur frequently in the angiosperms. Secondary wind-pollinated flowers occur in all major clades with a predominance of insect pollination (e.g. *Thalictrum* in Ranunculaceae, *Sanguisorba* in Rosaceae, *Macleaya* in Papaveraceae, some *Acer* in Sapindaceae, some *Erica* in Ericaceae, *Fraxinus* in Oleaceae, *Xanthium* in Asteraceae, *Theligonum* in Rubiaceae, *Leucadendron* in Proteaceae). Larger, predominantly wind-pollinated clades include Fagales and Poales. Wind-pollinated clades sometimes include species that have reverted to insect pollination (e.g. *Trochodendron* in Trochodendraceae, *Castanea* in Fagaceae, Buxaceae).

Specific elaborations of petals and/or staminodes are clearly linked with pollination syndromes and are triggered by the kind of pollinators that evolved with the flowers. Secondary stamen and carpel increases are widespread in the core eudicots, and are linked with the potential for higher pollen or ovule supply. Very often, there is a close mechanical correlation between petals and stamens, or stamens and style, in the release of pollen or the protection of the anthers. Different elaborations on the petals are also directly linked to the pollinator (e.g. the building of landing platforms, nectar containers and nectar guides: see Endress and Matthews, 2006b).

Flowers are morphologically highly dynamic entities with a potential for evolution reaching far beyond our preconceived ideas of the limitations of floral evolution. This flexibility is closely linked with what is available at the organ level.

1.2   Floral organs

1.2.1   Perianth

> If treated in isolation, there is no character combination which could stringently prove an organ's nature as a petal or sepal. (Endress, 1994: 26)

The perianth is the envelope of sterile leaves enclosing the fertile organs of the flower. The perianth is either differentiated into sepals and petals, or undifferentiated (perigone or tepals). In the latter case the perianth can be green (sepaloid) or pigmented (petaloid). A distinction between sepals and petals is applicable only when two different series of perianth parts are found. In cases with more than two whorls the transition between sepals and petals can be progressive with blurred limits.

In some cases there is unclear distinction between the perianth and surrounding bracts. In others, bracts can be variably associated with the flower, often in the form of an epicalyx, or sometimes as a fused cap (calyptra). Bracts can be distinguished from the perianth by presence of axillary buds (never in floral organs) and differences of plastochron (see p. 28; transition of a decussate to spiral phyllotaxis: Buzgo, Soltis and Soltis, 2004). Endress (2003a) suggested that bracts should be considered as phyllomes with a lower complexity than tepals. However, I believe that a distinction between bracts and tepals is sometimes impossible to make, given the existence of intermediate organs and the easy incorporation of bracts in the flower. Inclusion of bracts at the base of the flower makes the perianth bipartite, as in Magnoliaceae. Some taxa have transitional stages between bracts and tepals (in German called *Höchblätter*), as in Myrothamnaceae and some Ranunculaceae.

The distinction between sepals and petals is not always clear. If only a single whorl of perianth parts is present, it is sometimes difficult to categorize members of this whorl as sepals or petals, because one whorl may have been lost. A distinction is sometimes made between primary apetaly (as in basal angiosperms with tepals and no distinction between perianth parts) and secondary apetaly (apopetaly: Weberling, 1989), in cases where evidence exists that petals have been present and have been lost during evolution. Sepals are petaloid in several families or can have a mixed nature (partly green and pigmented; e.g. *Impatiens, Polygala*). One of the reasons for this variability is that the perianth can change function at different stages of the development of the flower. In general, the calyx tends to protect inner organs and is photosynthesizing. Later, it can become attractive for the dispersal of fruits (e.g. in *Physalis*). The main purpose of petals is to attract pollinators but they can become transformed into protective organs or dispersal units (e.g. in *Coriaria*).

The origin of the perianth has been discussed in several textbooks and papers, with various interpretations for the origin of petals: either from stamens, from bracts, from both structures, or as something totally new (see Ronse De Craene, 2007, 2008 for a recent review). In general, recent evidence from morphological and evo-devo studies suggests that petals in the majority of angiosperms are derived from bract-like structures and are homologous to the sepals in the flower. Contrary to a general assumption, cases where petals are unequivocally derived from stamens are rare in the angiosperms. In other groups, petals have variously evolved by insertion of bracts in the confines of the flower and their differentiation into two functional whorls.

An undifferentiated perianth was reconstructed as ancestral for the angiosperms in recent phylogenetic studies (Soltis *et al.*, 2005). In basal angiosperms, attraction and protection are combined with pigmentation of the entire perianth. A differentiated perianth has evolved independently several times, at least once at the base of core eudicots (Ronse De Craene, 2008).

Undifferentiated perianth (tepals or perigone)

An undifferentiated perianth tends to be concentrated in the basal angiosperms and monocots, and is usually associated with spiral or trimerous flowers. A distinction needs to be made between a primary undifferentiated and a secondary undifferentiated perianth. In the first case, tepals have evolved from bracts that became associated with reproductive organs and have acquired secondary functions of protection and attraction of pollinators. This kind of perianth is usually spiral, with a gradual differentiation from outer bract-like to inner petaloid tepals (e.g. Austrobaileyaceae, Calycanthaceae). Alternatively, two trimerous perianth whorls of several monocots are undifferentiated and petaloid. However, the switch between sepals and petals can be easy in these cases (Ronse De Craene, 2007). A secondarily undifferentiated perianth arises by loss of either the calyx (e.g. Santalaceae) or the corolla (e.g. Cunoniaceae) and should be referred to as a reduction.

Calyx

Sepals have a spiral initiation sequence with rapid growth, a broad base, three vascular traces and an acuminate (pointed) tip. The homology of sepals with leaves is based on similar anatomy as well as on several characteristics such as the presence of stipules and stomata, and sepals are often compared to the petiole of a leaf due to their broad shape (Guédès, 1979). On the contrary, petals often arise nearly simultaneously and have a delayed growth. They have a narrow base with only a single vascular trace and the tip is bifid or emarginated. Characteristics are more closely comparable to stamens than to leaves (Ronse De Craene, 2007).

Sepals can be fused together (gamosepaly). This fusion is often congenital at their margins, leaving free calyx teeth. Sepals are more often persistent than deciduous; very often, they increase in size after pollination and function in fruit dispersal. Reduction of sepals or their transformation into small scales or bristles is occasionally found in some families (e.g. in Asteraceae, Caprifoliaceae). The calyx may also vanish completely (e.g. in Santalaceae), leaving a single petal whorl combining attraction with protection of the flower.

Corolla

The corolla, or petals, represents the inner perianth whorl of the flower and is usually pigmented (petaloid). The number of petal whorls can be high, as in Annonaceae or Berberidaceae (up to four whorls). Petals can sometimes be highly distinctive, with a claw (a narrow base, compared with a broadened base in sepals), or they can be indistinct from sepals.

The corolla can be highly elaborate by development of ventral appendages, fimbriate margins or extended tips. Petals are often trilobed, with a protruding middle lobe (e.g. Saxifragaceae) or extensive lateral lobes (e.g. Elaeocarpaceae). Bilobed petals can be formed by reduction of the central lobe (Endress and Matthews, 2006b).

In some families, petals are highly elaborate in relation to specific pollination mechanisms. For example, Byttnerioideae (Malvaceae) have petals as inverted spoons, with a hood-like base and an extended apex. More examples were given by Endress and Matthews (2006b), who suggested that there is a correlation between the smaller size or reduction of petals and elaborations on the petal surface. However, petal reduction or loss is linked with other factors, such as petaloidy of sepals and the development of a hypanthium (Ronse De Craene, 2008). Fusion of petals (sympetaly) is a frequent phenomenon in angiosperms that occurs independently or is closely linked with the evolution of the androecium. Petal tubes appear to regulate access to flowers by a variable extent of development.

1.2.2   Androecium

The androecium consists of stamens, which make up the male part of the flower. Stamens are relatively uniform. They mostly consist of four pollen sacs arranged in two lateral thecae grouped in an anther that is linked by a connective to a filament. The orientation of anthers can be inward (introrse), lateral (latrorse) or outward (extrorse). Stamens can be basally connected into a tube (e.g. Meliaceae, Malvaceae) or fused with the petals in a common tube (e.g. Caricaceae, Rubiaceae). Anthers may become laterally connivent or fuse postgenitally (e.g. Asteraceae). The number and position of stamens appear to be the most variable in the flower compared with other floral organs, but this variation is never randomized and relatively conservative in angiosperms. Flowers with more than two stamen whorls are rare and largely restricted to basal angiosperms and some basal monocots. In other instances they are the result of a secondary increase (see p. 10: e.g. Molluginaceae, Rosaceae).

When there is only a single whorl, stamens are either inserted opposite the sepals (haplostemony: e.g. Gentianaceae, Violaceae), or less frequently

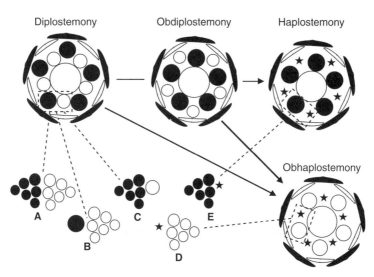

Fig. 1.1. Diagram representing possible changes in the androecium of a pentamerous flower with two stamen whorls. Black dots, antesepalous stamens; white dots antepetalous stamens; asterisk, lost stamen position or staminode. A, increase of antesepalous and antepetalous stamens; B, increase of antepetalous stamens only; C, increase of antesepalous stamens only; D, sterilization or loss of antesepalous stamens and increase of antepetalous stamens; E, sterilization or loss of antepetalous stamens and increase of antesepalous stamens.

opposite the petals (obhaplostemony: e.g. Vitaceae, Primulaceae). When two stamen whorls are present, they arise separately and are often spatially separated (Fig. 1.1). The number of stamens is ten in pentamerous flowers or eight in tetramerous flowers, a common pattern among rosids. Diplostemony is the condition where the outer stamen whorl is situated opposite the sepals and the inner whorl opposite the petals (e.g. Coriariaceae, Burseraceae). Obdiplostemony is the opposite condition, with outer stamens opposite the petals (e.g. Geraniaceae, Saxifragaceae). The distinction between diplostemony and obdiplostemony is often the result of developmental constraints. Obdiplostemony is caused by the pressure of developing carpels resulting in more outward orientation and shift of the antepetalous stamens (e.g. Ericaceae, Rutaceae: Leins, 1964a; Eckert, 1966; Ronse De Craene and Smets, 1995b). In some groups of plants, the antepetalous stamens arise before the antesepalous stamens; this inverted development is often linked with weakening or sterilization of the antesepalous stamens (e.g. Malvaceae). Obdiplostemony is often a first step towards evolution of (ob)haplostemony. This pattern may be related to a conversion of one whorl into staminodes.

Staminodes

Staminodial structures are defined as sterile stamens. The main criterion in recognizing staminodial structures is that of position, as any emergences in the flower can arbitrarily be called staminodial (Leins and Erbar, 2007). Rarely, the presence of sterile anthers is an indication of a former function as stamens. Staminodes are widely scattered in the angiosperms (e.g. Walker-Larsen and Harder, 2000; Ronse De Craene and Smets, 2001a), and are the result of sterilization affecting a complete whorl of stamens, or a variable number of stamens within a whorl. Sterilization of stamens has evolved frequently in angiosperms; for example, it affects 72% of all families of rosids (Walker-Larsen and Harder, 2000). In cases where an entire stamen whorl becomes obsolete, stamen remnants may still be identified in flowers, but they will eventually disappear and this process is irreversible. However, in many instances loss of fertility was accompanied by acquisition of novel functions not performed by stamens, such as differentiation into food-bodies (*Calycanthus*), osmophores (*Austrobaileya*), building of nectar containers (e.g. Loasaceae), enhanced optical attraction (Theophrastaceae, Eupomatiaceae), differentiation into nectaries (e.g. *Helleborus* in Ranunculaceae) or a combination of these (*Ranunculus*, *Aquilegia*). This process offers new evolutionary potential in many groups of plants. Walker-Larsen and Harder (2000) argued that sterilization of stamens within a whorl is evolutionarily reversible, such as in Lamiales, where the adaxial staminode can be restored to a fully fertile stamen by a reversal to polysymmetry. However, there is a threshold beyond which such a reversal is impossible.

The presence of staminodes is important in floral diagrams as it indicates evolutionary transitions in stamen numbers or changes in flower configurations.

Polyandry

In cases where the number of stamens is higher than double the number of petals or sepals the androecium is polyandrous. However, various non-homologous forms of polyandry exist in the angiosperms. Polyandry can be primary, by initiation of a high number of stamens in whorls or following the spiral initiation of a flower, or secondary (complex) by division of primary (common) primordia into secondary stamens (Fig. 1.1; Ronse De Craene and Smets, 1987, 1992a). In the past a fundamental distinction was made between many and few stamens, the former being interpreted as ancestral. However, the mere distinction between many and few stamens is simplistic and is not reflected in the development of polyandry.

Primary polyandry ('true polyandry') implies that a high number of stamen primordia arise in a specific sequence in the flower, whereas each primordium develops into a single stamen. Primary polyandry can be spiral or whorled, depending on the merism of the flower (see p. 30). Trimerous or dimerous flowers with multistaminate androecia in several whorls often show an alternation of paired and unpaired stamens (e.g. Papaveraceae). With primary polyandry, initiation of stamens is usually centripetal, though it can be reversed to a centrifugal development (e.g. Alismatales) due to spatial constraints.

In the case of secondary polyandry, the number of stamens is increased by developmental multiplication of stamen primordia. This development is fundamentally different from spiral or whorled polyandry, which is characterized by genesis of one stamen per primordium. A stamen primordium can divide into a pair or higher number of stamens, which may initiate in different directions (centrifugally, laterally or centripetally), depending on spatial constraints (Ronse De Craene and Smets, 1987, 1992a). The development of complex polyandry proceeds from one or two whorls of stamens (Fig.1.1). In some cases, the boundaries of the initial stamen primordium (primary or complex primordium) can be identified and the secondary stamens appear fused into groups or fascicles. The stamens remain united at the base and they have a common vascular supply (trunk bundles). Secondary polyandry can be linked with diplostemony, when one whorl remains simple while the other is multiplied (e.g. *Reaumuria* in Tamaricaceae: Ronse De Craene, 1990), or when the other whorl becomes staminodial (e.g. several Malvaceae). Very often, only a single initial whorl is present in the flower (e.g. *Hypericum*). Polyandry becomes much more complicated when the identity of separate primordia is lost by development of a ring primordium (ring wall). Stamen groups may still be discernible on the ring (e.g. Malvaceae), or a high number of stamens arise in girdles without clear boundaries between stamen groups. The advantage of a ring primordium is that a very high number of small stamens can develop, considerably increasing the pollen load. The direction of stamen development is often centrifugal and can be ongoing when the carpels are initiated. The growth of the receptacle or hypanthium can influence the extent of the secondary increase of stamens and the direction of development (see p. 20; Ronse De Craene and Smets, 1991a, 1992a).

The general pattern of the androecium in the angiosperms probably evolved from a moderate number of stamens arranged in a spiral (e.g. *Amborella*). Different evolutionary lines led to an increase or reduction in stamen number, with several transitional steps. An outline of this evolutionary progression is shown in Fig. 1.2 (Ronse De Craene and Smets, 1998a). As stamen number decreases and positions become fixed, it is possible to characterize specific 'types' of androecia. From primary polyandrous flowers, further increases or reductions can evolve, leading to much higher or lower numbers of stamens. In the case of whorled polyandry (polycycly), the number of whorls

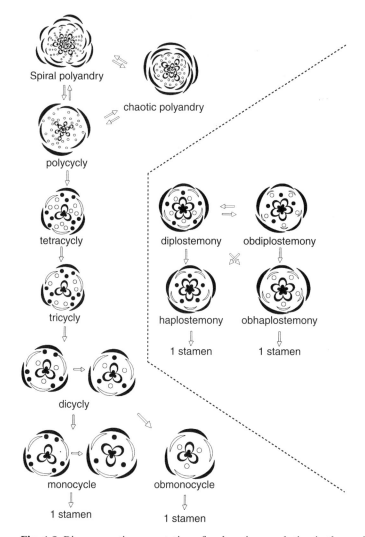

Fig. 1.2. Diagrammatic presentation of androecium evolution in the angiosperms (from Ronse De Craene and Smets, 1998a – modified, with permission of *Plant Biology*). The terminology used refers to different convergent origins of flowers with lower stamen numbers. Dotted line gives separation between polymery (left) and oligomery (right). For lower stamen numbers: black dots, antesepalous stamens; white dots, antepetalous stamens.

can be variable or decrease to two or one. These cases were described as dicycly or monocycly because they have a different origin from diplostemony and haplostemony (Ronse De Craene and Smets, 1987, 1998a). The two-whorled androecium of monocots is described as dicyclic and is not homologous with diplostemonous androecia. Even in groups with primary polyandry, the number of stamens can be secondarily increased, by extending the ontogenetic spiral on a longer cone, or by development of much smaller stamen primordia (e.g. Annonaceae, Papaveraceae). However, in most cases secondary polyandry develops from an androecium with an initially low number of stamens.

In the case of secondary polyandry it is assumed that these stamen primordia were initially single stamens with a specific position in the flower. Taxa with a secondary stamen increase are nested in clades with diplostemonous or haplostemonous androecia. The number of stamens in the flower appears to be highly flexible, with multiplications arising frequently in the angiosperms, especially core eudicots, as a means to increase the amount of pollen on offer to pollinators.

Paired stamens

Stamens in double position are widespread in the flowering plants. They have a different origin which is phylogenetically important, either (1) as the result of a transition of a spiral to whorls (Magnoliales), (2) a shift of alternisepalous stamens to the middle of the sepals, as in the Caryophyllales, or (3) a doubling of primordia, mainly in the core eudicots (a process called *dédoublement*: Ronse De Craene and Smets, 1993).

1. Stamen pairs are frequently found in flowers with primary polyandry. The transition between broad perianth parts and much smaller stamen primordia causes a shift of the first stamens in pairs between the perianth lobes (see p. 29; Erbar and Leins, 1981; Leins and Erbar, 2007). In the case of a whorled initiation, stamens tend to alternate in whorls of paired and unpaired stamens. A paired stamen arrangement is still found in cases where the flower is whorled with fewer stamens (e.g. Cabombaceae: Ronse De Craene and Smets, 1996a, 1998a).

2. With the loss of petals, stamens can shift from an alternisepalous position towards the middle of the sepals. This leads to a paired arrangement of stamens as found in Polygonaceae or Phytolaccaceae.

3. A stamen can become halved or split in two, leading to 15 stamens in a diplostemonous flower or to ten stamens in a haplostemonous flower. In some cases this division can be observed during the development of stamen primordia (e.g. *Theobroma*, Malvaceae), but more often the division is not visible and a pair arises in the position of single stamens (e.g. *Peganum*: Ronse De Craene, De Laet and Smets, 1996). The doubling can be sporadic without specific location (e.g. some Lythraceae: Tobe, Graham and Raven, 1998; Ronse De Craene and Smets, 1996a), or it can characterize a genus (e.g. *Peganum, Aristotelia*).

It can affect the antepetalous stamens only (e.g. Elaeocarpaceae), the antesepalous stamens (e.g. Cunoniaceae) or more rarely both whorls (e.g. Lythraceae). A list of families with double stamens is given in Ronse De Craene and Smets (1996a). Doubling of the antepetalous stamens is often linked with obdiplostemony; the doubling of stamens probably results in the retardation and smaller size of the stamen pair.

Contrary to dédoublement, a division of anthers in two half anthers is a process that does not increase the pollen load, although the number of organs is doubled. Halving of stamens occurs infrequently in angiosperms, but is found in *Adoxa* (Adoxaceae: Roels and Smets, 1994) and Malvaceae (van Heel, 1966) where it is superposed on stamen multiplication from primary primordia. The opposite process, i.e. the fusion of half-anthers into a single unit, occurs in *Hypecoum* (Papaveraceae: Ronse De Craene and Smets, 1992b).

The difference between doubling of stamens or shifts in stamen position has fundamental significance although this is often overlooked in studies of flowers. Cases with 15 stamens in a pentamerous flower can be confusing as they may be caused by dédoublement of one whorl in a diplostemonous flower (e.g. *Kirengeshoma* in Hydrangeaceae: Roels, Ronse De Craene and Smets, 1997; *Peganum* in Nitrariaceae: Ronse De Craene, De Laet and Smets, 1996), or by division of single stamens into triplets in haplostemonous flowers (*Nitraria*: Ronse De Craene and Smets, 1996a). The floral vasculature does not help in solving this confusion.

1.2.3   Gynoecium

The gynoecium represents the female part of the flower and is made up of a single or several carpels (megasporophylls) arranged in a single or several whorls, or in a spiral. While this is less obvious for stamens, carpels are comparable to leaves in being dorsiventrally flattened and having a midrib; they enclose a single to several ovules (which develop into seeds) attached on a marginal or submarginal proliferation of tissue called placenta (Fig. 1.3). The placenta is occasionally laminar-diffuse, in cases where ovules are scattered over the inner margin of the carpel (e.g. Nymphaeaceae, Butomaceae).

A major distinction exists between an apocarpous gynoecium (or choricarpous, with all carpels free) and a syncarpous gynoecium (with carpels fused to a variable degree). Apocarpous gynoecia seldom have a well-differentiated style with stigma. More often there is a slit-like opening filled with secretion to guide pollen tubes to the ovules.

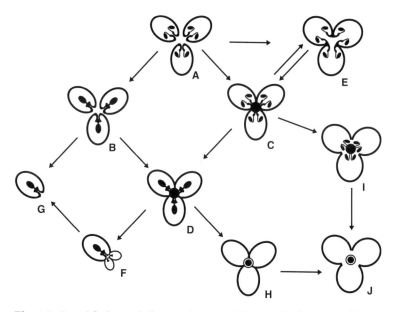

Fig. 1.3. Carpel fusion and placentation types (shown with three carpels).
A, apocarpous with marginal placentation; B, apocarpous with axile placentation;
C, syncarpous, axile placentation with paired ovules; D, axile placentation with
single ovules; E, parietal placentation; F, pseudomonomerous; G, single carpel;
H, basal or apical placentation with single ovule; I, free-central placentation;
J, basal or free-central placentation with single ovule. Arrows show possible direction
of evolution.

In syncarpous gynoecia there is usually a distinction between an ovule-
bearing zone (the ovary with a differentiation of a synascidiate and a symplicate
zone) and styles with stigmas. Pollen is collected on the stigma and a pollen tube
grows through the stylar tissue to reach the ovules. Styles may be separate and
connect to individual carpels, or be basally fused with apical style branches
(stylodes), or there is a common style with single apical stigma connecting all
carpels. When carpels are laterally fused there is a common zone shared by all
carpels (compitum). This has the selective advantage for the flower to control
fertilization much more efficiently by selecting for the fittest pollen and spreading pollen tubes equally over all carpels (Endress, Jenny and Fallen, 1983; Erbar,
1998). Stigmas can reflect the number of carpels in syncarpous gynoecia, or they
are undivided. The position of styles and/or stigma lobes can be opposite the
carpels (carinal), or alternating with the carpels (commissural). Styles are generally terminal, but become occasionally basal (gynobasic styles) by the
increased growth of the abaxial side of the carpels so that the style appears to
emerge from the base of the ovary (e.g. Lamiaceae, Chrysobalanaceae).

The number of carpels in apocarpous gynoecia can be highly fluctuating with a strong tendency for a stabilization to a single carpel. In syncarpous gynoecia the number of carpels is relatively stable and generally fluctuates between two and five. In some groups there is a tendency for one to two carpels to become sterilized with the retention of a single fertile carpel (pseudomonomery; see p. 17). There are different degrees of fusion of carpels: (1) carpels may arise separately and only fuse postgenitally (frequently in monocots), (2) carpels may arise separately but are soon unified by a ring meristem (e.g. Rosaceae) and (3) carpels arise on a ring meristem (congenital fusion, e.g. Primulaceae).

There is a general evolutionary pattern from apocarpy to syncarpy (Fig. 1.3). More than 80% of gynoecia are syncarpous, 10% are apocarpous and 10% are monocarpellate (Endress, 1994). Apocarpous gynoecia are generally found in the basal groups of angiosperms and magnoliids with a high incidence of monocarpellate gynoecia (derived by reduction of carpels except one). Basal groups have carpels with an ascidiate or plicate shape, or a combination of the two during their development (Endress, 1994; Endress and Igersheim, 1997, 2000a). Syncarpous gynoecia may rarely become secondarily apocarpous and this is usually associated with postgenital fusion of the stylar parts. This evolved independently at least in three different orders of core eudicots but also in monocots (Endress, Jenny and Fallen, 1983). Secondary apocarpy has the advantage of economy if only one carpel is fertile and there is no need to develop all carpels in fruit. In some basal angiosperms an external compitum is formed by other parts of the flower such as the floral receptacle (e.g. *Cananga* in Annonaceae, *Illicium* in Illiciaceae, *Tambourissa* in Monimiaceae: Endress and Igersheim, 2000a; Leins and Erbar, 2000, 2007). The central apex is usually used up in the development of the gynoecium. Only in some basal groups does it remain prominent (e.g. Illiciaceae, Trimeniaceae).

Placentation

The lateral walls of individual carpels may develop in the ovary as septa or divisions, separating individual locules or cells (multilocular ovary), or septa may not develop and there is only a single cell or locule (unilocular). The main types of placentation in the angiosperms are axile or parietal. In cases where septa are fully formed, placentation often tends to be axile with ovules clustered in a central position (Fig. 1.3C,D). During development of carpels the entire carpel margin may contribute to the formation of a septum, developing marginal placentae with ovules.

Disappearance or breakdown of septa leads to a free-central placentation (Fig. 1.3I; occasionally called a paracarpous gynoecium). This can be progressive during development (e.g. Caryophyllaceae), or septa may vanish altogether with occasional remnants present in some species (e.g. Primulaceae). In some groups the septa are restricted to the lower portion of the ovary (basal septum), the upper part of the ovary (apical septum) or both. Reduction of the septum may lead to a basal or apical placentation (Fig. 1.3H,J). A parietal placentation develops by contraction of the septa to the periphery (Fig. 1.3E). Intermediate conditions exist with septa reaching halfway into the ovary (e.g. *Papaver*, Papaveraceae), or septa do not develop at all and the placenta develops on the ovary wall. Secondary partitions (false septa) may occasionally arise late within the locule, either on areas of the carpel wall that are not linked to carpel margins (e.g. Linaceae), or from the placenta (e.g. Brassicaceae). The development of false septa has a function of separating developing ovules in partitions (e.g. Lamiaceae). False septa can evolve into real partitions and can ultimately increase the number of carpels (Ronse De Craene and Smets, 1998b).

Changes between different placentations are frequent and can go in different directions (Fig. 1.3). Ovules are often arranged in two series on the placenta (linear arrangement); they may also spread out in double series or without clear order (diffuse arrangement). By expansion of the placenta in the locule and increase of the number of ovules parietal or axile placentae can become protruding-diffuse. By contraction of the locular space two series can merge in a single line. There is much variation in the number of ovules, ranging from a very high number (e.g. Orchidaceae) to a pair or just a single ovule per carpel (e.g. Anacardiaceae). Ovules are attached to the placenta by a funiculus. The ovules are generally curved (anatropous) with the opening (micropyle) downwards by bending of the funiculus, more rarely erect (orthotropous or atropous) or curved around the funiculus (campylotropous).

The transition of pollen from the style to the ovule via the placenta is made possible by secretion or bridging devices: this can be an obturator as an extension of the placenta (e.g. Euphorbiaceae, Sapindaceae), protrusions of the funiculus (e.g. Anacardiaceae), extension of the style (e.g. Plumbaginaceae) or apical extensions of the placental column in ovaries with free-central placentation (e.g. Myrsinaceae). The gynoecium can be isomerous with the petals, which means that the number of carpels is the same as the petal whorl. In that case carpels are situated opposite the sepals or petals. Very often, the number of carpels is lower (three or two) and their position tends to be influenced by the position of the flower on the inflorescence (Ronse De Craene and Smets, 1998b).

Pseudomonomerous gynoecia

In several groups of core eudicots and commelinid monocots the gynoecium is reduced to a single fertile carpel. A pseudomonomerous gynoecium can be defined as a gynoecium that consists of more than one carpel of which only one is fertile and fully developed while the other(s) are empty or absent at maturity. Reduction series can easily be constructed in some groups of plants, such as Restionaceae, ranging from three fertile carpels to a single median carpel (Ronse De Craene, Linder and Smets, 2002; Fig. 1.3F), or 'Urticales' with a series running between different families from two carpels to a single one (Bechtel, 1921).

Pseudomonomerous gynoecia were variably defined in the past. Guédès (1979) defined pseudomonomery as a pluricarpellate ovary with a single ovule. The ovule can be on the margins of the sterile carpel or is basal. Eckardt (1937) and Weberling (1989) interpreted it as a pluricarpellate ovary with rudimentary carpels. The fertile carpel can have more than one ovule. The latter interpretation is the correct one, as there are different degrees of sterilization of carpels and ovules possible.

Pluricarpellate gynoecia

As for the androecium the number of carpels can be considerably increased relative to the other whorls of the flower. However, the extent of increase is more limited than for the androecium, because the gynoecium occupies a distal position in the flower with little space for extension. A possible cause for polymery is duplication of carpels through the building of false septa (e.g. Linaceae). The number of carpels is rarely increased by proliferation from complex primordia (e.g. *Kitaibelia, Malope* in Malvaceae: van Heel, 1995). A higher number of carpels in a whorl tends to distort the whorled arrangement and is often correlated with a polymerization of the androecium or increased merism of the entire flower (p. 31; e.g. Araliaceae, *Actinidia, Citrus*). Carpel and stamen numbers tend to be increased in a lateral girdle in most cases. *Tupidanthus* (Araliaceae) represents an extreme in lateral polymerization resulting in a convolute flower (Sokoloff *et al.*, 2007). In some taxa a second whorl of carpels is initiated (e.g. *Pavonia* and *Urena* in Malvaceae, *Punica* in Lythraceae: Ronse De Craene and Smets, 1998b). In Rosaceae, the swelling of the receptacle permits the development of several tiers of uniovulate carpels (Kania, 1973). Also in basal groups extension of the cone-shaped receptacle can lead to an increase in the number of carpels. This is clearly visible in some Ranunculaceae (e.g. *Myosurus, Ranunculus*), or Alismataceae.

1.3   The floral axis and receptacle

The floral receptacle (axis or torus) is the central part of the flower on which floral parts are inserted and is homologous to the stem. The floral receptacle can be inconspicuous or strongly developed. A conical shape

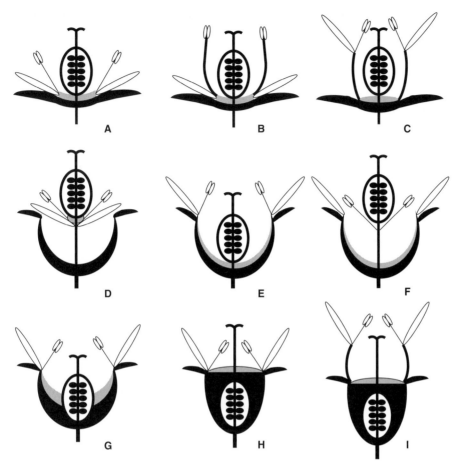

Fig. 1.4. Hypanthium types and ovary position. Nectary shown by grey area. Sepal lobes and hypanthium shown in black; petals shown in white. A, flower with superior ovary and without hypanthium; B, flower with superior ovary and with stamen tube; C, flower with superior ovary and with stamen-petal tube; D, flower with sepal tube and anthophore; E, flower with cup-like hypanthium; F, flower with cup-like hypanthium and androgynophore; G, flower with half-inferior ovary and hypanthium; H, flower with inferior ovary; I, flower with inferior ovary and stamen-petal tube.

is characteristic for some basal angiosperms, such as Magnoliaceae or Ranunculaceae, while a cup-shaped receptacle (hypanthium) is more wide-spread. Flowers are hypogynous with superior ovary, in cases where the peri-anth and stamens are clearly inserted below the gynoecium (Fig. 1.4A–F). In cases where the receptacle expands in a tube or a cup by intercalary meristem-atic growth, a hypanthium is formed which may place the perianth and stamens at a higher level than the gynoecium (Fig. 1.4C,E). During development of the flower the gynoecium can sink within the receptacle by intercalary growth occuring at the level of the gynoecium. The flower becomes perigynous (with a half-inferior ovary: Fig. 1.4G) or epigynous (with an inferior ovary: Fig. 1.4H,I) as a result. Differences between superior and inferior ovaries are only a matter of degree. A large part of the twentieth century was dominated by the controversy between protagonists of the appendicular (the American School) versus the axile origin of the receptacle (German School). This fruitless discussion was centred on the question of whether the hypanthium represents the congenital fusion of the bases of different organs, or the result of receptacular growth. Transitions, however, can go in both directions with the possibility of reversals. Examples of changes in the position of the ovary are cited in Gustafsson and Albert (1999), Endress (2003a) and Costello and Motley (2004).

A hypanthium can have the same colour as the perianth, leading to it being mistaken for a calyx or corolla tube (e.g. Grossulariaceae, Tropaeolaceae, Onagraceae). The presence of a hypanthium is often accompanied by petaloid sepals and a tendency for reduction and loss of petals, especially in rosids (Ronse De Craene, 2008). Depending on the level of intercalary growth, calyx, corolla and stamens can become connected by a hypanthial tube (e.g. Myrtaceae: Fig. 1.4E), or only calyx and corolla (e.g. Cucurbitaceae: Fig. 1.4F). Stamen-petal tubes are a specific kind of hypanthium arising by formation of a ring meristem at the base of corolla and stamen lobes (Fig. 1.4C,I; see p. 34).

The development of a hypanthium has huge influence on the growth of organs, for example by polarizing the direction of stamen increase (see p. 11), or by influencing a shift in position of the ovary from superior to inferior, and the direction of ovule development on the placentae. Anthophores (with an extension lifting ovary, petals and stamens above the calyx whorl: e.g. *Lychnis, Silene* in Caryophyllaceae: Fig. 1.4D), androphores (Euphorbiaceae), androgynophores (e.g. Passifloraceae: Fig. 1.4F) or gynophores (e.g. Capparaceae) are all special cases of extensive receptacular growth. Initiation of a hypanthium can start at different stages of development of flowers, influencing floral shape and position of organs. Development of an inferior ovary can be seen as a progressive evolutionary process, where the onset of hypanthial growth is pushed progressively earlier in the floral development. An early development of a hypanthial rim is characteristic for several asterids and is linked with a rapid development of a stamen-petal tube and an inferior ovary (early sympetaly sensu Erbar, 1991; see p. 34). The development of a hypanthium has an important ecological function in flowers in regulating the access to nectaries, separating pollen from stigmas (avoiding selfing) and protecting the inner parts of the flower.

1.4   Floral nectaries

Nectaries consist of glandular tissue that secretes sugar-containing fluids. They can arise anywhere in the flower and function as a major attractant to pollinators. Nectaries do not make up an organ category of their own but are specific tissues not necessarily confined to particular floral organs. They rarely comprise whole organs (staminodial structures or carpellodes), more often appendages situated on the receptacle or on floral organs. There were several attempts to classify floral nectaries, either on shape, position, histology and function (reviewed in Pacini, Nepi and Vesprini, 2003 and Bernardello, 2007). Vogel (1977, 1998c) distinguished three types of nectary based on their histological properties: mesenchymatous (usually a disc with underlying nectar tissue and secretion through transformed stomata), epithelial (secretion through the epidermis) and trichomatic nectaries (through transformed hairs). Smets (1986, 1988) distinguished between *nectaria caduca*, occurring on accessory structures in the flower (stamens and petals), and *nectaria persistentia*, occurring on parts of the flower that persist at fruiting (e.g. sepals, receptacle, ovary). The latter comprises two main types: disc nectaries common to core eudicots and septal nectaries of the monocots.

Nectaries are frequently associated with specialized compartments (nectar containers or spurs) that are independent of the nectar glands (e.g. Loasaceae, *Viola, Aconitum, Linaria*) or perform both functions (e.g. *Impatiens, Tropaeolum, Aquilegia*). In other instances the nectaries are inconspicuous and present as an epithelial layer on the hypanthium (e.g. Myrtaceae), thickenings at the base of the stamens (several Caryophyllales, including Polygonaceae, Geraniaceae), trichomes (e.g. Cucurbitaceae) or septal (or gynopleural) nectaries arising where septa are postgenitally formed in the monocots. Disc nectaries, although common in eudicots, are absent from monocots. Some authors discourage the use of the word 'disc' for nectaries, as it is indiscrimately employed for many non-homologous structures in flowers (see Bernardello, 2007). However, the use of 'disc' reflects a structure in the flower with a clearly identifiable shape. Contrary to a disc, nectaries can also occur as distinct glandular appendages or scales, occasionally resembling staminodes. Disc nectaries are receptacular in origin, but can undergo important shifts in the flower, usually in a centripetal direction (e.g. Polygonaceae: Ronse De Craene and Smets, 1991b). The shift can run concomitantly with invagination of the ovary: from disc nectary to gynoecial and further to stylar nectary.

The origin of nectaries is to be sought in the vegetative parts of the plant, in the form of transformed hydathodes or water pores (Vogel, 1998c; Smets *et al.*, 2000). Outside the angiosperms nectaries are found in some ferns and

Gymnosperms (Pacini, Nepi and Vesprini, 2003). Association of nectaries with pollination evolved in *Ephedra* and *Welwitschia* (Gnetales, Gymnosperms) and flowers of angiosperms. Some groups of plants have developed elaiophores or oil-secreting glands instead of nectar glands (e.g. *Diascia* in Scrophulariaceae, calyx glands of Malpighiaceae). In some angiosperms with strongly reduced flowers that have switched from wind pollination to insect pollination, nectary tissue may develop outside the flower and become associated with the reproductive organs. In *Salix* bracteoles of the flower have been transformed into nectar-secreting tissue. The inflorescence of *Euphorbia* (cyathium) is a cup of bract origin with marginal glands. Pseudonectaries represent a special category of structures that imitate nectar drops to attract pollinators (Endress, 1994; Endress and Matthews, 2006b). They can occur on stamens (e.g. *Memecylon*, Melastomataceae), staminodes (e.g. *Parnassia*, Parnassiaceae) or petals (e.g. *Lopezia*, Onagraceae; *Gillbea*, Cunoniaceae: Endress, 1994).

1.5   Relationship of flowers with inflorescences

1.5.1   Terminal and lateral flowers

Flowers are seldom solitary. They are mostly grouped in inflorescences, which can be highly complex. Several attempts were made to morphologically interpret inflorescences, mostly with emphasis on describing inflorescences typologically, and more rarely developmentally (see Weberling, 1989; Tucker, 1999a). However, the complexity of the terminology can be overwhelming and without clear introduction one is easily lost. Moreover, the distinction between flowers and inflorescences can sometimes be difficult or almost impossible to make. Inflorescences can be basically subdivided in two categories, polytelic (racemose or monopodial) versus monotelic (cymose or sympodial) inflorescences (Tucker, 1999a). A polytelic inflorescence has one principal growing axis producing flowers acropetally and is indeterminate. Lateral branches are a reiteration of the main branches. Indeterminate inflorescences are theoretically indefinite but they eventually decline in activity. A monotelic inflorescence is determinate, with the principal axis developing in a flower and a secondary axis developing basipetalous in its axil. The distinction between polytelic and monotelic inflorescences is not always clear. A raceme can evolve into a cyme and vice versa, the former by a dominance of apical flowers or terminalization (grouping of terminal pseudanthia into an apical flower), the latter by loss of apical dominance (loss of a terminal flower, called truncation) (Sokoloff, Rudall and Remizowa, 2006). This change, though quite frequent, is triggered by genetic shifts and can be the cause of evolutionary transitions from one type of inflorescence to another. A racemose inflorescence can produce a terminal flower opening before the other flowers in a number of families (e.g. Eames, 1961) or resume vegetative growth after developing flowers (e.g. *Callistemon*, Myrtaceae). Lateral flowers can also shift in a terminal position, in cases where an apical residuum is lost. Alternatively, a pseudanthium is formed by the loose grouping of carpels and stamens, in cases where the space for flower inception is too limited or flower identity breaks down (e.g. *Potamogeton, Piper, Triglochin*: Buzgo et al., 2006; Sokoloff, Rudall and Remizowa, 2006). It can be difficult to distinguish flowers from inflorescences because of the aggregation of highly reduced flowers (e.g. *Cercidiphyllum*: Endress, 1986).

More complex inflorescences can be derived from monotelic and polytelic inflorescences, with several lateral branches repeating the patterning of the main branch (e.g. panicles, umbels, compound capitula). However, it remains essential to link these inflorescences to their mode of initiation to recognize their initial form, which is either polytelic or monotelic, and this can best be achieved by concentrating on the subunits of inflorescences.

1.5.2   Pseudanthia

The definition of a pseudanthium differs among authors. The traditional definition implies that pseudanthia are inflorescences that mimic flowers (e.g. Eames, 1961; Weberling, 1989; Endress, 1994). For Rudall and Bateman (2003) a pseudanthium can be something in between, neither a true flower nor a true inflorescence. Two kinds of pseudanthia can be identified: those that retain the identity of individual flowers and those in which flower identity is lost (Sokoloff, Rudall and Remizowa, 2006). The first case is often associated with small flowers grouped in compact inflorescences, either with a functional division between outer sterile attractive flowers and inner fertile flowers (e.g. Hydrangeaceae with sterile outer flowers), or a contribution of bracts external to the inflorescence (e.g. Cornaceae with showy bracts or Marcgraviaceae with cup-like nectariferous bracts). More examples are presented by Weberling (1989) and Classen-Bockhoff (1990).

Small, highly reduced flowers can also be aggregated in inflorescences. In this case it is far more difficult to distinguish inflorescence from flowers, as one can be fooled by flower-like cyathia of *Euphorbia* (Prenner and Rudall, 2008). As mentioned earlier, some flowers have probably evolved from pseudanthia, as shown by Sokoloff, Rudall and Remizowa (2006) for the alismatids Potamogetonaceae and Cymodoceaceae.

1.5.3   *Bracts and bracteoles*

The transition between leaves and flowers is usually intermediated by bracts. These are appendages that can be leaf-like, or are often much smaller, sometimes intermediate between leaves and perianth parts. Bracts (pherophylls) and bracteoles (prophylls) are not floral organs, although they may become closely associated with flowers and secondarily included as part of the perianth (Ronse De Craene, 2007). Most eudicot flowers have one subtending bract and a pair of lateral bracteoles. In the monocots there is usually one abaxial bract and one adaxial bracteole (Arber, 1925). Numbers of bracts can be much higher without differentiation of bracteoles, or bracts and bracteoles can be secondarily lost. It could be argued that a distinction between bracts and bracteoles is artificial. In compound monotelic inflorescences with bracts and bracteoles, the bracteoles of the main flowers act as bracts of the lateral flowers, and this is continued with the further initiation of more flowers. A clear distinction between bracts and bracteoles is therefore impossible. For descriptive purposes one can distinguish between first-order bract, second-order bract, etc.

Bracts and bracteoles are usually well delimited from flowers. However, especially in more basal groups the transition between bracts and floral parts is progressive. Bracteoles can regulate the transition from a decussate arrangement of the vegetative leaves to a spiral phyllotaxis in the flower (Eichler, 1875; Prenner, 2004a; Ronse De Craene, 2008; see p. 28). The first sepals tend to be arranged pairwise in alternation with the bracteoles. When bracteoles become lost, the first sepals tend to occupy the empty lateral position of lost bracteoles. In some monocots the bracteole is transversal, not adaxial, and on the same level as the first sepal (Remizowa, Sokoloff and Kondo, 2008). Bracts are occasionally showy, attractive organs simulating a perianth (e.g. *Cornus*), or contrasting with the flowers in colour (e.g. *Mussaenda* in Rubiaceae; *Melampyrum* in Orobanchaceae). Bracts are sometimes strongly associated with the flower. They can occur high on the pedicel, close to the flower (e.g. Phytolaccaceae, Berberidopsidaceae) or may become part of the flower, enclosing sepals as a secondary calyx (e.g. *Aextoxicon* in Aextoxicaceae, *Afzelia* in Leguminosae). This implies that in some cases bracts can have contributed phylogenetically to the differentiation of the perianth in calyx and corolla by spatial shifts (Albert, Gustafsson and Di Laurenzio, 1998; Ronse De Craene, 2008).

1.5.4   *Epicalyx and calyculus*

Appendages may occur below the calyx as an extra whorl of sepals or as smaller structures. They are described as the epicalyx (calicle). Epicalyx members can be leaf-like, sometimes indistinguishable from sepals (e.g. *Potentilla, Malva*) or bristle-like (e.g. *Agrimonia, Neurada*). Origins of the epicalyx are varied and often not homologous, either derived from bracts or bracteoles that became closely associated with the flower (e.g. Dirachmaceae, Malvaceae, Caprifoliaceae, Passifloraceae, Convolvulaceae), as stipules of the sepals (Rosaceae) or as commissural emergences without clear homology (e.g. Lythraceae) (Pluys, 2002). In Rosaceae the two outer sepals bear stipules, and the third sepal has only a single stipule, which is missing in the inner sepals (Trimbacher, 1989).

A calyculus is a narrow rim of tissue found in some Santalales (e.g. Loranthaceae, Olacaceae). It was described as a reduced calyx by most authors. However, ontogenetic evidence points to its derivation from lateral bracteoles (Wanntorp and Ronse De Craene, 2009). Epicalyx and calyculus are often confused but are comparable in referring to bracteolar organs.

1.6   Symmetry and orientation of flowers

Floral symmetry is one of the main structural factors affecting the Bauplan of the flower besides merism (see p. 30). The symmetry of flowers is the result of the initiation of floral organs and their subsequent growth and differentiation (Endress, 1994, 1999; Tucker, 1999b). A regular floral symmetry results from equal growth of organs within whorls. Such flowers have a radially symmetrical plan (polysymmetry or actinomorphy) and are accessible to insects from all directions. The flower can be divided from any orientation in two equal halves relative to the floral axis. In cases where organs develop unequally within a whorl, one or two sides of the flower can become differentiated from the other sides. If two sides develop differently, a disymmetric flower with two lines of symmetry is formed. This is relatively rare and can be found in Fumarioideae of Papaveraceae or in Brassicaceae. The unequal development of one side of the flower leads to monosymmetry or zygomorphy (bilateral symmetry), with only one possibility to divide the flower in two equal halves. This pattern of development is widespread among angiosperms and is aimed at a specific access for pollinators. Most monosymmetric flowers have their symmetry line running along a median line from the axis to the bract. Transversal monosymmetry is much rarer (e.g. *Corydalis* in Papaveraceae), while oblique monosymmetry (with the symmetry line neither transversal nor radial) characterizes several families (e.g. Moringaceae, Sapindaceae, Tropaeolaceae). Monosymmetry affects flowers by degrees and is an evolutionary progressive development. The timing of the onset of monosymmetry is fundamental, and can be either manifested at early organ initiation, at organ growth and enlargement or as a late differentiation of organs (Tucker, 1999b). Bract and inflorescence axis act as two opposing gradients in shaping the floral bud during development (Endress, 1999). As a result, flowers may be monosymmetric in early stages of development but can become regular at maturity. Other examples were described by Ronse De Craene *et al.* (2001) for Melianthaceae and by Olson (2003) for Moringaceae.

There are different degrees in the extent of monosymmetry. Endress (1999) distinguished between constitutional, positional and reduced monosymmetry. Positional monosymmetry causes a subtle displacement of organs and differential growth. The orientation of stamens and style and a size difference of the petals between adaxial and abaxial side of the flower may lead to weak monosymmetry (e.g. *Epilobium* in Onagraceae, *Gladiolus* in Iridaceae). Constitutional monosymmetry (structural zygomorphy *sensu* Rudall and Bateman, 2004) implies that monosymmetry is strengthened by elaboration, reduction or suppression of organs that may occur simultaneously in a same flower (e.g. Orchidaceae, Scrophulariaceae). Petals arise sequentially and develop a different size, while stamens may be reduced or lost. Monosymmetry may also be caused by extreme reduction. Some families have flowers reduced to a single carpel and one to few unilateral stamens (e.g. Chloranthaceae, Lacistemataceae, Callitrichaceae). Such flowers are monosymmetric because most organs are lost. However, they may regroup in pseudanthial structures (see p. 23).

In strongly monosymmetric flowers, initiation of organs can be unidirectional and associated with suppression or loss of some organs (one side of the flower develops more extensively at the expense of the other side, e.g. *Bauhinia, Cassia, Senna* in Leguminosae: Tucker, 1984, 1996, 2003a). This development can be accompanied by shifts in function between the anterior and posterior side of flowers (e.g. in the androecium of Commelinaceae). Tucker (1997, 1999b) demonstrated a strong correlation between the extent of monosymmetry and the developmental stage at which it is first apparent. This correlation also affects major taxonomic groups: in groups with a majority of polysymmetric taxa, monosymmetric flowers tend to differentiate late in floral development (e.g. *Tropaeolum* in Brassicales: Ronse De Craene and Smets, 2001b), while groups which are predominantly monosymmetric have unidirectional development and zygomorphic characteristics much earlier in floral development (e.g. Lamiales: Endress, 1999). Monosymmetry can affect a single organ in groups, which are mainly polysymmetric. For example, a functional differentation between different kinds of stamens (heteranthy: differentiation of feeding stamens and pollen stamens) often leads to monosymmetry and a reduction of stamens, or also to highly monosymmetric flowers with elaborate androecia (e.g. Lecythidaceae, Commelinaceae). Rudall and Bateman (2004) argued that two patterns of derivation of monosymmetry are operating in the more advanced groups of monocots, one affecting the adaxial side (with suppression of one to five stamens: e.g. Zingiberaceae) and the other the abaxial side of the flower (with suppression of the outer abaxial stamen: e.g. Musaceae). Basal groups of monocots are more variable in degree of merism and apocarpy.

In a few groups of plants (e.g. Cannaceae, Marantaceae, Caprifoliaceae, Vochysiaceae, some Leguminosae), flowers can be asymmetric. These flowers have no clear symmetry line, but are derived from monosymmetry (Endress, 1999). Their rarity may indicate that the pollinators are not eager for such flowers, or that there are too many developmental constraints. However, these flowers appear often grouped in pairs with mirror-image flowers (enantiostyly: Marantaceae, *Wachendorfia* in Haemodoraceae, *Senna* in Leguminosae), or are small and arranged in compound inflorescences that appear regular (e.g. Caprifoliaceae).

Flower shape and symmetry are directly linked to pollination. Monosymmetric flowers can be subdivided in two kinds: flag-flowers and lip-flowers (Endress, 1994). Flag-flowers are sternotribic (pollination by the underside of the visitor through stamens that are exposed and curved up: e.g. Capparaceae, Caesalpinioideae); lip-flowers have a lower lip functioning as a landing platform and are nototribic (pollination by the back of the insect; the stamens are usually concealed by the fused upper petals: e.g. Lamiales). The initial number and orientation of flower parts constrain the way a flower is differentiated into adaxial and abaxial parts (Donoghue, Ree and Baum, 1998). The ubiquitous presence of sepal two in adaxial position restricts the way two-lipped flowers are orientated (e.g. Lamiales, *Pelargonium* in Geraniaceae). Petals are generally arranged as an upper and lower lip in Lamiales, with three possible orientations of the petals: a 2:3 orientation has two adaxial and three abaxial petals; a 1:4 orientation has one adaxial and four abaxial petals; a 0:5 orientation has all petals moved to the abaxial side. Such orientation appears rigid and is guided by the position of the sepals. Only in cases where sepal two is in abaxial position is an inverted arrangement possible (Donoghue, Ree and Baum, 1998; Endress, 1999). The orientation of the flower is also responsible for the reduction or loss of the adaxial stamen in many Lamiales. This is linked with the curvature of anthers and style in the upper part of the flower where they connect with the back of the pollinators and where an adaxial stamen would be in the way (Endress, 1994, 1999).

Although most evolutionary patterns indicate that monosymmetric flowers are derived from polysymmetric flowers, there are indications of a reversed evolution. There are several cases of polysymmetric flowers within mainly monosymmetric groups (e.g. Lamiales, Caprifoliaceae). Structural evidence indicates that fusion of posterior petals is linked with the loss of the adaxial stamen. This transition is linked with a switch to tetramery and appears to have arisen at least nine times in the asterids (e.g. Ronse De Craene and Smets, 1994; Donoghue, Ree and Baum, 1998; Endress, 1999; Bello *et al.*, 2004; Soltis *et al.*, 2005).

1.7   Phyllotaxis

1.7.1   *Whorls and spirals*

Flowers can have organs in spirals or in whorls similar to leaves. Spirals and whorls are regulated by phyllotaxis, which is the pattern of initiation of organs on vegetative stems and flowers. Guiding principles in floral phyllotaxis are the divergence angle (the angle between two organs arising in succession) and the plastochron (the time interval between the initiation of two successive organs). In spiral flowers, organs arise in regular plastochrons with an equal divergence angle regulated by an inhibition zone emitted by the floral apex. In whorled flowers organs arise in pulses with unequal plastochron and unequal divergence angles. Within a whorl the plastochron is very short or approaching zero, while it is longer between different whorls (Endress, 1987; Endress and Doyle, 2007). The sequence of initiation can be detected from contact spirals (parastichies) and contact lines (orthostichies). Parastichies and orthostichies can be mathematically described by the Fibonacci series, reflecting the number of organs formed within a sequence. In spiral flowers there are several sets of parastichies with a particular number of organs (5,8,13, …) and no orthostichies, as is clearly visible on a pine cone or sunflower head. Whorled flowers have two equal parastichies running in opposite directions and several orthostichies. As a result change from one organ category to another (e.g. outer to inner tepals) is progressive in spiral flowers and different floral organ categories can be bridged by intermediate organs. In whorled flowers the transition between different organ categories tends to be abrupt. For reviews on phyllotaxis the reader is referred to Endress (1987) and Endress and Doyle (2007).

Changes in phyllotaxis are very common in flowering plants and happen at the transition of vegetative organs to flowers (e.g. bracteoles to sepals), but also between different organ categories (sepals and petals) and are dependent on a change in the size of the floral organs. As will be shown in this book these shifts have important consequences for the arrangement of organs in flowers.

Phyllotaxis is best discussed in a phylogenetic context. A major evolutionary step in flowers is the transition from spiral to whorled flowers. Spiral phyllotaxis is considered ancestral in the angiosperms, as it is found in several basal angiosperms including *Amborella* (Ronse De Craene, Soltis and Soltis, 2003; Endress and Doyle, 2007). Erbar and Leins (e.g. Erbar and Leins, 1981, 1983; Leins and Erbar, 2007) demonstrated that in spiral flowers such as Magnoliaceae a switch to a whorled arrangement can result from an increase in size of the perianth parts. The transition between much larger tepals and small stamens leads to an interruption in the continuous plastochron. Outer stamens tend to arise simultaneously as pairs in alternation with the inner perianth parts. Inner stamens and carpels may still be initiated in a spiral sequence, or organs tend to be arranged in alternating whorls repeating a sequence of pairs and non-paired organs (Ronse De Craene and Smets, 1993, 1994, 1998a). The paired arrangement of outer stamens can be retained while the number of parts is considerably reduced (Ronse De Craene and Smets, 1993). Instead, Endress (1987, 1994) argued that stamens have been doubled as the result of a shortened plastochron and narrow organ shape (see p. 13).

The acquisition of stable whorled flowers has been considered an essential breakthrough in floral evolution (e.g. Endress, 1987, 1990; Ronse De Craene, Soltis and Soltis, 2003; Zanis *et al.*, 2003; Endress and Doyle, 2007). Spiral flowers have a weak synorganization; the floral parts are loosely arranged with fluctuating organ number and little interaction between neighbouring organs. Whorled flowers have strong synorganization with close interaction between neighbouring organs; fusions and changes in symmetry are only possible with a concerted interaction between different organs in the flower. As a rule of thumb, phyllotaxis is labile in basal angiosperms and basal eudicots, contrary to core eudicots and monocots. In most eudicots spiral phyllotaxis is retained in the calyx (with four to five sepals arranged in a 1/2 or 2/5 phyllotaxis). The corolla often arises in a rapid helical sequence but a spiral phyllotaxis is not clear at maturity. In some early diverging eudicots (e.g. Nelumbonaceae) and core eudicots (e.g. Paeoniaceae, Clusiaceae, Theaceae) the petals have occasionally reverted to a spiral sequence (Ronse De Craene, 2007). Whorled phyllotaxis can revert to a spiral if constraints of synorganization disappear. The retardation or loss of petals may influence the phyllotaxis of stamens, as in Apiaceae or Brassicaceae (Erbar and Leins, 1997). If constraints disappear (e.g. by loss of a perianth), whorled flowers may revert to a spiral or unordered initiation (e.g. *Cercidiphyllum, Trochodendron, Achlys*: Endress, 1986, 1989).

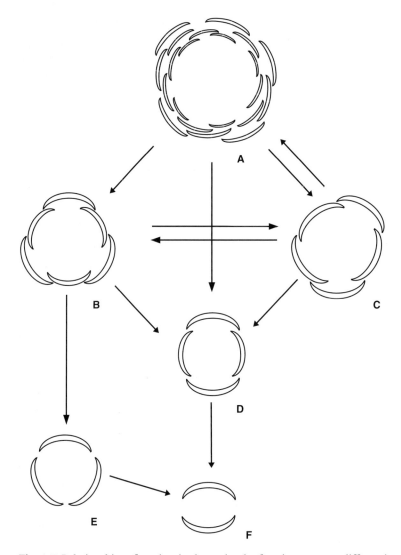

Fig. 1.5. Relationships of merism in the perianth of angiosperms; undifferentiated perianth shown. A, spiral with variable number of tepals; B, trimery/hexamery; C, pentamery; D, dimery/tetramery; E, trimery, single whorl; F, dimery, single whorl.

1.7.2   Merism

Merism (or merosity) refers to the number of parts per whorl in the flower (usually based on petals or perianth, taken as reference). Merism is either indefinite, but more commonly trimerous or dimerous (basal angiosperms, basal eudicots, monocots), or pentamerous and tetramerous (core eudicots) (Fig. 1.5). Merism is a major feature affecting the Bauplan of flowers besides symmetry and phyllotaxis. Origins of different merisms have been discussed in Endress (1987), Kubitzki (1987) and Ronse De Craene and Smets (1994).

Trimerous flowers (and by extension dimerous flowers) are widespread in basal angiosperms and monocots. A derivation from a spiral condition is the most plausible interpretation for their origin (p. 29). Dimery has arisen repeatedly by loss of a sector in each whorl of a trimerous flower, sometimes resembling a tetramerous flower, although the flower tends to be disymmetric. Origins of pentamery are not well understood, although five-merous flowers are found in the majority of angiosperms. There are at least five origins of pentamery, the major one affecting the core eudicots (Ronse De Craene, Soltis and Soltis, 2003). Postulated origins of pentamery are a derivation from trimerous flowers (by amalgamation of two whorls), from dimerous flowers (through apical expansion), or from spiral flowers (Fig. 1.5). Tetramerous flowers are further derived, occurring sporadically in five-merous groups or characterizing a whole family. It is understood that a single mutation or proportionate size differences can trigger a change to tetramery. Pentamerous flowers can occasionally evolve into hexamerous flowers. As for tetramery and dimery, hexamery and trimery can become superficially similar. The reason is that a continuous 2/5 initiation sequence of a pentamerous flower breaks down into a sequence of two whorls with a 1/3 initiation due to space constraints on the floral apex. Therefore, hexamery has often been confused with trimery. In hexamerous flowers, organs seldom arise simultaneously; more often there is an alternation of two sequential whorls, with the outer overlapping the inner whorl. This is effectively seen in *Manilkara* (Sapotaceae), Loranthaceae or Fagaceae. Polygonaceae is another example, where Rumiceae are hexamerous and appear trimerous, because the inner whorl behaves differently from the outer. Other merisms (seven- to ten-, 12-, 18-) are far less common. Although pentamerous flowers may show an intrinsic variation to higher numbers, these are not common and seldom stabilized. Certain genera have jumped to much higher merisms, such as 18-merous to 32-merous *Sempervivum* (Crassulaceae). These increases occasionally only affect part of the flower such as perianth and androecium, or androecium and gynoecium (e.g. Sapotaceae, Rhizophoraceae, Araliaceae). Apparently higher merisms are spatially less stable, as a greater surface is required for organ insertion.

Flowers are generally conservative in the presence of the same merism in most whorls. However, in some groups merism can fluctuate between different whorls (e.g. Winteraceae, Ranunculaceae, Sapotaceae). Merism is intricately linked to phyllotaxis, as whorled phyllotaxis is a prerequisite for stable merism. The acquisition of stable merism has led to the differentiation of the perianth into sepals and petals.

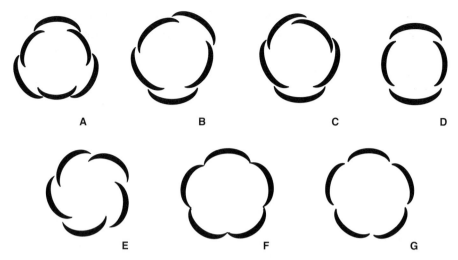

Fig. 1.6. Aestivation patterns of the perianth. A–E, imbricate: A, decussate-trimerous; B, quincuncial; C, cochleate; D, decussate-dimerous; E, contorted. F, valvate; G, apert.

1.7.3   Aestivation patterns

Aestivation is the relative arrangement of neighbouring perianth parts in bud and is usually applied to the petals, which are often the most obvious organs at maturity. Three types are generally recognized: imbricate (with margins overlapping: Fig. 1.6A–E), valvate (with margins touching: Fig. 1.6F) and apert (or open, with margins not touching: Fig. 1.6G). Aestivation is often – but not always – dependent on the initiation pattern of the perianth (Ronse De Craene, 2008). A sequential (spiral) initiation of organs will give rise to an imbricate aestivation (quincuncial or 2/5 pattern in pentamerous flowers: Fig. 1.6B), in cases where some organs are completely outside and some completely within. Spiral flowers generally have an imbricate perianth arrangement (Fig. 1.6A,D). Simultaneous initiation pattern of the petals leads to a valvate or imbricate (contorted or convolute) aestivation (Fig. 1.6E). Unidirectional initiation gives rise to an imbricate-cochleate pattern that can be ascending or descending (Fig. 1.6C). Apert aestivation is associated with retarded organs. In trimerous or hexamerous flowers, the outer perianth whorl can be imbricate, while the inner is apert (e.g. *Manilkara* in Sapotaceae), or the opposite occurs (e.g. *Monanthotaxis* in Annonaceae). Aestivation is also correlated with the symmetry and the variable function of sepals and petals in flowers (Endress, 1994).

Despite different origins, aestivation has systematic significance in several families of flowering plants and can be emphasized in floral diagrams.

Aestivation of sepals is generally imbricate (2/5 aestivation), reflecting their spiral sequence, especially in rosids and caryophyllids. Contorted sepal aestivation is rare or absent. A valvate calyx is rare in pentamerous flowers, but when present is often accompanied by an epicalyx (e.g. Rosaceae, Malvaceae). For petals, aestivation patterns can be variable, but can often be consistently contorted or valvate in some clades (e.g. Santalales, Malvales; Ronse De Craene, 2008).

1.8   Fusion of floral parts

A major requirement for fusion between organs in flowers is a whorled phyllotaxis. Therefore, basal angiosperms with spiral flowers rarely have fused organs. Fusion may occur independently in different organ whorls, such as sepals (synsepaly), tepals (syntepaly), petals (sympetaly), stamens (monadelphy) or carpels (syncarpy). In other cases different organ categories may be fused. Anthers and styles can be fused into a gynostemium by congenital fusion (e.g. Orchidaceae, Aristolochiaceae) or gynostegium by postgenital fusion (e.g. Apocynaceae). Fusions of organs are often confused with the development of a hypanthium (see p. 20) as the limits between fused organs and a floral cup may not be visible.

1.8.1   *Congenital and postgenital fusion*

In the literature a distinction is commonly made between *congenital fusion* versus *postgenital fusion*. However, this distinction is confusing. What is described as congenital fusion is difficult to ascribe as a fusion. Endress (1994: 10) defined congenital fusion as: 'a common base of several organs develops as a ringwall which may eventually form a tube'. Morphological surfaces cannot be detected at the site of congenital coherence (Endress, 2006). Sattler (1977, 1978) described congenital fusion as interprimordial growth or zonal growth. Whenever a meristematic zone situated below primordia starts to grow, a tube is formed, be it a sepal tube, a petal tube, a stamen tube or a stamen-petal tube (Fig. 1.4). All sorts of tubes are to be considered as extensions of the hypanthium in those cases where zonal growth leads to the formation of a common ring. In the same way an inferior ovary is to be seen as the result of extensive hypanthial growth (Fig. 1.4H,I; cf. Leins and Erbar, 2007). Fusion should be restricted only to observable developmental processes.

Postgenital fusion is a real process of fusion and arises by marginal adhesion (false or pseudosympetaly: e.g. petals of *Napoleonaea, Correa, Oxalis, Pittosporum*; anthers of Asteraceae). This can fool us in believing that petals are truly fused. In some cases holes appear between the congenitally fused parts and the postgenitally fused section (fenestrations; e.g. *Paederia*: Puff and Igersheim, 1991). The development of a calyptra also belongs to this category: fusion affects the upper part of the corolla, which is dropped as a whole (e.g. *Eucalyptus*; *Vitis*). Similar developmental processes are involved in fusions of the calyx, androecium and gynoecium.

1.8.2   *Sympetaly and common stamen-petal tubes*

The development of sympetaly (or syntepaly in monocots) is related to pollination strategies restricting rewards to long-tongued pollinators, so that nectaries are hidden deep in the flower and ovaries are better protected. The development of a stamen-petal tube is the most common form of sympetaly, but has been often misinterpreted in the past. Two processes are clearly involved in the development of sympetaly: (1) growth of a common meristematic zone below the petal and stamen lobes, lifting petals and stamens on a common tube (stamen-petal tube), and (2) lateral coalescence of petal lobes, possibly linked with a postgenital fusion with stamen bases (corolla tube) (Erbar, 1991). The two processes are independent phenomena that occur in combination, although both are often indiscriminately described as a 'petal tube', and stamens as 'epipetalous stamens'. However, the common stamen-petal tube is to be interpreted as some sort of hypanthium and the variable combination of these two growth processes gives highly different results (see also evidence in Sattler, 1977, 1978; Nishino, 1983a,b). The petal tube proper can only exist above the insertion of the stamens. Erbar (1991) considered the two processes as strictly independent by concentrating on the corolla tube *sensu stricto*. She made a distinction between *early sympetaly* (in cases where the corolla tube arises before or together with the petal lobes) and *late sympetaly* (in cases where the corolla lobes are initially free but become connected by meristem fusion). In my opinion a strict distinction between two processes cannot be made. Early sympetaly tends to be correlated with the formation of an early depression in the floral bud; stamens arise on the inner slope of the depression with petals on top, and petals tend to grow rapidly. Late sympetaly is linked with the formation of a convex or flat plateau with stamens on top and petals on the outer slope; petals tend to grow more slowly (e.g. Roels and Smets, 1994, 1996; Ronse De Craene and Smets, 2000). Erbar also recognized transitional stages (e.g. Acanthaceae, Asclepiadaceae), intermediate between early and late sympetaly, which appear doubtful. There is an obvious correlation between the position of the ovary and early or late sympetaly. For example, most Gentianales have late sympetaly and superior ovaries, while Rubiaceae have early sympetaly and inferior ovaries. Secondary loss of sympetaly occurs in several 'sympetalous' families by the failure of a common tube to develop (see Suessenguth, 1938; Montiniaceae: Ronse De Craene, Linder and Smets, 2000; Rubiaceae: Ronse De Craene and Smets, 2000; Plantaginaceae: Hufford, 1995). In some cases the corolla tube can be partially split (e.g. Goodeniaceae, Lobelioideae of Campanulaceae, Haemodoraceae).

2

Floral diagrams

2.1   Definition and significance of floral diagrams

A floral diagram is a schematic cross-section, preferably through a young flower in which all the individual organs or elements of the flower are projected into one plane. A floral diagram is the best way to show the number and topological properties of floral organs. Different parts of the flower are represented by clear symbols, and the spaces between organs are an approximate reflection of the distance between the organs. Fusions between different parts are shown by connecting lines, and the orientation of the flower by placing reference points in relation to the axis of the inflorescence and the subtending bracts. Floral diagrams have two major attributes: (1) the information that can be retrieved from a good floral diagram is immense and replaces extensive descriptions or even drawings, and (2) they facilitate a whole-scale comparison of floral structures across the angiosperms. An inconvenience is that it is concentrated on a limited number of characters at the expense of others that do not fit on the diagrams. Endress (2008a) discussed the benefits of accurate drawings over detailed descriptions. A floral diagram is a clearer way to convey information, as a synthesis of the details of flowers.

There are different ways to create floral diagrams. These range from a simplified drawing of the position of organs in the flower in relation to each other, to a very complex representation of details that cannot always be seen without a lens. Complexity and simplicity of flowers will undoubtedly be reflected in their floral diagrams, but a large amount of additional information can be presented in a clear way on the drawing. This information can include developmental evidence (sequence of initiation), details of organ morphology (such as the insertion of anthers, major vascular bundles, obvious hairs) or obvious appendages that characterize certain flowers and have much significance for comparative morphology (e.g. nectaries, spurs). It can be difficult to represent organs that are small and occur in large numbers in the flower (such as many stamens – in some flowers ranging from 100 to 1000), and often no attempt is made to represent these numbers accurately. Information on the development of organs can facilitate their representation on floral diagrams, as a multitude of stamens can be derived from a few primary primordia. Floral diagrams can be used to represent the typical Bauplan of the flower of a certain family, order, genus or species, and this has much significance for didactic and research purposes.

There is no rule as to how floral diagrams should be built. All floral diagrams imply a certain interpretation, even if efforts are made to depict the flower as accurately as possible. In this book a specific code of colours and shapes is used to depict different organs (see Table 3.1 below, p. 52).

2.2   Types of floral diagrams

Depending on what needs to be emphasized, different kinds of floral diagrams can be constructed.

Empirical floral diagrams represent the spatial relationships of organs in flowers without any interpretation of intervening changes or evolution. The representations can be simple, depicting the organs and their topological relationship in the flower, or more complex, showing other less obvious structures that a botanist wants to emphasize (e.g. presence of appendages on stamens or petals, nectaries, the orientation of organs). For a floral diagram to be useful special attention needs to be paid to the orientation of the flower relative to the inflorescence. Orientation of flowers is often overlooked.

Developmental floral diagrams are mainly based on information that has been obtained through painstaking microscopic studies on the development of flowers. They can be combined with empirical floral diagrams. Developmental floral diagrams can represent specific stages of the development of a flower that need to be emphasized and often necessitate more than one diagram in some cases. An illustration of this is *Tropaeolum*, which shows a change in flower orientation during development of the flower (Ronse De Craene and Smets, 2001b): young flowers have an oblique monosymmetry. At maturity, the floral symmetry becomes median because of the development of a large spur on the adaxial side of the flower, completely changing the visual configuration (Fig. 10.24). A similar change in orientation is caused by resupination of flowers (e.g. Orchidaceae, Balsaminaceae). By a torsion of the pedicel, flowers can turn around for about 180° or less, as shown by drawing a curved arrow on the floral diagram. The sequence of initiation of whorls or organs can be shown by numbers and can more easily be combined with an empirical diagram. Developmental diagrams can depict the groupwise arrangement of stamens, something that cannot be seen in adult flowers. Mature flowers with many stamens often look deceptively similar and it is difficult or sometimes impossible to know the basic androecium configuration. Examples of developmental diagrams are given in Leins and Erbar (2007). Floral diagrams are useful in describing differences in patterns of development of the androecium. In this book details of the sequence of development of stamens will be shown for polyandrous taxa. Phyllotaxis can be shown, either by a sequence of numbers, or by the aestivation of sepals and petals.

Theoretical (or hypothetical) floral diagrams usually contain information that is not visible externally, certainly not in mature flowers. These include organs that have been reduced or lost but for which there is evidence that they were present in an ancestral flower. Clear examples where this can be used are Poaceae or orchids (Figs. 6.7, 6.23), depicting missing perianth parts or stamens with a star. In some cases hypothetical and developmental diagrams can be combined. An example of this is *Melianthus* (Melianthaceae). Mature flowers have four petals only, but during development a fifth petal is initiated but growth aborts at mid-development and no trace is visible at maturity (Ronse De Craene *et al.*, 2001). The sister genus *Bersama* has five petals, indicating that five petals were present in ancestral flowers of *Melianthus*. Presence of the fifth petal can be shown by an asterisk on the floral diagram (Fig. 10.10B). Even without developmental evidence floral diagrams can be used to present hypotheses in a clear way.

2.3   Floral diagrams and floral formulae

Another widely used method to describe flowers succinctly is the use of floral formulae (Table 2.1). This provides information on the kind and number of organs, type of symmetry, presence of fusions and level of ovary. However, it lacks detail of position of organs. The way to use floral formulae differs between the European and American tradition, but they tend to convey the same information. More extensive detail can be given for fluctuations within a whorl. This was recently done by Prenner and Klitgaard (2008) and will be used in a similar way in this book in addition to floral diagrams. A floral formula can be given for a whole family or for a given species. In the first case numbers shown will give an approximate figure of variation, which is unhelpful in cases of high variation. For each individual species that was drawn, a floral formula is given following details given in Table 2.1. If the family is variable in merism and number of organs, a general formula is also given, only referring to the number of floral organs.

Table 2.1. *Floral formulae: symbols used in this book*

Symmetry	
✳	Polysymmetric (actinomorphic)
↓↑←→↖↗↙↘	monosymmetric (zygomorphic; orientation of arrow corresponds to orientation of the flower)
↺	spiral
↔	disymmetric
↯	asymmetric
Floral organs	
P	perigon (no differentiation between calyx and corolla)
K	calyx (sepals)
C	corolla (petals)
A	androecium (stamens)
G	gynoecium (carpels): superior (G̲); half-inferior (-G̲-); inferior (Ğ)
A°	staminode (sterile stamen)
G°	pistillode (sterile carpel)
[...]	fusion between whorls of different organs
(...)	fusion within a whorl or between the same organs
+	more than one whorl can be distinguished
:	there is a clear morphological difference within a whorl
∞	number or organs is numerous or indefinite
-	refers to a variable number within a whorl

Examples

↓ K(5) [C(5) A4:1°] G̲(2) *Penstemon fruticosus* (Lamiaceae)

✳ K4 C4 A4+4 G̲(4) *Ruta graveolens* (Rutaceae)

↯ K3 [C3 A1°-3°+½:2°] Ğ(3) *Canna edulis* (Cannaceae)

2.4   Problems of three-dimensional complexity

Floral diagrams are by essence two-dimensional representations of flowers. The structural information that is provided cannot allow for the description of differences that occur from the base to the top of the flower. In bilabiate flowers the lower part tends to be regular, while the upper part including the stamen-petal tube is highly monosymmetric. Inferior or half-inferior ovaries are taxonomically important and can be problematic to represent on a floral diagram. A hypanthium, lifting stamens, petals and sepals, is found in several major groups and presents the same problems as inferior ovaries (Fig. 2.1).

However, there are ways to solve the problem of three-dimensional complexity, either by superposing two diagrams at different levels of the flower (as a sequence of sections: Fig. 2.1A) or by use of simple symbols. If upper and lower parts of the flower are markedly different (e.g. *Lonicera*, *Cuphea*), this can be

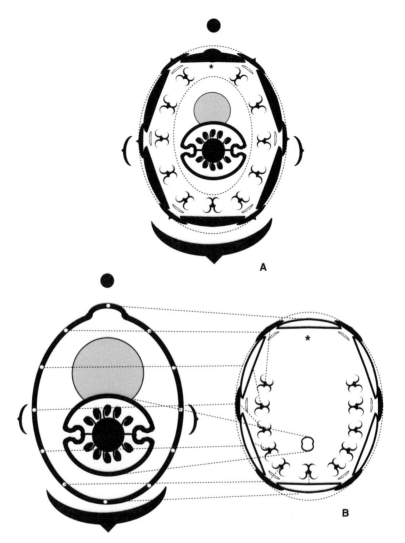

Fig. 2.1. Two ways of accommodating three dimensions in a bidimensional floral diagram for *Cuphea micropetala* (Lythraceae). A, compression of flower in one plane with the presentation of the basal ovary at the same level as the stamens inserted at the top of the hypanthium (broken circles); B, presentations of two diagrams at different levels, the figure on the left representing the level of the ovary and nectary; the figure on the right shows the attachment of the stamens (broken line) and their position in the flower. White dots represent the vasculature of perianth and stamens.

shown by connecting two diagrams at different levels, similar to transverse sections in the flower (Fig. 2.1B). An arrow can also show the orientation of symmetry in the upper part of the flower, especially when monosymmetry is weak and caused by the orientation of a single organ (e.g. the style in *Exacum*,

Gentianaceae). Presence of a hypanthium can be shown by depicting two broken lines to delimit the area covered by the hypanthium. The same could be applied for petal-stamen tubes, which are connected by common growth, although this is not practical because one often cannot differentiate between different sections of the tube. In this book stamen-petal tubes have been depiced as a set of stamens attached to the fused petals. Eichler (1875, 1878) did not represent stamen-petal tubes in his floral diagrams, avoiding the controversy. The position of inferior or half-inferior ovaries can be shown by small grey triangles, next to the illustration of the gynoecium (Table 3.1). Variations in the ovary occur from top to bottom: parietal at top (symplicate zone) and axile at bottom of the ovary (synascidiate zone). The floral diagram will show the placentation at the section where the highest number of ovules is attached.

Receptacular spurs, pouches or appendages represent often striking structures that cannot be ignored in floral diagrams. However, their origin is highly diverse, developing on a receptacular hypanthium (e.g. *Tropaeolum*), on a stamen-petal tube (e.g. *Diascia*) or at the base of the ovary (e.g. *Pelargonium*). The presence of spurs can be shown by drawing them with a broken line below the area where they are attached.

2.5 Representation of accessory structures

Restricting floral diagrams to a representation of the major organs of the flower limits their use considerably. Only the major positional and structural elements can be shown. However, several species have flowers with elaborate structures that cannot be overlooked. Accessory structures (in addition to the four main organ categories) can have different origins and should be included in a floral diagram whenever possible. A few will be presented below.

Organ appendages

Petals and stamens, or more rarely sepals and carpels, can have obvious appendages in the form of awns, tufts of hairs, flaps, wings, etc. These appendages are usually secondary elaborations on the floral organs, arising at a late stage of development, although they may occasionally represent whole organs such as staminodes. Representing these obvious features in floral diagrams helps to convey information that is potentially significant.

Petal appendages are tissue proliferations of the ventral surface of the petal. They have been variously interpreted as ligules, a corona or scales, but their interpretation often remains unclear or contradictory in the absence of developmental evidence. Appendages on petals have a function in pollination (e.g. nectar production, storage and cover, attraction, landing platforms) and are correlated with the architecture of flowers (Endress and Matthews, 2006b). If the appendages are clearly linked to the ventral surface of the petals, they are usually referred to as part of the petal. However, interpretations can be variable if the appendage is only loosely linked to the petal or not at all (e.g. the corona of *Narcissus*, Amaryllidaceae). A hollow space may be present between petal blade and appendage that can act as a nectary (e.g. Ranunculaceae). Leins and Erbar (2007) considered ventral appendages as evidence of sterile ventral pollen sacs and as an indication of a staminodial origin, at least in Caryophyllaceae and Ranunculaceae.

A true spur arises as a ventral invagination leaving an appendage to the dorsal side of the petal (e.g. *Aquilegia*) or hypanthium (e.g. *Tropaeolum*). Spurs arise late in the development of flowers. They either combine storage and production of nectar (e.g. *Tropaeolum*) or only storage (*Corydalis, Delphinium, Aconitum, Viola*). Spurs can be part of a transformed organ (petal: Ranunculaceae) or more commonly a hypanthial outgrowth. Some taxa have spurs that are not visible externally (internal spurs: e.g. *Pelargonium*, Chrysobalanaceae). Bernardello (2007) gave an overview of nectar spurs (see p. 21).

By invagination of the petal blade an abaxial cavity can form on the back of the petal, similar to the development of an inverted spur. The invagination restricts entry to the flowers and is common in some asterids (e.g. Boraginaceae, Plantaginaceae, Orobanchaceae, Gentianaceae).

Corona

A corona is a highly obvious outgrowth, usually of receptacular origin, but also arising as part of the perianth or the stamens. The corona can be in one or several series, is often showy and resembles the perianth (e.g. paracorolla of Amaryllidaceae: *Narcissus*) or can be highly elaborate (e.g. Passifloraceae, Apocynaceae). In some cases the corona is staminodial in nature, such as in *Napoleonaea* (Lecythidaceae: Ronse De Craene, unpubl. data) and *Pachynema* (Dilleniaceae: Endress and Matthews, 2006b). In *Narcissus*, the corona is tubular and is connected to the perianth tube; in other Amaryllidaceae, such as *Pancratium* or *Hymenocallis* the corona is an outgrowth of the filaments (Guédès, 1979).

Pseudostaminodes

In some cases it is difficult to separate true staminodes (as derived from stamens) from receptacular or petal appendages that can have a similar morphology and function (pseudostaminodes: Ronse De Craene and Smets, 2001a). Examples are the stalked glands at the base of the stamens in several Laurales, glandular emergences between the stamens in *Francoa* or *Greyia* (Melianthaceae: Ronse De Craene and Smets, 1999b), antepetalous appendages in *Brexia* (Celastraceae: Edgell, 2004), the interstaminal appendages between the fused stamen bases of Amaranthaceae (Eliasson, 1988) or scales on the petals of several asterids (Ronse De Craene and Smets, 2001a). These appendages may be shown on floral diagrams, but they need to be distinguished from staminodes.

Nectaries

The presence of nectary tissue is often ignored in floral diagrams. However, it represents an important element of the flower that can occupy a considerable space (e.g. Rhamnaceae. Celastraceae) and cannot easily be ignored. The nectaries, as well as any area of glandular tissue, have been presented on the floral diagrams in all cases where they are discernable (grey colour). Eichler (1875, 1878) did not include nectaries in his floral diagrams, except in cases where they are very prominent.

2.6   Floral heteromorphism and unisexual flowers

Flowers of a given species tend to be homogenous in most cases. However, there are instances for flowers of the same species to develop in different shapes and forms. Heteromorphism can be occasional, depending on external factors such as the availability of pollinators, nutrients or the time of the season, or it can be inherent to a species. An occasional heteromorphism is dependent on nutrients, which affect the number of parts in certain flowers. This phenomenon is frequent in annual species with variable organ numbers (e.g. *Nigella* in Ranunculaceae; *Papaver* in Papaveraceae), and was experimentally studied at the beginning of the twentieth century (e.g. Murbeck, 1912).

Within an inflorescence, merism of flowers may fluctuate between a terminal tetramerous flower and lateral pentamerous flowers (e.g. *Ruta, Adoxa*), or lateral trimerous flowers and a terminal pentamerous flower (e.g. *Berberis*) (Eichler, 1875; Rudall and Bateman, 2003). In families with umbel-shaped or capitate inflorescences such as Apiaceae, Hydrangeaceae and Asteraceae, marginal flowers may differ from central flowers, as they gain bigger petals at the expense of fertility. The occurrence of cleistogamous flowers (flowers that do not open at anthesis, against chasmogamous flowers that do open) is a temporary phenomenon that may also be a cause for an important heteromorphism, especially in cases where pollinators affect the structure of the flower. Cleistogamous flowers often have a different number of floral parts, such as reduction of petals, nectaries, stamens, pollen sac numbers per anther, carpels and ovules

(Endress, 1994). This heteromorphism must be taken into account in producing floral diagrams.

Dimorphic species have two morphologically distinct kinds of flowers, which are different from heteromorphism or intrinsic variation found within a species. The most important cause for dimorphism is unisexuality, which can be expressed in different degrees (Mitchell and Diggle, 2005). Heterostyly is a first step in the transition to unisexual flowers. Mitchell and Diggle (2005) distinguished between two types of unisexual flowers. Type I represents unisexual flowers with rudiments of the other sex. In this case, unisexuality is a late-developmental phenomenon, in cases where anthers or ovules fail to develop or reach maturity (e.g. Restionaceae, Sapindaceae). Very often, staminodes or carpellodes remain present in the mature flower. In other cases, sterility occurs much earlier in the floral development. Type II represents flowers that are unisexual from inception. The earlier abortion occurs in the development through a process called heterochrony (see p. 48), the greater the difference between staminate and pistillate morphs (e.g. Euphorbiaceae, Caricaceae). Unisexual flowers can be derived in different ways from bisexual flowers, by organ abortion, loss of organs or homeosis. The degree of reduction of one or the other gender in flowers affects the outcome of floral diagrams to a great extent. In some cases the reduced organ takes over different functions, such as the pistillodial nectary of *Carica* (Ronse De Craene and Smets, 1999a) or *Buxus* (Von Balthazar and Endress, 2002b), or carpels are replaced by organs of the other sex through homeosis (such as the three inner stamens in *Jatropha* of Euphorbiaceae).

In wind-pollinated flowers there is a whole range of adaptations including presence of unisexuality (see p. 5). The loss of the other gender is accompanied by an absence or reduction of the perianth, and various associations with bracts. This leads to highly simplified (reduced) flowers, often limited to the androecium and carpels, with an important dimorphism between the genders (e.g. Betulaceae, Casuarinaceae).

In this book two floral diagrams are presented where differences between genders are more pronounced (type II flowers). If unisexual flowers are only slightly different, this will be mentioned in the text.

2.7   Floral development and floral diagrams

Meristems and developmental constraints

Floral development (as well as vegetative development) is determined by a growing apical meristem. The molecular basis for floral development is being increasingly studied and different models have been proposed (see p. 48). Genetic control of the floral apex influences the nature of organs as well as their position in the flower. Phyllotaxis of flowers is influenced by the size of organ primordia and an apical inhibition zone, which determines the position of subsequent organs. The result is a clear pattern of organ distribution. For floral development two independent processes regulated by two different kinds of genes are fundamental in guiding the position of primordia in the flower and the differentiation of primordia into organs. Different developmental programmes that are seemingly contradictory can be superposed in flowers. For example, in Malvaceae there are common stamen-petal primordia reflecting a reduction of petals; at the same time there is centrifugal multiplication of the stamens (Ronse De Craene and Smets, 1995b).

An important concept in floral morphology is Hofmeister's rule. The rule formulated by Wilhelm Hofmeister in 1868 insists that a subsequent primordium arises as far as possible from a primordium already formed on the apex (cf. Kirchoff, 2000). In flowers a whorl should always alternate with the previous one. While this rule holds for the majority of flowering plants, there are several exceptions. The main reason is that organs invariably arise where there is sufficient space for their initiation, and this can be variable during the initiation of the flower because of various influences such as shape of the apex and the position of previously formed organs (Kirchoff, 2000). In regular core eudicot flowers, whorls are usually alternate, and the antepetalous whorl is most prone to be lost. However, obhaplostemonous flowers are not that rare. They are usually linked with retarded petals, often arising from common stamen-petal primordia. Stamen and petal function as a single unit in that case, adapting to Hofmeister's rule. Interestingly, carpel number and position can have an influence on the number of stamens. Trimerous carpels may affect the number of stamens by reducing space for their initiation (e.g. *Hibbertia, Hypericum, Mollugo*). Alternatively, carpel proliferations (e.g. *Tupidanthus*, Araliaceae) are linked to an increase of stamen numbers, even as carpels are initiated after the stamens. This means that spatial pattern formation of carpels is initiated before the onset of stamen primordia.

The developmental process will strongly determine the outcome of flower structure and arrangement of organs in mature flowers. Taxonomic groups that are largely monosymmetric will have a much earlier onset of monosymmetry, already starting during the development of the perianth. In groups where monosymmetry is a less current phenomenon, organs arise as in symmetric flowers and the onset of zygomorphy happens shortly before maturity (see Tucker, 1997).

Position of organs

Floral organs develop in a specific sequence and position characteristic for a species, genus or family. With five sepals in core eudicots, sepals one and three are in an abaxial position and sepal two is in an adaxial position. This position is mainly regulated by transversally placed bracteoles, forcing the first floral organs into a median position. Loss of bracteoles may lead to displacement of the first-formed sepals in a lateral position (see p. 24). In tetramerous flowers, sepals are generally positioned in median and transversal positions, and petals in a diagonal position.

In most pentamerous core eudicots, the orientation of one petal is abaxial, two are lateral and two are adaxial. There are few exceptions that are not the result of a resupination of the flower enumerated by Eichler (1878), of which the Leguminosae is the most important representative. The position of the petals corresponds more or less with the position of the alternating sepals (two anterior, two lateral and one posterior). This abaxial–adaxial duality is very stable and is usually maintained whenever flowers have divergent petal sizes in monosymmetric or bilabiate flowers. The two adaxial petals may converge and fuse into one unit (this can be the origin of tetramerous flowers in some cases), while the anterior petal can become much larger or become associated with the laterals in a large unit (lower lip). In Faboideae, the two anterior petals become associated as the keel. Orientation of the perianth is variable in monocots and depends on the position of the bracteole (Stevens, 2001 onwards). Most monocots have their median outer tepals in abaxial position and the inner median in adaxial position. Loss of the bracteole can lead to a switch of 180° (e.g. *Dioscorea, Allium, Agave, Lilium*). In alismatids, a similar situation is found in *Acorus*, Juncaginaceae, Potamogetonaceae and *Tofieldia*, where the switch in position can be mediated by uncertain homology of bracts and tepals (Buzgo, 2001; Remizowa and Sokoloff, 2003). In *Metanarthecium* the bracteole is transversal, with the first tepal in transversal position (Remizowa, Sokoloff and Kondo, 2008).

Stamens and carpels usually have a stable position. This has been discussed extensively for the androecium, where loss of whorls can result in highly different configurations (e.g. Ronse De Craene and Smets, 1987, 1993, 1995b, 1998a). Position of carpels in eudicots is regulated by available space and number of stamen whorls. In cases where carpels are isomerous with the other whorls, they can either be in antepetalous or antesepalous position, and this may occasionally fluctuate in a family (e.g. Malvaceae, Rosaceae, Rutaceae). When fewer than five, the carpels can have specific orientations. With three carpels, two are often adaxial and the odd abaxial in the core eudicots, rarely the opposite (Ronse De Craene and Smets, 1998b), while it is the reverse in most monocots. With two carpels, position is mostly median, more rarely transversal. An oblique position is rare and is found in Solanaceae (Eichler, 1875), and occasionally in Capparaceae (Ronse De Craene and Smets, 1997a).

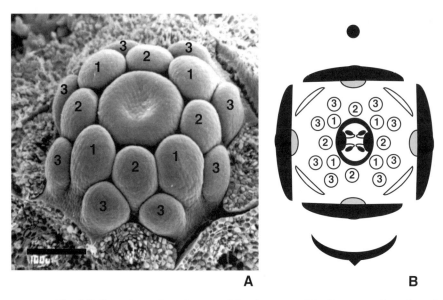

Fig. 2.2. *Capparis yco.* A, early stage of development of the flower; sepals and petals removed (from Ronse De Craene and Smets, 1997a – modified, with permission of Botanische Jahrbücher für Systematik, www.schweizerbart.de). B, corresponding floral diagram; numbers represent the sequence of initiation of the stamen whorls.

The principle of variable proportions

The principle of variable proportions was formulated by Troll (1956) and implies that the shape of structures can be modulated along a gradation; an organ can progressively be transformed into a different structure by a change of size and differential growth processes (Ronse De Craene and Smets, 1991e; Ronse De Craene, De Laet and Smets, 1996). Flower development is largely guided by spatio-physical factors, such as an inverse relationship between the size of organs and the apical meristem. The larger the apical meristem the more organs can be produced, if they are smaller in size. The same applies for secondary primordia on primary primordia (see p. 11). Complex androecia can differentiate into a relatively large number of stamens (Fig. 2.2), or much higher numbers if the size of individual stamens is decreased. In Annonaceae the number of stamens is indirectly related to the size of stamen primordia and the extent of development of the floral apex (e.g. *Annona*: Leins and Erbar, 1996; Ronse De Craene, Soltis and Soltis, 2003).

Reduction and loss of organs

Flowers as we observe them are the result of a long evolutionary process, in which changes have progressively taken place. In many cases with reduced numbers of organs in a whorl, it is assumed that some organs were lost during evolution. Guédès (1979) refers to this condition as 'aborted' versus 'unborn', while Tucker (1988b, 1997) refers to 'suppressed' versus 'lost' organs. Sometimes this assumption can be verified by a study of floral development showing that a primordium is initiated, but is subsequently arrested in growth and has completely vanished at maturity (suppression). Such an organ can be described as an ephemeral or 'phantom organ' that appears during a short period of development. Heterochrony describes the process by which the onset of initiation is shifted in time. As a result, an organ can become progressively aborted at an earlier stage until it fails to initiate (e.g. Li and Johnston, 2000; Mitchell and Diggle, 2005). Various examples are found in Angiosperms, such as stamens in Leguminosae (*Bauhinia, Saraca*: Tucker, 1988a, 1997, 2000c), staminodes in pistillate Restionaceae (Ronse De Craene, Linder and Smets, 2002), petals in Melianthaceae (*Melianthus*: Ronse De Craene *et al.*, 2001) and Polygalaceae (*Polygala*: Prenner, 2004b) and stamens and petals in Brassicaceae (*Lepidium*: Bowman and Smyth, 1998). Organ abortion can occur at different stages of development. Loss of organs implies that certain organs were present in ancestral flowers but that they were lost in the descendant lineages. A clear example is the orchids with one–three functional stamens and one to two staminodes, derived from six stamens in two whorls (Rudall and Bateman, 2003; Stützel, 2006).

2.8   Evolutionary developmental genetics and floral diagrams

The approach to morphology has dramatically changed by increased emphasis on the genetic basis of organ and flower development. The discovery that several genes are responsible for the flower expression and the development of floral organs has opened new areas in research that have strongly interacted with a classical approach to morphology. Flower development is the result of a cascade of expressions of different genes, transforming the inflorescence meristem into flowers and consequently leading to expression of different floral organs (Glover, 2007). Floral diagrams are ultimately an illustration of the morphological expression of genes responsible for flower induction and differentiation. At every level of flower differentiation, different genes are active in controlling developmental events.

The ABC model of flower development was elaborated in the early 1990s (Coen and Meyerowitz, 1991). The identity of organs in each whorl is determined by a combination of three classes of identity genes: A genes are responsible for sepal expression, A+B function determines petal identity, B+C function specifies stamens and C function determines carpel development. The model has been more recently expanded with D and E genes with specific roles in the development of flowers; D genes are responsible for ovule and placenta development, while E genes tend to control the development of all organs except the sepals (e.g. Theissen *et al.*, 2002; Glover, 2007). A and C genes are mutually antagonistic in their expression. The genetic model is based on the study of several mutants of core eudicot model genera *Arabidopsis* and *Antirrhinum*, two highly evolved and derived organisms. This evidence appears to be much more fluid in other groups of plants and cannot be applied to monocots or basal angiosperms in the same way as for core eudicots, with an easy transition of gene expressions across the organ boundaries and the evolution of numerous paralogous substitutions in the latter groups (e.g. Kramer, Di Stilio and Schlüter, 2003; Kim *et al.*, 2005; Ronse De Craene, 2007).

The definition of whorls is different from morphology, as describing a domain in which a single type of organ is produced in one or more concentric circles (Leyser and Day, 2003; Glover, 2007). Leins and Erbar (2007) rightly criticized the unfortunate use of 'organ whorls' in evo-devo, where it is better to refer to 'organ categories'. The androecium or perianth can consist of more than one whorl and this is not accounted for. 'Asymmetry' has also been used in the sense of 'monosymmetry', clashing with the morphological definition of asymmetry (Endress, 1999).

The evolution of gene expression in the angiosperms is highly complex and can be applied to understand major evolutionary processes in flowers. Genes have undergone several duplications in their history, as is clear for the B function genes *APETALA3* (*AP3*) and *PISTILATA* (*PI*) (Kramer *et al.*, 2006, reviewed in Ronse De Craene, 2007). The origin of flowers has also been tested through an understanding of gene expressions (Theissen *et al.*, 2002).

There are interesting parallelisms in the evolution of floral structures in the angiosperms and their gene counterparts. Basal angiosperms with spiral flowers and an undifferentiated perianth have a broader expression pattern of A and B genes than the core eudicots (e.g. Kim *et al.*, 2005). In the core eudicots there is a stricter compartmentalization of gene expression due to greater synorganization. This can occasionally break down and affect the development of the whole flower. Gene expression studies have also been applied to monocots to understand the homology of the perianth, especially Poaceae to understand the nature of lemma, palea and lodicules, but in most cases the answers are not straightforward or are more complex than initially expected (Ronse De Craene, 2007).

Changes in the floral Bauplan are also regulated by genes. It is generally recognized that genes such as *CYCLOIDEA* (*CYC*) and *DICHOTOMA* (*DICH*) are responsible for the change in symmetry from regular to bilateral flowers (e.g. Coen *et al.*, 1995; Citerne, Moeller and Cronk, 2000; Leyser and Day, 2003; Citerne, Pennington and Cronk, 2006; Busch and Zachgo, 2007). The expression of *CYC* leads to the abortion of the adaxial stamen while *DICH* affects the adaxial petals. Changes in expression patterns can lead to dramatic shifts in flower morphology, such as secondary polysymmetry in *Cadia purpurea* of Leguminosae (Citerne, Pennington and Cronk, 2006) or abortion of lateral stamens in *Mohavia* of Plantaginaceae (Hileman, Kramer and Baum, 2003). The pattern of genetic mutation is thought to cause a reversal to polysymmetric flowers in some asterids. Donoghue, Ree and Baum (1998) suggested that a return to polysymmetry could be either triggered by loss-of-function *CYC* mutants or by structural fusion of the posterior petals linked with a loss of stamens. The former possibility, based on cases of *Antirrhinum*, implies that the adaxial side of the flower develops like the abaxial side and leads to mutants with six petals (Luo *et al.*, 1996). This evolution may be responsible for regular four- to six-merous flowers of *Sibthorpia* (Scrophulariaceae) or *Ramonda* (Gesneriaceae).

Although evo-devo can give clear indications of the nature of floral organs, an understanding of structural morphology remains essential. Pollinator coevolution is probably the principal trigger to changes in symmetry, and this is not clarified by evo-devo studies. The fact that B genes are sometimes expressed in sepals is not an indication of a petal nature, but of petaloidy (Ronse De Craene, 2007). Gene expression can be extremely complex and future genetic studies will need to be closely linked with any morphological study. Floral diagrams are a way to convey morphological information that can be inspirational for further evo-devo research.

3

Floral diagrams used in this book

Table 3.1 summarizes the symbols for the floral diagrams used in this book. When constructing floral diagrams it is essential to orientate the flowers correctly (Fig. 3.1). Arrangement of individual organs and their orientation can only be accurately compared if a common reference point is being used. The main axis relative to the flower is shown by a black dot. The main axis of the inflorescence is shown by a crossed circle. Flowers are conventionally depicted with the axis on top and the main subtending bract below along the median line. Bracteoles are placed more or less in transversal position depending on their orientation relative to the main axis. To differentiate bracts and bracteoles from the perianth, a small triangle is placed on the abaxial side of the bract(eole). A bract or bracteole that is lost or early deciduous is shown in white with a broken outline. A large straight arrow represents the main direction of monosymmetry, when present. If flowers are resupinate, this is shown by a curved arrow.

The distinction between sepals, petals and tepals is based on the presence versus absence of petaloidy, as the differentiation between sepals and petals is sometimes unclear (see p. 6). White curves represent pigmented (petaloid) organs, without distinction between sepals, petals or tepals, as these represent homologous organs in most cases (cf. Ronse De Craene, 2007, 2008). Absence of petaloidy is shown by black arcs, which is mostly present in the green sepals but occasionally in the petals (e.g. *Callistemon citrinus*: Fig. 10.13A). Stamens are represented as a cross-section through the anther, with occasional details reflecting variation of size and shape. In case many stamens are formed in a flower, stamens are represented by open circles, occasionally with the sequence of initiation shown by numbers. Staminodes are represented by a circle with a black dot, or occasionally by a blackened anther (representing a sterile anther). Staminodial petals are shown by an arc with a black central dot. The ovary is

Table 3.1

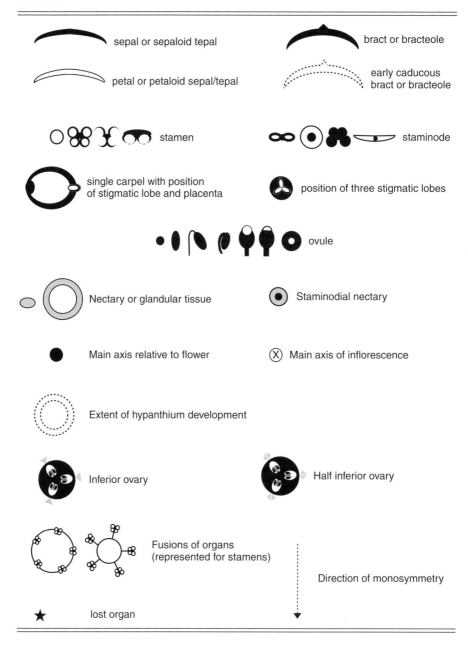

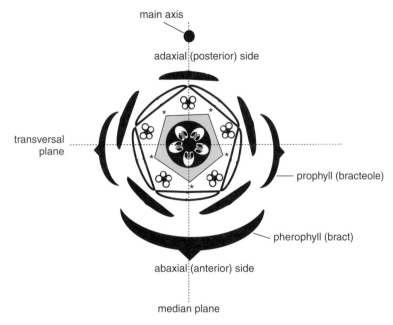

Fig. 3.1. Representation of an idealized floral diagram.

represented as a cross-section at the level of attachment of the ovules. When possible, the position of the styles is shown, as these may be opposite the carpels (carinal) or alternate (commissural). Where the ovary is inferior or half-inferior this is shown by grey triangles inserted on the periphery of the ovary. In order to convey extra information, two drawings are occasionally combined, showing a diagram at the level of the style and a diagram of the inferior ovary separately. Carpellodes are depicted by blackened or empty carpels. Secretory tissue is represented in grey, wherever it occurs in the flower. Fusions between organs are shown by full connecting lines. The presence of a cup-like hypanthium is shown by two broken lines. Specific appendages of organs, such as conspicuous hairs or lobes, are shown on the diagram when they are conspicuous. In cases where there is abundant evidence to conclude that organs were present but were lost through abortion, their position is shown by an asterisk.

In most cases the floral diagrams were prepared based on living or pickled material. In cases where no flowers were available, several photographs, drawings and literature descriptions were used that were as representative as possible.

PART II   FLORAL DIAGRAMS IN THE MAJOR CLADES

4

Systematic significance of floral diagrams

4.1 Floral diagrams and molecular phylogeny

4.1.1 *Molecular and morphological characters*

Since the early 1990s molecular systematics has dramatically changed the approach to studying relations of plants and led to major changes in the classification of plant groups. Premolecular classifications such as those of Cronquist (1981), Thorne (1992), Takhtajan (1997) and to a lesser extent Dahlgren (1975, 1983), were mostly intuitive, with specifically selected characters considered to be more important than others. Especially for flowers, certain characters considered as important were shown to be mere convergences by the molecular phylogenies. For example, families sharing three carpels with parietal placentation grouped in an order Violales or Parietales *sensu* Engler were shown to belong to three different lineages. Other comparable earlier associations of families include the 'Contortae' (based on contorted petal aestivation), 'Sympetalae' (taxa with stamen-petal tubes) or 'Rhoeadales' (Papaveraceae and Brassicaceae) and were largely used in the book of Eichler (1875, 1878) and the Englerian systems.

Molecular phylogenies have proposed several shifts in relationships, some of them predicted by other characters (e.g. the link between Salicaceae and former Flacourtiaceae), others unexpected (e.g. the circumscription of Proteales) or apparently questionable because of superficial morphological similarities (e.g. the separation of Oxalidaceae and Geraniaceae), and a few controversial (e.g. Anisophyllaceae in Cucurbitales). Two major milestones, the Angiosperm Phylogeny Group or APG I (1998) and APG II (2003), have increasingly brought stability to the system with most major groupings firmly supported by a wide range of genes. A third publication (APG III) is due to appear at the time of the publication of this book. More recently, a number of studies have improved understanding of phylogenetic relationships on a large scale and increased the predictive value of inferred relationships of major plant groups (e.g. Moore *et al.*, 1997; Chase *et al.*, 2005; Jansen *et al.*, 2007; Brockington *et al.*, 2009; Wang *et al.*, 2009).

Nevertheless, morphology of flowers remains essential, even more important than before in understanding affinities of plant families. Morphological characters belong to the most versatile category of phylogenetically useful information. Compared with molecular characters, morphology has far fewer characters, but most characters are phylogenetically informative. Morphology informs on phylogenetic relationships but also on function, and through function it is a reflection of natural selection (Bateman, Hilton and Rudall, 2006). A major drawback of floral morphology is the occurrence of convergences towards similar adaptations to specific pollinators, or a secondary simplification due to changes towards wind pollination. The evolution of floral characters is shaped by ecology (flower–pollinator interactions) but also by developmental constraints and potential. Endress (2006) pointed to the importance of key-innovations (e.g. sympetaly, spurs, syncarpy), leading to bursts of diversification in flowers. The interaction with pollinators or their absence will have major influence in shaping flowers. Floral adaptations range from mechanisms to avoid selfing (dichogamy, herkogamy, dicliny) to specific adaptations to pollination (wind, water or animal).

However, molecular data are equally riddled with problems and constraints (reviewed in Bateman, Hilton and Rudall, 2006). The study of morphological characters becomes increasingly important in the framework of a stable molecular phylogeny. Although morphological characters tend to improve the support for phylogenetic trees when used in combination with molecular characters, a combination of morphological characters also has great predictive value, giving indications about principal characteristics of taxa. Morphological characters are increasingly mapped on phylogenetic trees, as an indication of evolution (e.g. Doyle and Endress, 2000; Ronse De Craene, Soltis and Soltis, 2003; Zanis *et al.*, 2003; Soltis *et al.*, 2005; Doyle, 2008; Endress and Doyle, 2009). As discussed in Ronse De Craene (2008), this is not without problems because of the uncertainty about the presence of characters in ancestral groups.

The changes in the relationships of families proposed by molecular classifications have sometimes put the value of morphological characters in doubt, but they have also opened new challenges and emphasized the value of previously unrecognized characters. As flowers remain a prime tool in the identification of plants next to vegetative parts, the understanding of floral structures is essential, and this can be achieved through the use of floral diagrams. In this book

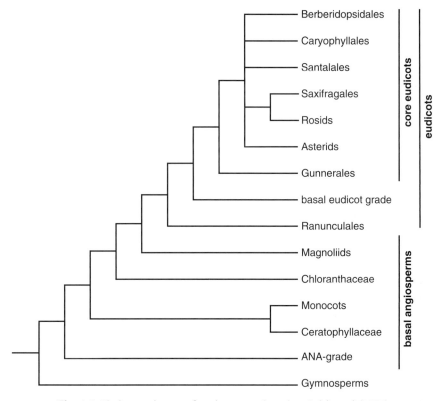

Fig. 4.1. Phylogenetic tree of angiosperms, based on Soltis *et al.* (2005).

efforts are made to demonstrate how floral diagrams can help in understanding relationships between or within families, but also to clarify the evolutionary shifts that have shaped floral structures. Floral diagrams will be discussed within the system of APG II (2003) and some more recent publications (e.g. Soltis *et al.*, 2005). At the same time the existence of common characters or cryptic apomorphies for major groups will be demonstrated based on floral diagrams.

4.1.2   The Angiosperm Phylogeny Group classification

Figure 4.1 shows a phylogenetic tree with the major clades of angiosperms. Angiosperms are the sister group of the gymnosperms. There was much recent discussion about the nearest gymnosperm group, with contrary opinions based on molecular characters and fossil evidence. As it is not within the scope of this book to discuss the origin of angiosperms, the reader is referred to recent updates in Soltis *et al.* (2005), Bateman, Hilton and Rudall (2006) and Doyle (2008).

Although the relationships within major clades are not always fully supported, the foundations of the molecular classification are sound and are not expected to change radically in the future. The angiosperms consist of two major clades, the basal angiosperms, including monocots, and the eudicots. Although a main morphological character is the presence of pollen with a single aperture in the former (monosulcate) and pollen with three apertures in the latter (tricolpate), differences of flowers are not clear-cut, with a gradual transition between basal groups and core eudicots through an intermediate basal eudicot grade. Monocots are nested within basal angiosperms or are sister to eudicots, rejecting the previous concept of a clear distinction between dicotyledonous versus monocotyledonous Angiosperms. The core eudicots, representing 75% of the angiosperms (Soltis *et al.*, 2003) consist of the well-supported larger clades caryophyllids, rosids and asterids, but relationships among the major lineages are still unresolved.

Family delimitation

For a number of large families (e.g. Scrophulariaceae, Malvaceae, Flacourtiaceae, Euphorbiaceae, Alliaceae) delimitation has changed considerably compared with premolecular classifications in an attempt to reflect phylogenetic relationships. These changes were implemented rather inconsistently, with some families being heavily split up (e.g. Scrophulariaceae) and others grouped into megafamilies (e.g. Malvaceae). This discrepancy reflects the inherent tendencies of botanists to be splitters or lumpers, but also reflects a distinction between theorists, who want a phylogenetic classification at all costs, or pragmatists, who favour a system that is phylogenetically sound and at the same time practical.

Some families can be broadly circumscribed and are represented by a clear set of vegetative and floral morphological characters, and this is reflected in their taxonomic circumscription (e.g. Malvaceae: Alverson *et al.*, 1998, 1999; Bayer *et al.*, 1999; Caprifoliaceae: Donoghue, Bell and Winkworth, 2003), while the split-up of some larger families (e.g. Scrophulariaceae: Olmstead *et al.*, 2001) has increased the difficulty in identifying subunits with distinctive morphological characters. The primuloid clade is a good example of the challenge to accommodate practicality with phylogenetic accuracy. The paraphyletic Primulaceae can be split into several smaller clades. Källersjö, Bergqvist and Anderberg (2000) proposed the recognition of four families, Maesaceae, Theophrastaceae, Myrsinaceae and Primulaceae, although there are indications that these groupings lack obvious homogenous characters with transitional genera (e.g. *Samolus*, *Ardisiandra*, *Coris*) that can arguably be recognized as separate families. I agree with the statement of Källersjö, Bergqvist and Anderberg (2000: 1339) that 'The dilemma of phylogenetic analyses including molecular data is that although the results may be robust, and hence our best estimate of the evolutionary relationships, they may indicate a close relationship between taxa which may be very different morphologically.' However, I believe practical reasons need to be a major consideration in building classifications. A concept of a larger superfamily Primulaceae with a number of subfamilies makes more sense, as all taxa are characterized by a syndrome of common characters, including sympetaly, obhaplostemony and a free-central placentation. Another example of the difficulty in delimiting families is the relationship between Brassicaceae and Capparaceae. Capparaceae appears to be paraphyletic on molecular evidence, with subfamily Cleomoideae sister to Brassicaceae. Placing all Capparaceae in an expanded Brassicaceae would be phylogenetically correct but would obscure the many clear morphological characters. Therefore, Hall, Sytsma and Iltis (2002) recommended recognition of three well-supported monophyletic families, Capparaceae, Cleomaceae and Brassicaceae.

Several larger orders are phylogenetically well supported but remain unpractical, with almost no sound morphological synapomorphies (e.g. Asparagales, Ericales, Brassicales, Malpighiales, Polygonales). There is a need to break up these large entities into workable units. This challenge is to be covered in this book by describing smaller clades within the orders, whenever there is a possibility to do so.

4.1.3   *Fossil flowers and floral diagrams*

The last decades have seen a tremendous advance in the discovery of fossil angiosperms, and coupled with a molecular clock it has been possible to identify the age of several angiosperm families (e.g. Dilcher, 2000; Soltis and Soltis, 2004; Friis, Pedersen and Crane, 2006; Friis, Pedersen and Schönenberger, 2006; Crepet, 2008; Wang *et al.*, 2009).

The earliest angiosperms date back to the Early Cretaceous (ca. 130 million years ago). The fossil record is remarkably concordant with floral evolution postulated for extant angiosperms. Earliest flowers tend to be perianthless or have a simple perianth resembling extant basal angiosperms and early diverging eudicots, while there is a broad-scale diversification of flower structures in mid-Cretaceous floras when the major clades become recognizable, including the core eudicot rosids and asterids. Radially symmetrical flowers with calyx and corolla are widespread. It is only in the Tertiary that large groups with floral tubes and monosymmetric flowers became abundant. Wind pollination arose several times and appears to be derived from perianth-bearing ancestors. The fossil diversity is huge and our understanding is increasing rapidly by constant new discoveries and by comparing fossil flowers with extant angiosperms (e.g. Friis, Pedersen and Crane, 2006; Endress, 2008b). The discovery of the early diverging *Archaefructus* and the recognition of several taxa with reduced floral structure (e.g. Ceratophyllaceae, Chloranthaceae, Hydatellaceae) have indicated an early diversification of angiosperms mainly in an aquatic environment (Doyle, 2008; Endress and Doyle, 2009).

Floral diagrams play an important part in representing our understanding of structures from fossil flowers, which have to be reconstructed from compressions or charcoalified remains. Fossil flowers become increasingly incorporated in phylogenetic studies and evidence from these flowers can be neatly summarized in floral diagrams. Several studies have incorporated floral diagrams in their analyses of fossil flowers (e.g. Friis, 1984; Crane, Friis and Pedersen, 1994; Gandolfo, Nixon and Crepet, 1998; Schönenberger and Friis, 2001; Schönenberger *et al.*, 2001; Friis, Pedersen and Crane, 2006). These floral diagrams make comparisons with extant flowers much easier to interpret evolutionary patterns.

4.2   Overview of floral diagrams in the major clades of flowering plants

The overview of families and their floral diagrams cannot be complete. Even Eichler did not cover all families, despite the fact that his overview was more extensive than mine. Attempts were made to include as much diversity as possible as well as to include all major orders recognized by APG II (2003). Missing from the list are some minor orders such as Canellales and Chloranthales (basal angiosperms), Crossosomatales and Zygophyllales (rosids) and Aquifoliales (asterids), as well as some families within larger orders. Chapters 5–11 cover the diversity of floral diagrams in the major clades of Angiosperms.

5

Basal angiosperms: the ascent of flowers

The basal angiosperms are a grade comprising the main part of Magnoliidae *sensu* Cronquist (1981) minus Ranunculales, which belong to the basal eudicots. The monocots are nested within the basal Angiosperms. Figure 5.1 summarizes the phylogenetic relationships of the main families and orders according to Soltis *et al.* (2005). The position of Chloranthaceae and Ceratophyllaceae is debatable according to recent phylogenies, depending on emphasis on molecular or morphological analyses (see e.g. Moore *et al.*, 1997; Jansen *et al.*, 2007; Endress and Doyle, 2009).

5.1   The ANA-grade: Amborellales, Austrobaileyales, Nymphaeales

The ANA-grade (formerly ANITA-grade: Qui *et al.*, 1999) groups the basalmost lineages of the Angiosperms. Among these taxa the monotypic *Amborella trichopoda* has recently been identified on molecular evidence as the basalmost extant angiosperm (Soltis *et al.*, 2005; Jansen *et al.*, 2007). Since then, morphological studies have confirmed the basal nature of these early-diverging lineages (see e.g. Endress, 2001; Ronse De Craene, Soltis and Soltis, 2003; Buzgo, Soltis and Soltis, 2004; Endress and Doyle, 2009). Most taxa of the ANA-grade share spiral (or whorled) flowers with a morphological continuum between bracts and perianth, indefinite number of organs and a central receptacular residue. However, perianth initiation and differentiation is variable in the grade. Endress (2008c) argued that a

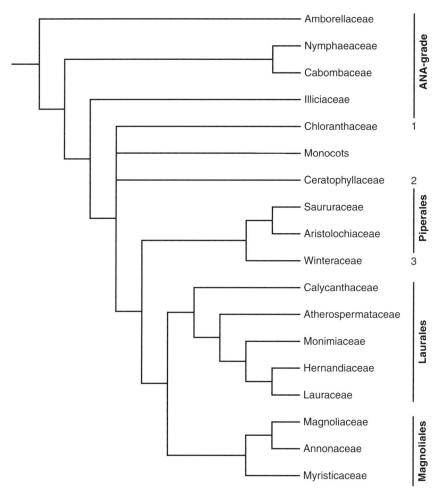

Fig. 5.1. Phylogenetic tree of basal angiosperms, based on Soltis *et al.* (2005).
1. Chloranthales, 2. Ceratophyllales, 3. Canellales.

multistep gradation from bracts to inner tepals may be secondary, as this is absent from *Amborella*. Some characters appear derived and have evolved independently, such as unisexual flowers and adaptations to specific pollinators.

Amborellales

Amborellaceae

Fig. 5.2A,B. *Amborella trichopoda* Baill., based on Endress and Igersheim (2000b) and Buzgo, Soltis and Soltis (2004)

Fig. 5.2. *Amborella trichopoda* (Amborellaceae): A. staminate inflorescence; B. pistillate flower. Numbers give the sequence of initiation of the perianth.

Staminate: ♀P(6)9–11(15) A(8)12–21(22) G0
Pistillate: ♀P7–8 A(0)1°-2° G̲(4)5(6)

Flowers have several characters considered to be ancestral in basal angiosperms: spiral flowers with undifferentiated perianth and no clear transition between bracts and perianth, variable number of floral parts and orthotropous ovules (Ronse De Craene, Soltis and Soltis, 2003). However, some characters appear to be advanced: unisexual flowers (with remnants of the other gender in pistillate flowers) and a floral cup (hypanthium).

The differentiation into vegetative shoot and flower is progressive with terminal flowers arranged in short cymose inflorescences. Two bracteoles are closely connected with the flower and occur at the transition of a decussate to a spiral phyllotaxis in the first two tepals (shown by numbers in Fig. 5.2A,B: Buzgo, Soltis and Soltis, 2004). In pistillate flowers staminodes resemble fertile stamens with sterile anthers. Both staminate and pistillate flowers have a central pyramidal extension probably representing a sterile apex (Endress and Igersheim, 2000b). At anthesis the inner tepals are larger and reflexed.

Nymphaeales

The order contains the three families Cabombaceae, Nymphaeaceae, and Hydatellaceae. The latter was recently moved from an unsettled position in monocots to a basal position as sister to Nymphaeales and this has important consequences for the interpretation of the earliest-diverging flowers (Rudall *et al.*, 2007). Flowers of Hydatellaceae (a single genus *Trithuria* with 12 species) have simple, occasionally unisexual flowers that are highly different from other Nymphaeales and could represent a reduction.

Compared with other early diverging angiosperms, flowers of Nymphaeaceae show important evolutionary patterns, such as a transition to whorled phyllotaxis, differentiation of sepals and petals with outer peri-anth parts larger than inner, development of a hypanthium, presence of nectaries and a secondary increase of floral parts (e.g. Ronse De Craene, Soltis and Soltis, 2003; Endress, 2008c).

Cabombaceae

Fig. 5.3. *Brasenia schreberi* Gmel., based on Richardson (1969) and Ito (1986a)

Fig. 5.3. *Brasenia schreberi* (Cabombaceae), solitary flower.

✳P3+3 A 6+6+3+3+6+6 G̲3+6+3
General formula: ✳P4–8* A3–36 G(1-)2–18
*or K(2-)3(-4) C(2-)3(-4) in *Cabomba*

The trimerous flowers are axillary or extra-axillary and solitary. In *Brasenia* flowers are dull purple and wind-pollinated without differentiation of sepals and petals, while *Cabomba* is insect-pollinated with a differentiation of sepals and petals, and nectar secreted on the petal margins through localized glandular cells (nectarioles: Vogel, 1998a; Endress, 2008c). Petals are retarded in development compared with sepals, which is unusual for Nymphaeales (Endress, 2001; Schneider, Tucker and Williamson, 2003).

Stamens are numerous in *Brasenia* with an arrangement of trimerous whorls of stamens in double and single positions (Fig. 5.3); carpels arise in one to three trimerous whorls (Richardson, 1969; Ronse De Craene and Smets, 1993). Only six stamens are found in *Cabomba* (rarely three), arising simultaneously and alternating with three carpels opposite the petals (Endress, 2001). The six stamens of *Cabomba* correspond to the three outer stamen pairs found in *Brasenia* and the lower stamen number is probably derived.

Carpels are strongly ascidiate and contain one to three ovules in variable position (laminar-diffuse placentation). In *Brasenia* one to two ovules are attached to the dorsal side of the carpel in a single row (Ito, 1986a).

Nymphaeaceae

Fig. 5.4. *Nymphaea alba* L.

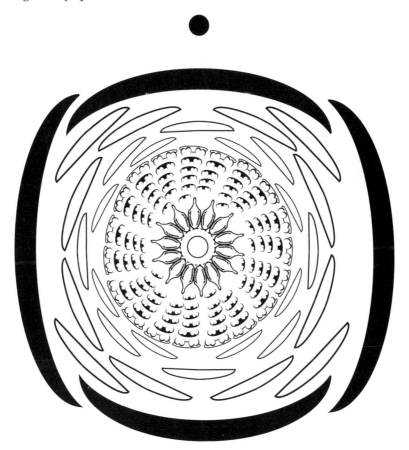

Fig. 5.4. *Nymphaea alba* (Nymphaeaceae). Central receptacular residue shown by circle.

$$\text{*K2+2 C4+8+4+4+8 A}\infty\,\underline{\text{G14–18}}$$
$$\text{General formula: *K4–6(-12) C0–70 A14–200 G3–35}$$

Only *Nuphar* has flowers associated with an abaxial bract. In other genera a bract is lacking (Schneider, Tucker and Williamson, 2003). Endress and Doyle (2009) discussed various possibilities that the bract has been lost or incorporated in the calyx of other genera.

Merism is variable in the family, ranging from di-, tri-, (tetra-?) to pentamery, but the ancestral phyllotaxis is unclear (Endress, 2001). Flowers are large to very large with a well-developed hypanthium in *Victoria* and *Nymphaea*. The depression arises by sinking of the apex during floral development (Schneider, Tucker and Williamson, 2003). Flower initiation is whorled with an easy transition between different merisms. Endress (2001) interpreted the flowers of *Nymphaea* and *Victoria* as tetramerous as a result of the extension of the diameter of a trimerous flower. Ronse De Craene, Soltis and Soltis (2003) suggested that tetramery is equivalent to dimery in Nymphaeaceae, because organs appear in two sequences within a whorl. The whorled phyllotaxis is lost in the inner parts of the androecium, probably by excessive crowding (e.g. *Nuphar*: Endress, 2001). The perianth is weakly differentiated into sepals and petals, although the distinction is unclear. The outer petals alternate with the sepals in all genera and subsequent whorls contain the same or double that number. Petals intergrade progressively with stamens through intermediate (staminodia-like) structures (Hiepko, 1965). Stamens are flattened with elongated pollen sacs. Outer stamens are petaloid and become progressively narrower towards the centre of the flower. The number of stamens is usually high to very high, with closely packed parastichies. Carpels are grouped in one whorl at the base of the hypanthial slope and are laterally and abaxially embedded in the receptacle. The stigmatic slit extends on the adaxial margin of the carpels. In the centre of the flower of *Nymphaea* and *Victoria* there is a small knob-like apical residue. In *Nuphar* the apex remains dome-shaped. Ovules are formed on laminar-diffuse placentae. The genus *Ondinea* lacks petals or their presence is linked with higher stamen numbers (Williamson and Moseley, 1989). Nectaries are present in *Nuphar* only, inserted on inner tepals and are covered with nectar-secreting stomata (Endress, 2008c).

The large flowers of Nymphaeaceae (especially *Victoria*) were interpreted as being secondarily increased in size with the potential for an addition of petals and stamens (Schneider, 1976; Schneider, Tucker and Williamson, 2003).

Austrobaileyales

The small order contains three to four families (Austrobaileyaceae, Illiciaceae with Schisandraceae, Trimeniaceae).

All have spiral flowers with undifferentiated perianth and variable organ number in common (Endress, 2001). Bracts, outer and inner tepals overlap gradually on a spiral, but outer tepals are smaller than inner ones in bud, as in *Amborella* (Endress, 2008c). Stamens and carpels are numerous, except for Trimeniaceae with a single carpel. The difference between Illiciaceae and Schisandraceae is the unisexual flowers of the latter (e.g. *Kadsura*, *Schisandra*), while *Illicium* has bisexual flowers with carpels arranged in a whorl around a broad central vestigial apex. In all taxa the transition between different organ categories is gradual, especially in Austrobaileyaceae with inner staminodes (Endress, 1980c, 2001).

Bisexual flowers are proterogynous, a common character in the basal angiosperms.

Illiciaceae

Fig. 5.5. *Illicium simmonsii* Maxim.

Fig. 5.5. *Illicium simmonsii* (Illiciaceae). Central receptacular residue shown by circle.

⟳P14–19 A13–19 G12–13
General formula: ⟳P7–33 A4–50 G5–21 (based on Endress, 2001)

Subterminal flowers are borne on short shoots enclosed by decussately arranged bracts. Transition between outer, inner tepals and stamens is progressive, but staminodes are lacking. Carpels are plicate with ventral-median ovule. They appear to be inserted in a whorl but arise in helical sequence (Endress, 2001).

In the related Schisandraceae staminate and pistillate flowers lack any residue of the other gender and the flower appears spiral throughout with a progressive transition from bracts to tepals (Endress, 2001: Figs. 5–6). However, intermediate bisexual flowers were reported in some taxa (Endress, 2001).

5.2   Magnoliales

The order is highly diverse and contains six families. Flowers range from indeterminate spiral flowers with many parts (e.g. Himantandraceae, Degeneriaceae, Eupomatiaceae) to trimerous whorled flowers (at least in the perianth: Magnoliaceae, Annonaceae) to reduced whorled flowers (e.g. Myristicaceae).

Eupomatiaceae have a single bract forming a circular calyptra and no perianth (Endress, 2003b). Corresponding calyptras are found in Magnoliaceae and Himantandraceae, but are less elaborate. The presence of a hypanthium with proterogynous flowers in some families and internal staminodes is similar to Calycanthaceae (Laurales), but inner staminodes are also found in other Magnoliales where they play a role in attraction. Nectaries are absent, except for inner tepals of some Annonaceae (Endress, 1994).

Myristicaceae

Fig. 5.6A,B. *Myristica fragrans* Houtt., based on Armstrong and Tucker (1986)

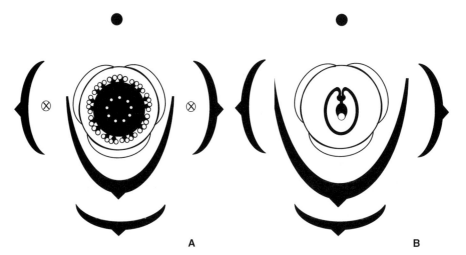

Fig. 5.6. *Myristica fragrans* (Myristicaceae): A. staminate flower; B. pistillate flower. White dots represent the vascular bundles of the stamens.

Staminate flowers: ✱P(3) A3+3+3(+3)* G0
* Occasionally with an additional whorl
Pistillate flowers: ✱P(3) A0 G̲1
General formula:staminate: ✱P(2)3(5) A2–40 G0; pistillate (P(2)3(5) G1

Myristicaceae differ from other Magnoliales by unisexual flowers with a single trimerous (rarely two-parted) perianth. The perianth is often partially to completely connate. Pistillate flowers have a single carpel with one basal anatropous ovule and are homogeneous throughout the family. Staminate flowers show much variation within the family. All stamens are fused into a synandrium of 2–60 anthers (Armstrong and Wilson, 1978; Sauquet, 2003). Anthers are extrorse and sessile, as a crown on a column, sometimes resembling an inverted cone (*Knema*), or they are occasionally stalked (*Maloutchia*). Each pair of bisporangiate lobes and a connecting vascular bundle are equivalent to a stamen (Armstrong and Wilson, 1978; Fig. 5.6A). Some Myristicaceae have three anthers only. Because of the phylogenetically derived position of the genus, Sauquet (2003) argued that the androecium of *Maloutchia* represents a reversal to an ancestral stamen phyllotaxis with several secondary increases of anthers, and that filaments have become secondarily free. However, due to the complexity of the fused androecium, there is little indication of the ancestral condition of Myristicaceae. In *Myristica* staminate and pistillate flowers arise terminally on short shoots and are enclosed by a bract and a single abaxial bracteole (Fig. 5.6A,B). Armstrong and Tucker (1986) demonstrated that the staminal column in *Myristica* is derived from the receptacle and not from fused filaments, as previously suggested. They also showed that stamens arise sequentially in trimerous girdles, starting with three stamens alternating with the tepals and forming 7–11 (or more) stamens depending on the size of the apex. Anthers develop four elongated pollen sacs with transversal septations. The general appearance of staminate flowers of several Myristicaceae is very similar to staminate *Nepenthes* or Canellaceae and represents a remarkable convergence.

Magnoliaceae

Fig. 5.7. *Magnolia paenetalauma* Dandy, based on Xu and Rudall (2006)

✱K3 C3+3 A∞ G̲∞
General formula: ✱K3 C6–21 A(3–6)-∞ G(2–4)-∞

For a long time Magnoliaceae were considered as the prototype for the ancestral flower in the angiosperms (e.g. Eames, 1961). Nowadays the flower is

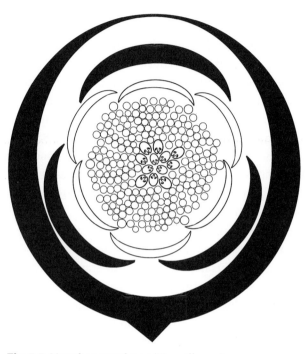

Fig. 5.7. *Magnolia paenetalauma* (Magnoliaceae).

considered as relatively specialized (Nooteboom, 1993; Endress, 1994). The reason for the status of primitiveness is the often conically developed receptacle with many spirally inserted stamens and carpels.

Flowers of the seven genera of Magnoliaceae are enclosed by one or more circular bracts, which are shed at anthesis (Nooteboom, 1993). The perianth is usually trimerous and differentiated in an outer whorl of three tepals (sepals) and an inner part of two trimerous (petal) whorls, rarely in only two trimerous whorls (e.g. *Michelia figo*: Hiepko, 1965, contrary to Eichler's [1878] opinion). In some species the number of tepals is higher with spiral phyllotaxis (e.g. *Magnolia stellata*, *Michelia champaca*). The tepals have sometimes been interpreted as transformed bracts (e.g. Baillon, 1868b; Eames, 1961), although ontogenetic evidence indicates that they belong to the flower and that they are differentiated into sepals and petals (Xu, 2006; Xu and Rudall, 2006). Erbar and Leins (1981, 1994) interpreted the flower as a transitional stage in the shift from spiral to whorled phyllotaxis. The transition between large perianth parts and androecium is abrupt and leads to a disruption of the spiral phyllotaxis. The initiation of the androecium does not follow a regular spiral, contrary to the claims of Erbar and Leins (1981, 1994), at least for some *Magnolia* species. Outer stamens tend to arise in flushes opposite the outer petal whorl with at least an indication of stamen pairs (Xu, 2006; Xu and Rudall, 2006), or clearly in pairs (see figures of *Liriodendron tulipifera* and *Magnolia denudata* in Erbar and Leins, 1981, 1994). The stamens have no distinct filament (except in *Liriodendron*), but consist of a laminate organ with embedded anther tissue. Staminodes are absent, although outer stamens can be morphologically transitional with the perianth (e.g. *M. stellata*: Hiepko, 1965). The gynoecium is apocarpous and each carpel produces two to several ovules on laminar-diffuse placentae. Carpels often have a petiole-like stipe (gynophore). Interestingly, the floral apex grows continuously during development, in a way similar to racemose inflorescences, providing space for more carpels (Xu and Rudall, 2006). Some genera have fewer than ten carpels (reduced two to four in pairs in *Michelia montana*), while numbers can reach up to 90 carpels. The number of floral parts in Magnoliaceae is fluid and the higher numbers could be secondarily derived by an extension of the receptacle.

Annonaceae

Fig. 5.8A. *Artabotrys hexapetalus* (L.f.) Bhandari, based on Leins and Erbar (1996)

✶K3 C3+3 A6+∞ G̲14–20

Fig. 5.8B. *Monanthotaxis whytei* (Stapf) Verdc., based on Ronse De Craene and Smets (1990a)

✶K3 C3+3 A6°+3°+6°+6+3 G9+9+9
General formula: ✶K(2)3(4) C3+3(-12) or C3+0 A(3–6)∞ with outer or inner staminodes G1,2–10,∞

Annonaceae are the most diverse family of Magnoliales. Flowers are generally grouped in cymes, or they are solitary. A hypanthium is present in *Xylopia* (Steinecke, 1993). Flowers are mostly trimerous with three perianth whorls: three outer sepaloid or petaloid tepals (sepals) and two inner whorls of petaloid tepals (petals). The inner tepal whorl may be reduced and apparently absent (e.g. *Annona*), or the calyx may be reduced to a rim (e.g. *Fenerivia*). More plasticity is found in *Ambavia* with supernumerary petals and variations between dimery and trimery (Deroin and Le Thomas, 1989). Petals are imbricate or valvate and are occasionally fused. Inner petals may be erect and form a closed nuptial chamber linked to beetle pollination. The androecium consists either of three or six stamens, or multiples of three (12–24) (e.g. *Orophea*, *Bocagea*, *Monanthotaxis*:

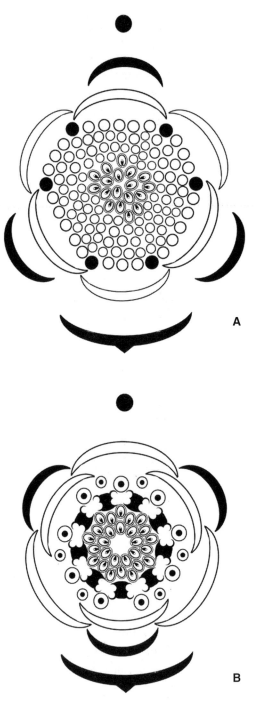

Fig. 5.8. Annonaceae: A. *Artabotrys hexapetalus*; B. *Monanthotaxis whytei*. Black dots in A represent outer stamen pairs.

Baillon, 1868c; Ronse De Craene and Smets, 1990a; Steinecke, 1993). The six outer stamens (or staminodes) tend to be inserted as three pairs in alternation with the petals (Fig. 5.8A,B). More often the androecium is highly polymerous (up to 2000 stamens in *Annona montana*) with a more or less whorled initiation, becoming spiral higher up on the receptacle (e.g. Leins and Erbar, 1996). In *Fenerivia*, there are six supernumerary petals corresponding in position and vasculature with outer stamen pairs (Deroin, 2007). In *Toussaintia* there are also supernumerary petals, although their relationship with stamens is less clear (Deroin, 2000). Outer stamens always alternate with petals as in flowers with few stamens. More stamens appear in flushes, not in a regular spiral, and form parastichies. Leins and Erbar argued that the number of stamens has been increased at the expense of their size.

Staminodes can be either inside or outside (Fig. 5.8B) fertile stamens. Anthers are usually extrorse, rarely latrorse and the connective is broadly developed, resembling a shield in some taxa. Pollen is released when stamen bases become detached from the receptacle. Carpels are usually numerous (rarely a single carpel) and apocarpous, but appear pseudosyncarpous in some genera by development of a basal collar of the receptacle connecting the carpels (Deroin, 1997). *Monodora* and *Isolona* are unilocular with parietal placentation. The origin of this gynoecium is controversial and was either seen as the result of a subdivision of a single carpel (Leins and Erbar, 1996), or the fusion of several carpels at their margin (Deroin, 1985, 1997). The latter interpretation is supported by the phylogeny, as *Monodora* and *Isolona* are nested within a clade with many apocarpous carpels. However, early developmental stages show the development of a single carpel primordium comparable to other apocarpous gynoecia (Leins and Erbar, 1996).

Diagrams of *Popowia* and *Bocagea* were shown by Baillon (1868c).

5.3   Laurales

The order consists of six families in two clades, sister to a basal Calycanthaceae (Renner, 1999).

Flowers are extremely variable with a floral phyllotaxis ranging from spiral to whorled. Endress and Doyle (2007) assumed that spiral phyllotaxis found in Magnoliales and Laurales is derived, because both groups are nested deep within clades that are whorled. While phyllotactic patterns can jump easily between whorls and spirals in basal angiosperms, it makes little sense to postulate a reversal in Magnoliales and Laurales, especially because families with spiral flowers are more basal in these orders, which makes repeated reversals unlikely. Endress and Doyle (2007) also considered chaotic phyllotaxis to be derived from whorled phyllotaxis by doubling of the position of organs leading to a loss of a regular phyllotaxis. Other factors probably lead to chaotic initiation patterns, such as the distortion of the floral apex (e.g. *Drimys* in Winteraceae), as paired stamens are part of a cyclization event that appeared several times within the basal angiosperms

Merism is variable, and mainly trimerous only in Lauraceae. There is a tendency for building flowers with several superposed stamen whorls (e.g. Monimiaceae, Lauraceae). Most Laurales have stamens accompanied by a pair of nectariferous glands on the filament. These were taken for staminodes in the past although there is no evidence for this (Endress, 1980b; Ronse De Craene and Smets, 2001a; Buzgo *et al.*, 2007). Staminodes occur frequently, either outside the stamens or as an inner whorl, and they are usually secretory. Dehiscence of anthers is mostly valvate by flaps and there are several patterns leading to the reduction of ventral or dorsal pollen sacs (Endress and Hufford, 1989). Monimiaceae is the most variable family, with evolutionary patterns ranging from spiral bisexual flowers with well-developed tepals to highly specialized unisexual and dimerous flowers (Endress, 1980b). Each carpel contains a single ovule in ventral median position, rarely two basal ovules (Endress and Igersheim, 1997). A hypanthium is characteristic for all families (least developed in Lauraceae) and is linked with half-inferior to inferior ovaries. The hypanthium can take huge proportions in some Monimiaceae, enclosing the carpels as a pseudo-fruit and forming a hyperstigma.

Calycanthaceae (including Idiospermaceae)

Fig. 5.9. *Calycanthus floridus* L., partly based on Staedler, Weston and Endress (2007)

$\circlearrowleft$P(26-)28(-29) A 13(-16) A°(16-)20(-22) G̲(25-)34(-35)
General formula: $\circlearrowleft$P15–40 A5–30 A°∞ G(1–3)∞

Organ numbers are much variable. In Fig. 5.9 the tepal initiation sequence is numbered (1–26), 13 stamens in one series and 16 staminodes in two series of eight.

Phyllotaxis of the four genera of Calycanthaceae was studied by Staedler, Weston and Endress (2007). They found a regular transition from the decussate vegetative phyllotaxis to the spiral flower in all genera. The different organ categories are bridged by intermediate organs, and within a group of organs there are distinct Fibonacci series of eight, 13 or 21 organs. This arrangement in distinct series can be seen as a preliminary phase in the transtion to whorled phyllotaxis found in other Laurales. Anthers are extrorse. Inner stamens are

Fig. 5.9. *Calycanthus floridus* (Calycanthaceae). Numbers refer to the initiation sequence of the perianth. Note the presence of food-bodies on inner tepals and stamens.

reduced into staminodes that play a role in closing the floral bud. Feeding tissue is present on staminodes, stamens and inner tepals of *Calycanthus* and *Sinocalycanthus*. Carpels occupy the bottom and sides of a deep floral hypanthium and are variable in number. Long filiform stylar appendages reach the top of the depression between the packed staminodes. Each carpel contains two ovules, but the upper one degenerates (Endress and Igersheim, 1997).

Tepals of *Chimonanthus* secrete nectar through scattered glandular cells, while *Calycanthus* lacks nectaries but offers food-bodies on the inner staminodes (Vogel, 1998a).

Lauraceae

Fig. 5.10. *Persea americana* Mill., based on Buzgo *et al.* (2007).

✳P3+3 A3+3+3+3° G1
General formula: ✳P4–6 A3–30 G1

Fig. 5.10. *Persea americana* (Lauraceae): partial inflorescence.

The perianth is made up of two inconspicuous, greenish-yellow to white tepal whorls held together in bud by trichomes. In a few species one or two outer stamen whorls are tepaloid in shape (e.g. *Eusideroxylon*), or the tepals are transformed into stamens (*Litsea*, *Lindera*: Rohwer, 1993a). The number of stamen whorls can be variable in the family, with a distinction between outer and inner whorls. Reductions to three stamens occur in a few species (*Silvaea*, *Endiandra*: Eichler, 1878). Up to five whorls can be found in *Umbellularia*, or the number of stamens can be higher (up to 32), probably in several whorls (Rohwer, 1993a). All stamens bear paired lateral appendages in only a few genera. In general only the third whorl bears two lateral nectary appendages and tends to have a different anther orientation from the outer whorls (Fig. 5.10). The inner whorl is reduced to small stub-like staminodes that can be secretory. Buzgo *et al.* (2007) interpreted the presence of the staminodes as the result of overlapping developmental signals between androecium and carpel. Anthers are either tetrasporangiate, opening with four valves, or disporangiate through the reduction of two pollen sacs (Ecklund, 2000). The single carpel bears one anatropous, pendent ovule. The orientation of the carpel is variable; Buzgo *et al.* (2007) showed it transversal on their diagram, while in Eichler (1878) and Rohwer (1993a) it is median descending.

Laurus among some other genera is unusual in having dimerous, unisexual flowers. These arise as a continuation of a subdecussate phyllotaxis found in several Lauraceae. The outer stamens alternate with the tepals (Eichler, 1878; Ronse De Craene and Smets, 1993).

Lauraceae appear early in the fossil record and are well documented, with a high variation of floral forms leading to extant taxa (e.g. Ecklund, 2000).

Hernandiaceae

Fig. 5.11A–C. *Hernandia nymphaeifolia* (C.Presl) Kubitzki, based on Endress and Lorence (2004)

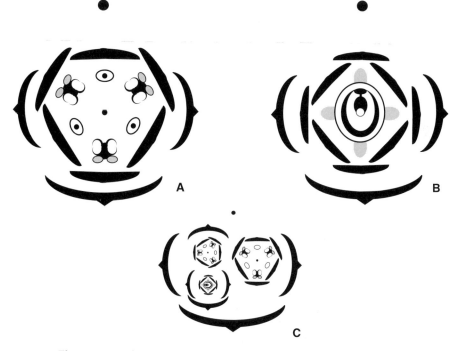

Fig. 5.11. Staminate (A) and pistillate (B) flower and (C) partial inflorescence of *Hernandia nymphaeifolia* (Hernandiaceae).

Staminate: ✳P3+3 A3+3° G0

Pistillate: ✳P4+4 A4° Ğ1

General formula: ✳P4–12 A3–5(-7) G1

The family is close to Lauraceae and shares similar flowers with undifferentiated perianth, anthers with flaps and basal nectaries, and a monocarpellate gynoecium. However, merism is more variable, ranging from tri- to tetra-, penta- or heptamery. Flowers are often unisexual and the ovary is inferior. The heptamerous perianth of *Gyrocarpus* is probably an amalgamation of two flowers with different merisms. Kubitzki (1969) provided several floral diagrams in his monograph of Hernandiaceae. In some species of *Hernandia* there is a tendency for the bracteoles to fuse into a cupular structure enclosing the fruit (Kubitzki, 1993). In *Hernandia nymphaeifolia* flowers are grouped in triads derived from a monochasial cyme. Staminate flowers are trimerous and pistillate flowers tetramerous (derived dimery). Four nectaries are present in pistillate flowers and are probably staminodial in origin, corresponding to three appendages in staminate flowers (Endress and Lorence, 2004).

Atherospermataceae

Fig. 5.12A–C. *Laurelia novae-zelandiae* A. Cunn., based on Sampson (1969)

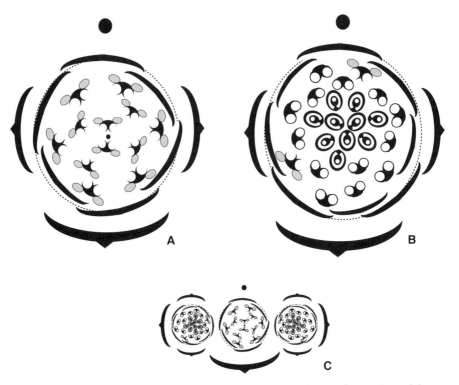

Fig. 5.12. *Laurelia novae-zelandii.* A. Staminate and B. bisexual flower; C. partial inflorescence.

Staminate: ∪P4–9A7–13 G0
Pistillate and bisexual: ∪P5–9 A0–6*A°13–24 G8–12
*only in bisexual flowers

Atherospermataceae used to be part of a traditionally circumscribed Monimiaceae (e.g. Philipson, 1993), which appears to be polyphyletic (Renner, 1999). Most inflorescences are dichasial cymes, often with three flowers or a single terminal flower. Flowers are bisexual or mostly unisexual, with frequent transitions of flowers.

The perianth is undifferentiated and bract-like and the transition between bracteoles and tepals is often unclear. In *Laurelia* the two outer tepals continue the decussate arrangement of the bracteoles (Sampson, 1969), while in *Atherosperma* there is a differentiation of two outer tepals enclosing eight 'petals'. In bisexual and pistillate flowers the stamens or non-functional stamens are accompanied by smaller inner staminodes (*Laurelia*: Sampson, 1969). Staminate flowers have no carpels. All stamens bear glandular appendages. The number of microsporangia is often reduced to two with valvate dehiscence.

Monimiaceae

Fig. 5.13A–C. *Ephippiandra myrtoidea* Decne, based on Lorence (1985) and Philipson (1993)

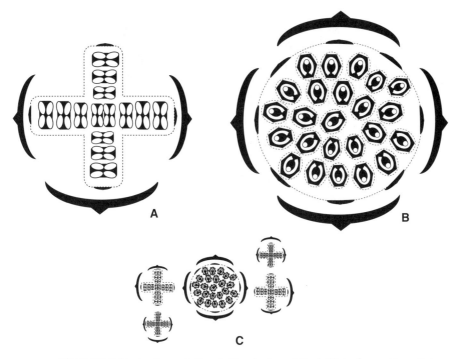

Fig. 5.13. *Ephippiandra myrtoidea.* A. Staminate and B. pistillate flower; C, partial inflorescence.

Staminate: ✱P2+2 A2+2+2+2+2+2+2+2 G0
Pistillate: ⟲P2+2+2+2 A0 G̲ 25–30
General formula: ✱/⟲P(0)2-∞ A9-∞ G̲ ∞

Flowers are bisexual to unisexual, often arranged in monoecious or dioecious dichasia. In *Ephippiandra* pistillate flowers are terminal and surrounded by axillary staminate flowers (Lorence, 1985; Fig. 5.13C). Flowers are spiral

(*Hortonia*, *Peumus*) or frequently dimerous, with decussate arrangement (*Tambourissa*, *Kibara*, *Ephippiandra*, etc.). The hypanthium is usually well developed and petaloid, flattened or deeply urceolate with apical pore, and often takes over the function of the strongly reduced perianth. In *Peumus* the hypanthium forms a flattened platform exposing the spirally arranged tepals and stamens. Paired nectaries are often lacking at the base of the filaments (present in *Peumus*).

Highly specialized flowers occur in *Tambourissa*, *Ephippiandra* or *Wilkiea*. In staminate flowers the number of stamens ranges from nine to 1800; carpel numbers range from a single carpel (e.g. *Xymalos*) and extend to 2000 in *Tambourissa* (Endress, 1980b; Endress and Igersheim, 1997). In some *Wilkiea* the number of stamens is reduced to four to six. Tiny tepals are arranged as several (generally five) decussate pairs around the apical pore or within the urceolate hypanthium. Dehiscence of the flower is by unequal fissuring of the cup forming a pseudoperianth (Endress, 1980b; Philipson, 1993). In *Epphippiandra* the fissuring is more regular, exposing decussate rows of anthers (Fig. 5.13A). In pistillate flowers the hypanthium is either open and cupulate (e.g. *Ephippiandra*: Fig. 5.13B) or urceolate. In the latter case the inner tepals are secretory and contribute to the formation of mucilage and a hyperstigma (a secretory pseudostigma or external compitum). While in most other genera with a floral cup the styles protrude out of the apical pore (also in Siparunaceae), the pistillate flowers have the carpels enclosed in the hypanthium, with the hyperstigma formed on the margin of the floral cup (Endress and Igersheim, 1997).

Tambourissa represents an extremely sophisticated evolution, with a correlation between an increased floral size and increased stamen and carpel numbers. The perianth is reduced or lost and its protective and attractive role is taken over by the hypanthium. In other genera there is a generalized pattern to a reduction of the size and number of carpels and the reduction of the size of the entrance to the floral cup, linked with complete enclosure of carpels, stamens and tepals (Endress, 1980b).

A floral diagram of *Hortonia* was presented in Endress (1980a).

The distinction between Monimiaceae, Siparunaceae and Atherospermataceae as suggested by the molecular phylogeny appears less convincing on a morphological basis, with several recurrent characters that represent floral 'tendencies' of the order.

5.4   Piperales

The order Piperales consists of five families, Aristolochiaceae, Hydnoraceae, Lactoridaceae, Piperaceae and Saururaceae. The relationship of Saururaceae with Piperaceae is very strong on a morphological and molecular

basis. Both families share perianthless flowers arranged in spicate inflorescences, and unidirectional floral development (e.g. Tucker, Douglas and
Liang, 1993; Jaramillo and Manos, 2001).

All families share a trimerous flower with simple perianth (except *Saruma*) or
the perianth is absent.

Aristolochiaceae

Fig. 5.14. *Asarum caudatum* Lindl., based on Leins and Erbar (1985)

Fig. 5.14. *Asarum caudatum* (Aristolochiaceae). Numbers give order of initiation of
stamens and staminodes.

✶K3 C3 [A3+3+6 Ğ3+3]
General formula: ✶/↓K3–4 C0–3 A5–40 G4–6

Aristolochiaceae is a highly specialized family of basal angiosperms, with
several evolutionary novelties in the flower culminating in the genus *Aristolochia*
(gynostemium, monosymmetry, tubular perianth, inferior ovary and syncarpy). In *Asarum* and *Saruma*, the flowers arise at the end of shoots with
distichous leaf arrangement and are enclosed by a bract that appears to arise

on the adaxial side (Tucker and Douglas, 1996: Fig. 7.2), although the flower could be terminal. *Saruma henryi* was considered to be the basal taxon, with well-developed petals. In other genera petals tend to become reduced or abort entirely. *Asarum* has either small petal appendages (e.g. *A. caudatum*) or these are mostly absent (e.g. *A. europaeum*), while *Aristolochia* or *Thottea* have no petals at all (Leins and Erbar, 1985; Leins, Erbar and van Heel, 1988; Kelly, 2001). The number of stamens ranges from six to 46. The most common arrangement of the androecium is whorled, with six stamens arising opposite the sepals before the additional initiation of six more stamens in two whorls in *Asarum* and *Saruma* (e.g. Leins and Erbar, 1985). The six stamens of *Aristolochia* correspond with the six first-formed stamens in other Aristolochiaceae. Some *Aristolochia* have a reduction to five stamens and five carpels (González and Stevenson, 2000b). *Thottea* is much more variable, with several stamen whorls of three, six, nine or 12 members, and inner staminodial structures in some species. Carpel number is also variable, ranging from three, six or nine, and there is often no correspondence between the number of carpels and stigmatic lobes. Carpels arise in two trimerous whorls in *Saruma* and *Asarum* when six in number (Leins and Erbar, 1985, 1995). It is possible that a secondary increase of stamen and carpel numbers has affected some species of *Thottea*, linked with an expansion of the floral meristem. Leins, Erbar and van Heel (1988: 369) attributed the increase of stamens in a whorl to 'a sudden phylogenetic change'. Stamens and carpels can occasionally proliferate up to 12 in subgenus *Pararistolochia* of *Aristolochia* (González and Stevenson, 2000b). Carpels are generally inferior, but can be superior in subgenus *Heterotropa* of *Asarum* (Kelly, 2001). The ovary contains two rows of ovules per carpel, with little to variable fusion of the carpels in *Saruma* and *Asarum*, and strong fusion in *Aristolochia*.

Aristolochia is the most derived genus, with a stable floral formula of K3 A6 G6. The perianth is highly diverse in structure, linked to a specific trap mechanism. Earlier theories, interpreting the perianth of *Aristolochia* as a single bract-like structure (such as a spathe in Araceae), were rejected by González and Stevenson (2000a), who showed that the perianth has a trimerous calyx-like nature. The development shows an early onset of monosymmetry in the development of the perianth linked to its curvature and the formation of a tube. Anthers are extrorse in Aristolochiaceae and the connectives often become fused with the upper part of the carpels in a gynostemium. In *Aristolochia* each gynostemium lobe is formed by extension of the commissural region of two carpels (González and Stevenson, 2000b); this corresponds with the alternating position of the stamens with the carpels.

Two possibilities exist for the evolution of the flower in Aristolochiaceae: the perianth is either basically bipartite, as in *Saruma*, and petals were progressively lost in other Aristolochiaceae, or the perianth is simple and trimerous, and petals were secondarily derived.

The phylogeny does not support a basal position of *Saruma* (Kelly and González, 2003). A simple trimerous perianth is also found in other Piperales, such as Hydnoraceae and Lactoridaceae. González and Stevenson (2000a) discussed the difference between sepals and petals in Aristolochiaceae, implying that petals are different structures. Vestigial structures in *Asarum* were also interpreted as petals or staminodes. Kelly (2001) reported the occasional presence of vestigial anther sacs in *Asarum*. This indicates that petals in Aristolochiaceae may be structures derived from staminodes, which is rare in angiosperms (see Ronse De Craene, Soltis and Soltis, 2003; Ronse De Craene, 2008). Further evidence comes from the genetic expression of B-class genes in Aristolochiaceae. Jaramillo and Kramer (2004) found the same gene expression in petals and stamens of *Saruma*, but a different expression in the perianth of *Aristolochia*. The expression of *AP3* and *PI* homologs in *Saruma* is also different from *Arabidopsis*, which may indicate a derivation from stamens.

Aristolochiaceae represent several plesiomorphic traits in the Piperales, despite their strong synorganization. The existence of three stamen pairs, trimerous simple perianth and multiple stamens links the family with other basal angiosperms. Lactoridaceae with trimerous flowers, a simple perianth and two whorls of stamens (Tucker and Douglas, 1996) has the same diagram as *Thottea tomentosa* (see Leins, Erbar and van Heel, 1988) and may represent the ancestral condition.

Saururaceae

Fig. 5.15. *Gymnotheca chinensis* Decne, based on Liang and Tucker (1989)

↔P0 A6 Ğ(4)
General formula: ↔/ ✳/↓P0 A3–6(8) G3–5

Piperaceae and Saururaceae share similar morphologies and are obviously closely related. The perianth is absent and is thought to precede the evolution of both families (Dahlgren, 1983). Flowers are insect-pollinated with pollen as the main reward, although some *Peperomia* secrete nectar from glandular cells on their subtending bract (Vogel, 1998a).

Inflorescences are spicate with each flower subtended by a single bract. In *Houttuynia* and *Gymnotheca* the basal involucral bracts are petaloid and the inflorescence mimics a flower. Flowers in Saururaceae are reduced, disymmetric (in cases where both sides are equally developed) or monosymmetric. Tucker, Douglas and Liang (1993) considered only the second possibility, which is reflected in the development of the flowers. The number of stamens in

Fig. 5.15. *Gymnotheca chinensis* (Saururaceae).

Saururaceae is six, except for *Houttuynia* with three stamens. Based on previous research, Tucker, Douglas and Liang (1993) provided floral diagrams to illustrate the variation of floral initiation among the five genera of Saururaceae and some Piperaceae. In the absence of a perianth, floral development is usually unidirectional, either from the abaxial to the adaxial side (e.g. *Anemopsis*, *Houttuynia*, *Piper*) or the opposite (e.g. *Gymnotheca*, *Zippelia*). Six stamens probably represent the basal condition, but linked with a unidirectional initiation of stamens the lateral stamens can arise on common primordia. *Anemopsis* forms three stamen pairs arising from common primordia. Single stamens occupy the same position in *Houttuynia*, which was interpreted by Tucker (1985) as the result of splitting, although the opposite possibility cannot be excluded (Jaramillo, Manos and Zimmer, 2004). As stamen pairs are common in other basal angiosperms, including Aristolochiaceae, there is some evidence that the ancestral androecium of Piperales was hexamerous. In Piperaceae the stamens have a comparable unidirectional initiation but are further reduced to two in some genera (usually the first-formed latero-anterior stamens) by terminal delation (Lei and Liang, 1998; Jaramillo and Manos, 2001). Jaramillo and Manos (2001) suggested that the loss of stamens in *Piper* is accompanied by closer packaging of flowers and larger anthers per flower.

The number of carpels ranges from four to three and the gynoecium is usually inferior and syncarpous through the development of a hypanthium (except in *Saururus*). Ovules are arranged in two whorls on parietal placentae. In *Peperomia* (Piperaceae) the ovary is reduced to a single median carpel.

Floral development is particularly labile in perianthless Piperales, leading to similar mature morphologies through different developmental pathways (see Tucker, Douglas and Liang, 1993; Jaramillo, Manos and Zimmer, 2004). The variability in stamen development is best considered to be the result of a gradual change in Piperales, not as independent origins as suggested by Jaramillo, Manos and Zimmer (2004). This may be a reflection of the loss of the perianth inducing higher lability in floral development (Endress, 1987, 1994).

6

Monocots: variation on a trimerous Bauplan

Figure 6.1 gives one recent phylogeny of monocots based on Chase *et al.* (2005). There is much uncertainty about the closest sister group of the monocots, which lies within the basal Angiosperms. Options, such as Ceratophyllaceae and Chloranthaceae, were discussed in Soltis *et al.* (2005) but are not resolved. Moore *et al.* (1997) suggested a sister group relationship of monocots and eudicots, including Ceratophyllaceae, but this was questioned by Endress and Doyle (2009), who associated monocots with magnoliids. The trimerous monocot floral formula (P3+3A3+3G3) is found in a number of basal angiosperms, but there is no certaintly of any morphological links and basal monocots have a more variable floral Bauplan.

Monocots consist of three major units: a basal Acorales–Alismatales grade, a lilioid grade and a higher commelinid clade.

6.1 The basal monocots: Acorales and Alismatales

The Alismatales contains 14 families and is the most diverse order of the monocots in organ number and floral morphs. While the formula P3+3A3+3G3 fits well with most other monocots, the number and development of organs is much more variable in the Alismatales. The link of Alismatales with *Acorus* is strong and the separation of an order Acorales is arbitrary, as some molecular studies place *Acorus* within Alismatales (see Buzgo *et al.*, 2006).

A bract tends to be present (sometimes included in the flower: *Acorus*) or is obviously reduced or absent (Araceae, Aponogetonaceae). Buzgo (2001) argued that a flower-subtending bract merged with the abaxial tepal in *Acorus* and some other Alismatales, forming a hybrid organ, contrary to Araceae or Potamogetonaceae where a bract is reduced or lost. The delimitation of bract

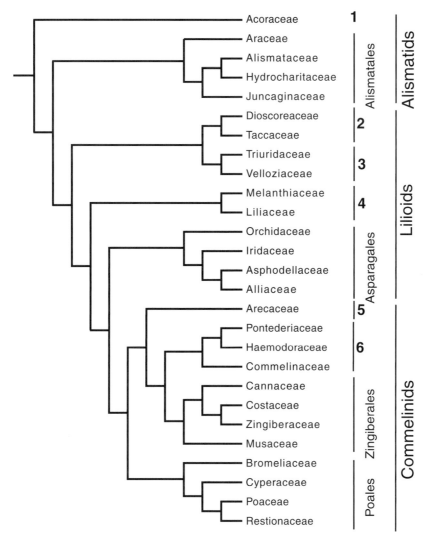

Fig. 6.1. Phylogenetic tree of the monocots, based on Chase *et al.* (2006). 1. Acorales; 2. Dioscoreales; 3. Pandanales; 4. Liliales; 5. Arecales; 6. Commelinales.

and original tepals is unclear in many Alismatales and *Acorus* (Remizowa and Sokoloff, 2003; Buzgo *et al.*, 2006). This corresponds with a reversed position of the tepals; in Araceae and Tofieldiaceae the median adaxial tepal is part of the outer whorl, while in *Acorus*, Juncaginaceae and Aponogetonaceae it is part of the inner whorl. In my opinion this arrangement is linked with the presence or absence of a bract, displacing the tepals in the most favourable position for their initiation. The arrangement of flowers in dense spikes is linked with the loss of the bracts and may lead to the impression of pseudanthial flowers (Buzgo *et al.*,

2006). Bracts and tepals are closely linked organs and bracts may invade the confines of the flower. Similar cases exist in the relationship between tepals and stamens in Juncaginaceae, where tepals were considered to be appendages of stamens because of their strong association (e.g. *Triglochin*: Dahlgren, Clifford and Yeo, 1985; Buzgo *et al.*, 2006).

Flowers are dimerous or trimerous, often arranged in spikes or compact inflorescences. The perianth consists of two whorls, with marked distinction between a sepaloid outer whorl and petaloid inner whorl in Alismataceae. Other families of Alismatales have small flowers with a tendency for reduction of floral parts (tepals are undifferentiated and in two whorls, seldom three to one or even absent: e.g. Araceae, Juncaginaceae, Potamogetonaceae), culminating in the highly adapted and reduced flowers of Zosteraceae and Cymodoceaceae. The presence of nectaries is variable, with occurrence of septal nectaries without settled position (often infralocular when present), or perigonal nectaries. The presence of septal nectaries is closely linked with the postgenital fusion of the carpels, which is most pronounced in the lower part of the ovary (e.g. Tofieldiaceae: Rudall, 2002; Remizowa and Sokoloff, 2003; Remizowa, Sokoloff and Rudall, 2006). It is assumed that apocarpous gynoecia are plesiomorphic in monocots because of significant differences between syncarpous gynoecia of early-diverging monocots (Remizowa, Sokoloff and Rudall, 2006). The gynoecium is variable in the order; it is often made up of numerous apocarpous carpels or two whorls of three (some Alismataceae, Aponogetonaceae), or three to one carpels (e.g. Araceae, Tofieldiaceae, Acoraceae).

Stamen number is variable, ranging from three to very numerous. Paired outer stamens are common in Alismataceae. Ronse De Craene and Smets (1995a) argued that a multistaminate, multiwhorled androecium represents the plesiomorphic condition in the monocots, which is also found in several basal angiosperms. Some taxa have a secondary multiplication linked with a diminution of the size of stamen primordia (e.g. Limnocharitaceae: Sattler and Singh, 1973, 1977), or more commonly a reduction to the two-whorled androecium. The presence of stamen pairs in the outer whorl of the androecium is probably plesiomorphic and is the result of a phylotactic shifting mediated by the trimerous perianth. Although *Acorus* has a two-whorled androecium, the hypothesis of a multistaminate ancestral androecium cannot be excluded.

Alismatales make up four subclades forming a polytomy (Buzgo *et al.*, 2006). The Alismataceae clade (five families) is characterized by relatively large pedicellate flowers subtended by bracts and a differentiation of calyx and corolla. In the clade of Aponogetonaceae there is a progressive reduction of flowers, with undifferentiated perianth or without tepals and a diffuse distinction between flowers and inflorescence, and bracts and tepals.

Acoraceae

Fig. 6.2. *Acorus calamus* L., based on Buzgo and Endress (2000)

Fig. 6.2. *Acorus calamus* (Acoraceae). Asterisk refers to lost abaxial outer tepal.

↑P2–3?+3 A3+3 G̲(3)

The monotypic family Acoraceae has become a major point of interest for systematists since it was placed at the base of monocots by molecular analyses. A morphological comparison with basal angiosperms (especially palaeoherbs) can help in understanding the evolution of the flower in the monocots. *Acorus* has typically trimerous monocot flowers and was interpreted as a prototype for all monocots (see Buzgo and Endress, 2000). However, other candidates for ancestral monocot flowers include members of Alismatales with polymerous flowers (Ronse De Craene and Smets, 1995a). Buzgo and Endress (2000) studied the floral structure and development of *Acorus* and found striking similarities with Piperales.

The flower of *Acorus* is unusual, as the abaxial tepal cannot be distinguished from a flower-subtending bract. In the past the tepal has either be interpreted as bract (e.g. Payer, 1857) or as part of the perianth (e.g. Sattler, 1973). The absence of a bract was the reason why *Acorus* had been included in Araceae in the past (e.g. Eichler, 1875; Dahlgren, Clifford and Yeo, 1985). However, early initiation of the first floral organ is reminiscent of a bract and takes over tepal features later on (Buzgo and Endress, 2000). Three possibilities exist: the abaxial tepal is lost and replaced by the bract that became displaced close to the flower; the organ is a precocious tepal and there is no bract; the organ is a complex hybrid organ of bract and tepal.

The first interpretation is to be favoured (Fig. 6.2), especially since the bract takes a larger size at maturity, covering the flower. The incursion of bracts in flowers is not rare in the Angiosperms and could have led to the loss of the abaxial tepal. A similarity exists with some Saururaceae (e.g. *Saururus, Gymnotheca*), with a showy bract displaced close to the flower (Buzgo and Endress, 2000).

Flowers are sessile and arranged in spikes. The flower of *Acorus* is monosymmetric with unidirectional initiation. The syncarpous ovary contains orthotropous ovules on apical placentae. Septal nectaries are absent, although non-secretory septal slits are present (Buzgo and Endress, 2000).

Hydrocharitaceae

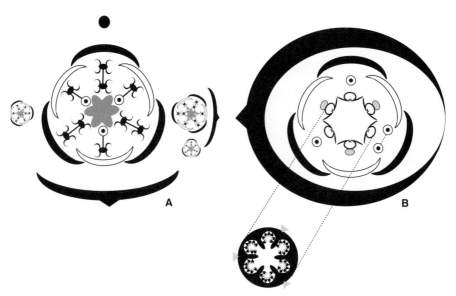

Fig. 6.3 *Hydrocharis morsus-ranae* (Hydrocharitaceae): A. partial staminate inflorescence, B. pistillate flower.

Fig. 6.3A,B. *Hydrocharis morsus-ranae* L., based on Scribailo and Posluszny (1985)

Staminate: ✳K3 C3 A3+3+3+3°(+3° +3°)* G0
*developing as a nectary

Pistillate: $*$K3 C3 A°3* $\breve{G}$3+3
*occasionally as a pair
General formula: $*$K(0)2–3 C0–3 A1–18 G0; $*$K(0)2–3 C0–3 A1°-18° G(2-)3–6(-20)

Contrary to other Alismatales, flowers are epigynous and mostly unisexual. Inflorescences are complex monochasia with well developed bracts enclosing several flowers. Flowers are variable, being either well developed (subf. Hydrocharitoideae) or strongly reduced as an adaptation to water pollination (e.g. Cox, 1998; Leins and Erbar, 2007). The perianth is either differentiated, rarely reduced to three to two, or absent.

Staminate flowers lack all traces of a pistillode, while pistillate flowers often possess staminodes. Pistillate flowers have carpels with carinal styles (Kaul, 1968; Cox, 1998).

The androecium often contains paired stamens (as an outer pair opposite sepals: *Stratiotes, Ottelia*: Eichler, 1875; Kaul, 1968) and inner stamens are often staminodial. Paired staminodes are often found in pistillate flowers while staminodes may be absent in staminate flowers (e.g. *Limnobium*). Staminate flowers of *Hydrocharis* have a central nectary, while pistillate flowers have nectaries as stylar appendages. Although the nectary of staminate flowers was interpreted as a pistillode, Scribailo and Posluszny (1985) argued that it is best interpreted as fused staminodes, based on developmental evidence, and the same applies to the nectaries of pistillate flowers. In *Stratiotes* a whorl of small appendages surrounds fertile stamens in staminate flowers, and carpels in pistillate flowers. These were interpreted as staminodes (Kaul, 1968; Cox, 1998), although Eichler (1875) argued against this in favour of receptacular emergences. In small staminate flowers number of stamens is often reduced to three in antesepalous position (e.g. *Vallisneria*), rarely one (*Maidenia*). In cases where there are only three perianth parts, stamens alternate with these (e.g. *Halophila*). In small pistillate flowers there are usually two or a single whorl of three carpels, mostly with staminodes.

The inferior ovary bears a protruding parietal placentation, with lobes corresponding to the number of carpels. Carpels can be multiwhorled (often as a multiple of three), and they are occasionally apocarpous (e.g. *Stratiotes*) despite being inferior. Kaul (1968) discussed the evolution of the gynoecium in the family as a progressive fusion of lateral carpel margins and their progressive reduction.

The related family Butomaceae has flowers with nine stamens (outer stamens in pairs) and six carpels with marginal placentation. The parietal placentation of Hydrocharitaceae can be seen as the result of a contraction of the placental areas to the carpel wall.

Alismataceae

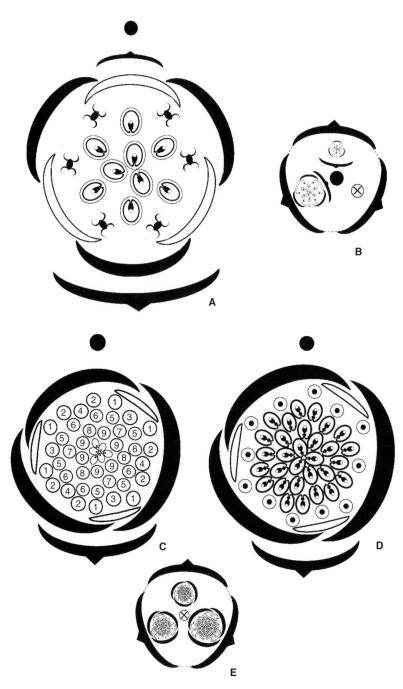

Fig. 6.4. Alismataceae: flower (A) and inflorescence (B) of *Luronium natans*; (C) staminate, (D) pistillate flower and (E) inflorescence of *Sagittaria lancifolia*.

Fig. 6.4A,B. *Luronium natans* based on Charlton (1999a)

∗K3 C3 A6 G̲3+3+3
General formula:∗K3 C3 A6–9-∞ G6-∞

Fig. 6.4C–E. *Sagittaria lancifolia* Willd., based on Singh and Sattler (1973)

Staminate: ∗K3 C3 A6+6+3+3+6+6+3+3+6 G̲°3+3
Pistillate: ∗K3 C3 A°6+3+3 G̲∞

Inflorescences often develop a succession of pseudowhorls interrupted by longer internodes (cf. Singh and Sattler, 1973; Charlton, 1999a, 2004). The pseudowhorls consist of three bracts with flowers (Fig. 6.4B) or lateral shoots (Fig. 6.4D). *Ranalisma* is exceptional in forming pseudo-terminal flowers on a sympodial system (Charlton, 1991). Flowers are generally bisexual to unisexual, and always trimerous. The perianth is differentiated into outer greenish sepals and inner caducous petals. Stamen number is variable, ranging from three (*Wiesneria*, *Caldesia*) to many in several whorls, originating centripetally. Six stamens in three pairs are found in *Alisma*, *Baldellia* and *Luronium*. Carpels are six to numerous, in one to several whorls. There is often a discrepancy between stamen and carpel numbers within a flower, with fewer stamens and higher carpel numbers (e.g. *Ranalisma*, *Alisma*, *Echinodorus*: Sattler and Singh, 1978; Charlton, 1991, 2004). The stamens and carpels are characteristically whorled with the outer stamens arranged as three pairs next to the petals, which are initially small primordia. *Wiesneria triandra* and *Caldesia parnassifolia* are exceptions with single antesepalous stamens (Charlton, 1999b, 2004). Sattler and Singh (1978) interpreted the association of the stamen pair and petal into common primordia as fundamental in the family and as the origin of a secondary stamen increase. Developmental evidence has shown that stamen pairs can be positionally associated with the petals but do not arise on common primordia (cf. Ronse De Craene and Smets, 1993, 1998a). Stamen pairs can form opposite sepals and petals in Alismatales, depending on the size of the latter. The initiation of the androecium in Alismataceae is interpreted as a retained plesiomorphy, with consecutive reductions or increases in complexity (cf. Ronse De Craene and Smets, 1995a). In *Sagittaria*, the numbers of stamen and pistil whorls, as well as the presence of staminodes and pistillodes in staminate and pistillate flowers, are widely variable. Outer stamens opposite the sepals can be paired (Fig. 6.4A) or single (e.g. *S. cuneata*: Singh and Sattler, 1977a), even fluctuating between different sepals (*S. montevidensis*: pers. obs.). Pistillode numbers can be much higher, correlated with their smaller size (e.g. *S. cuneata*). The pattern of development in *Sagittaria* is reminiscent of Nymphaeales, with alternations of single stamens and paired stamens but also of certain

Annonaceae with a secondary stamen and carpel increase. The apocarpous gynoecia bear a single (rarely two) basal ovule.

Echinodorus has perigonal nectaries. In *Sagittaria*, nectariferous cells are centred in cavities around the stamen bases and carpellodes (Smets *et al.*, 2000).

Araceae

Fig. 6.5. *Spathiphyllum patinii* (R. Hogg) N. E. Br.

Fig. 6.5. *Spathiphyllum patinii* (Araceae): A. flower, B. partial inflorescence.

✳P3+3 A3+3 G(3)
General formula: ✳P0/6 A4–6 G2–3(-∞)

Flowers are grouped in condensed inflorescences (a spadix) enclosed in a generally showy bract (spathe). The entire inflorescence functions as a pseudanthium and the inflorescence apex is often enlarged and functions as an osmophore. Flowers are closely packed, forming regular parastichies. Flowers are often unisexual on the same or different spadices. When monoecious, staminate flowers develop distally and pistillate flowers proximally, with an intermediate zone arising on a gradient in between (e.g. *Philodendron*: Barabé and Lacroix, 2000). Tepals are not petaloid and are often reduced or absent in derived groups. Lehmann and Sattler (1992) showed that in *Calla* the perianth was replaced by stamens through a process of homeosis (cf. *Macleaya* in Papaveraceae). Bracts are absent and stamens emerge at different levels and irregular intervals between tepals and gynoecium in *Spathiphyllum*. No distinct style is formed and stigmatic trichomes are formed on a triangular apical slit. Flowers generally lack nectaries. The number of carpels varies between one and several, and monocarpellate gynoecia may be pseudomonomerous. In *Spathiphyllum* three carpels bear two basally inserted collateral ovules embedded in mucilage.

A distinction is made between basal Araceae (proto-Araceae) and other groups. *Gymnostachys* and Orontioideae are both dimerous with extrorse anthers, with transitions from trimery to dimery in *Orontium* (Buzgo, 2001).

Juncaginaceae (incl. Lilaeaceae)

Fig. 6.6. *Triglochin maritima* L., based on Dahlgren, Clifford and Yeo (1985) and Buzgo *et al.* (2006)

Fig. 6.6. *Triglochin maritima* (Juncaginaceae).

✳P3+3 A3+3 G3+3°
General formula:✳P(0)1–6 A1–6 G1–6

Inflorescences are dense spikes and flowers are ebracteate, although limits between bract and first tepal are unclear (Buzgo *et al.*, 2006). Flowers have an extreme variation in perianth and stamen number, ranging from one to six (tri-, di- or monomerous), and tepals are generally small and attached as an appendage to the extrorse stamens. Tepals are occasionally interpreted as an appendicular extension of the stamen, as the inner tepal whorl is distinctly inserted inside the outer stamen whorl (see Dahlgren, Clifford and Yeo, 1985; Endress, 1995a). Carpel numbers range from one to six (in two whorls). When in two whorls, the inner is sterile in *Triglochin* (Fig. 6.6). Carpels are either free or basally fused to a variable extent, with a single basal ovule. No nectaries are present, as flowers are wind-pollinated. An extreme reduction occurs in *Lilaea* with unisexual flowers, a single stamen and carpel, and the perianth reduced to a bractlike organ. Related families such as Scheuchzeriaceae and Aponogetonaceae have six stamens and three carpels, but the perianth is often incomplete, ranging from six to one tepals. Flower development of these three families is highly similar (Singh and Sattler, 1977b; Buzgo *et al.* 2006). In Potamogetonaceae flowers are dimerous, but they have the same arrangement of tepals attached to extrorse stamens. The Scheuchzeriaceae clade shares a wind pollination syndrome (extending to water pollination in derived families). This explains the relative reduction of flowers and the confusion of flowers with inflorescences.

6.2   The lilioids: Asparagales, Dioscoreales, Pandanales, Liliales

Floral synapomorphies are rare as most flowers conform to the generalized monocot floral diagram. In cases where monosymmetry is present, this is rarely structural (except in Orchidaceae).

Asparagales

The order comprises approximately 24 families. Family delimitations are difficult as there are no obvious morphological characters. Floral diversity is highly diverse, with 'traditional' monocot Bauplan or with elaborate derivations. Ovaries are inferior to superior with several independent origins and reversals. Flowers are either polysymmetric to weakly or highly monosymmetric (Orchidaceae: Rudall and Bateman, 2004; *Gilliesia* in Alliaceae: Rudall *et al.*, 2002). Tepals are rarely spotted, and septal nectaries, ovaries with axile placentation and several ovules and single styles are generalized (Rudall, 2002). Stamens are usually in two whorls, rarely with some staminodial (some Alliaceae), the outer whorl staminodial (Themidae) or the inner whorl consistently absent (Iridaceae). A corona is often developed in Amaryllidaceae or Alliaceae.

In all Asparagales the median inner tepal is adaxial and carpels are opposite the outer tepals (Kocyan and Endress, 2001a,b).

'Lower' Asparagales have inferior ovaries, and it is assumed that superior ovaries common in 'higher' Asparagales were derived, although one would expect to find anatomical evidence for this as in Haemodoraceae (see Simpson, 1998a,b).

Orchidaceae

Fig. 6.7. *Oncidium altissimum* (Jacq.) Sw.

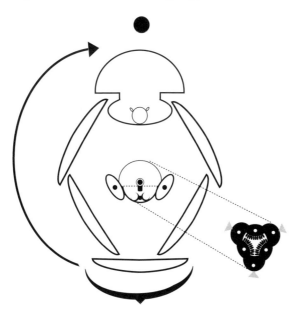

Fig. 6.7. *Oncidium altissimum* (Orchidaceae). Note appendage on labellum. The fertile stamen and stigma are covered by a hood (broken line). Asterisks refer to lost adaxial stamens.

↑P3+2:1 [A1+2° Ğ(3)]

Orchidaceae represents the most successful family of angiosperms, with species estimates ranging between 25 000 and 35 000. Despite this the general floral Bauplan is remarkably conservative. The Orchidaceae are an early diverging clade with strictly monosymmetric flowers, an inferior ovary, lack of septal nectaries, a strong differentiation of tepals and development of an adaxial labellum, and various reductions of the androecium to three, two or a single stamen, with staminodes connected with the style in a monosymmetric gynostemium (Endress, 1994; Kocyan and Endress, 2001a,b; Rudall and Bateman, 2004; Pabón-Mora and González, 2008). Flowers are generally grouped in racemes and subtended by a single bract. Flowers are resupinate at maturity, placing the labellum in abaxial position and stamens and stigma in adaxial position.

The labellum (adaxial tepal of the inner whorl) is mostly elaborate and different from the other tepals, which may be identical, or differentiated in an outer and inner whorl. In *Oncidium* the labellum bears a conspicuous ridged protuberance (callus). In some Orchidaceae, the labellum may revert to a structure identical to the other inner tepals (e.g. *Gymnadenia, Telipogon*: Pabón-Mora and González, 2008). Mandragón-Palomino and Theissen (2008) demonstrated that the labellum became progressively differentiated from other tepals by a distinctive *DEFICIENS* gene duplication in the evolution of the orchids. Basal Orchidaceae (Apostasioideae) have three adaxial fertile stamens (e.g. *Nieuwiedia*), occasionally with the middle one staminodial or missing (*Apostasia*) and flowers are weakly monosymmetric (Kocyan and Endress, 2001b). In other Orchidaceae the lateral stamens are staminodial (Fig. 6.7, subfamilies Orchidoideae, Epidendroideae), or the median stamen is staminodial (subfamily Cypripedioideae) (e.g. Graf, 1975; Dahlgren, Clifford and Yeo, 1985; Endress, 1994; Rudall and Bateman, 2004; Stützel, 2006). Basal Apostasioideae have a more or less axile placentation by postgenital fusion and primordia of abaxial stamens are initiated but soon become repressed (Kocyan and Endress, 2001b). These characteristics are lost in other Orchidaceae. The androecium is intimately fused with the style in a gynostemium, with the anther overtopping the stigma. Three commissural stigmatic lobes can be initiated and the adaxial lobe may develop as a protuberance (rostellum). Anthers have their pollen sacs confluent into two (basally) connected pollinia (Dahlgren, Clifford and Yeo, 1985). Pollinia are either exposed or covered by a hood (Fig. 6.7). The ovary is inferior and strongly ribbed, with parietal (marginal?) placentation and a high number of small ovules.

Contrary to most other Asparagales (except Hypoxidaceae with buzz-pollinated flowers) there are no septal nectaries in Orchidaceae, and Smets *et al.* (2000) linked this to the presence of inferior, unilocular ovaries. Adaptations to various pollinators have evolved in the family, including tepal nectaries, spurs and pollination by deceit (Bernardello, 2007).

Asphodelaceae

Fig. 6.8. *Aloe elgonica* Bullock

↓/✳P(3)+3 A3+3 G̲(3)

Flowers are grouped in terminal racemes or spikes and are subtended by a single bract. *Aloe* has monosymmetric flowers enhanced by curvature of the pedicel and compression of the lateral tepals. In *Aloe* outer tepals are fused up to the middle and variously connate to the free inner tepals, while certain genera have free tepals (*Asphodelus*). Stamens develop sequentially and are curved to the abaxial side. The inner stamens are longer and mature first in the species studied. The trimerous ovary has a simple style, and abundant nectar is produced by septal

Fig. 6.8. *Aloe elgonica* (Asphodelaceae).

nectaries. There are few morphological characters distinguishing the family from other Asparagales (Dahlgren, Clifford and Yeo, 1985).

Iridaceae

Fig. 6.9A. *Gladiolus communis* L. ssp. *byzantinus* (Miller) A. P. Ham.
↓[P3+3A3+0] Ğ(3)
Fig. 6.9B,C. *Sisyrinchium striatum* Sm.
✳ P3+3 A(3+0) Ğ(3)

Flowers are arranged in distichous spikes with flowers subtended by two unequal sheath-like bracts. All Iridaceae share an inferior ovary and loss of the inner stamen whorl. Stamens and carpels are opposite the outer perianth (carpels rarely alternating?). The flower of *Iris* is regular but functions as three floral units through the strong development of the three styles dividing the flower in three separate sections (see Stützel, 2006). The lower part of the flower is tubular (stamen-tepal tube) and stamens are free or fused. The inferior ovary has septal nectaries and three double rows of ovules. In subfamily Iridoideae nectaries are produced at the base of the tepals. The fruit is always a loculicidal capsule.

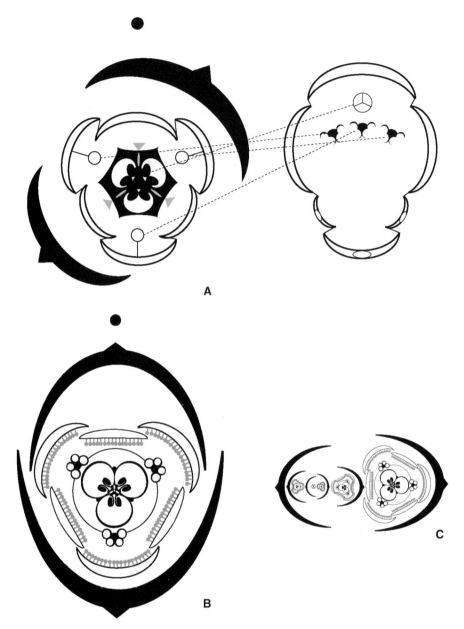

Fig. 6.9. Iridaceae: A. *Gladiolus communis* var. *byzantinus*; left, lower section of flower; right, upper section; B. *Sisyrinchium striatum*, flower; C. inflorescence. In B, note trichome nectaries on tepals.

Zygomorphic flowers of *Gladiolus* have a two-lipped perianth, with the lower lip often with a different pattern. The three stamens and stigma reach out at the adaxial side of the flower. Eichler (1875) wrongly depicted the zygomorphy as transversal in the flower. The basic construction of the flower is median mono-symmetric and is a late developmental event.

Alliaceae

Fig. 6.10. *Tulbaghia fragrans* Verdoorn

Fig. 6.10. *Tulbaghia fragrans* (Alliaceae). Note the corona with appendages surrounding the stamens.

✳ [P3+3 A3+3] G̲(3)

Flowers of Alliaceae and related families (Alliaceae *sensu lato*: Chase *et al.*, 2005) are in umbellate inflorescences enclosed by papery bracts. In *Tulbaghia*, a variable number of smaller bracts are basal to the flowers. Flowers are generally trimerous, with the typical monocot formula and diagram. Tepals are free or connected with the stamens in a hypanthial tube. A corona is occasionally present and associated with the tepal lobes (Fig. 6.10). Contrary to most monocots, styles are solid in the family. Ovules are in two rows. In *Allium* the number is reduced to two collateral ovules and the style is gynobasic. Septal nectaries are mostly present and well developed. Flowers are polysymmetric or weakly monosymmetric, except in *Gilliesia* and *Gethyum* (Rudall *et al.*, 2002), with a strong structural monosymmetry (reduction of the adaxial stamens and inner adaxial tepal).

Dioscoreales

The order contains a heterogenous assemblage of five families, including the mycotrophic families Thismiaceae and Burmanniaceae, which have always been a problem to place due to lack of chloroplast DNA (Merckx *et al.*, 2006).

The perianth is undifferentiated in Nartheciaceae and Dioscoreaceae, but well differentiated into sepals and petals in the other families. The androecium is reduced to three inner stamens in Burmanniaceae and some Thismiaceae, or to three outer in some *Dioscorea* (Caddick, Rudall and Wilkin, 2000). Septal nectaries are generally present (variously in Nartheciaceae, but not in Taccaceae, which lack nectaries). Dioscoreales share a number of similar characters (Caddick, Rudall and Wilkin, 2000): reflexed stamens with a prolonged connective (in genera such as *Stenomeris* and *Trichopus* of Dioscoreaceae connectives are long and lie over the ovary), an umbrella-shaped stigma and urceolate floral chambers formed by growth of a hypanthium. However, these characters are not found in Nartheciaceae (Remizowa, Sokoloff and Kondo, 2008).

Taccaceae

Fig. 6.11A,B. *Tacca palmatifida* Baker

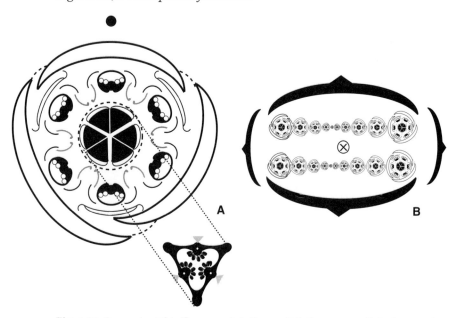

Fig. 6.11. *Tacca palmatifida* (Taccaceae): A. Flower, B. Inflorescence. Note the secretory (?) folds on the receptacle.

∗K3 C3 A3+3 Ǧ(3)

The inflorescence consists of paired cincinni surrounded by two pairs of bracts covering the hanging flowers. Like most Dioscoreales, Taccaceae have the monocot floral diagram. Flowers and inflorescences are well adapted to a specific pollination syndrome of gnats. Sepals and petals are differentiated;

sepals are broad and reflexed; petals are much smaller and erect. Stamens are characteristically coiled with broad connective, providing shelter and enclosing the anthers. Young flowers of *Tacca* show the thecae strongly curved over the filaments, as in Dioscoreaceae and *Saruma* (Aristolochiaceae). This similarity is later lost by curvature of the connective into a hood. The inferior ovary has three intrusive parietal placentae. Trichomes are situated in arcs on a hypanthium, but are reported not to represent perigonal nectaries (Dahlgren, Clifford and Yeo, 1985), and septal nectaries appear to be absent (cf. Caddick, Rudall and Wilkin, 2000 for other species), although their presence is reported in some species. Taccaceae and Thismiaceae appear to be sister groups in the phylogeny of Merckx *et al.* (2006).

Liliales

The order contains about 11 families, with or without a basal Petermaniaceae (Stevens, 2001 onwards). Liliales share a number of synapomorphies, such as absence of septal nectaries, absence of a hypanthium (free tepals) and the generalization of perigonal nectaries (Rudall, 2002). Tepals are often spotted and stamens are often extrorse. Ovaries are variously superior or inferior. Stigmatic lobes tend to be long and separate, in contrast to most Asparagales. The Liliales respect the common floral diagram of monocots without loss of stamen whorls (except *Scoliopus*), although merism can be variable (Melanthiaceae).

Liliaceae

Fig. 6.12A,B. *Tricyrtis puberula* Nakai & Kitag.

∗K3 C3 A3+3 G̲(3)

The perianth in *Tricyrtis* is differentiated in an outer whorl with basal pouches secreting nectar and inner tepals that are erect and have a dorsal crest. Both whorls are covered with spots, which is characteristic for the family. In *Scoliopus* the petals are small and linear and distinctive of the broader sepals, while the petals are broader in *Calochortus*. Most other Liliaceae lack the differentiation of outer and inner perianth parts. The androecium is two-whorled, except in *Scoliopus* where the androecium is reduced to three stamens opposite the sepals. Gynoecia are superior, with generally numerous ovules stacked in two rows on axile placentation. The family was variously circumscribed (e.g. Dahlgren, Clifford and Yeo, 1985; Stevens, 2001 onwards).

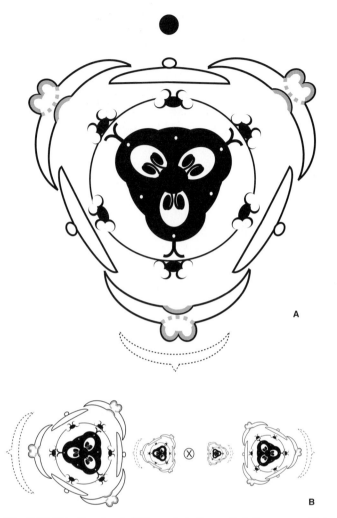

Fig. 6.12. *Tricyrtis puberula* (Liliaceae): A. flower; B. inflorescence. Note the position of branched styles and nectariferous pouches on outer tepals.

Melanthiaceae (incl. Trilliaceae)

Fig. 6.13. *Paris polyphylla* Sm. var. *thibetica* H. Hara

✳K 4–5 C4–5 A4–5+4–5 G(4–5)

The family is an assemblage of smaller units of mainly woodland plants. *Paris* is often placed in Trilliaceae (Dahlgren, Clifford and Yeo, 1985).

The merism of *Paris* is highly variable (up to 12-merous). Our material was either tetra- or pentamerous (Fig. 6.13). As in *Trillium*, flowers of *Paris*

Fig. 6.13. *Paris polyphylla* var. *thibetica* (Melanthiaceae). Note the small rounded petals.

are terminal without bracts and follow the pseudoverticillate leaves. There is a tendency for differentiation of sepals and petals in the family (e.g. *Trillium*) and the sepals are often persistent. Petals are occasionally thread-like (Fig. 6.13) or sometimes absent through suppression during development (*P. tetraphylla*; also in *Trillium apetalon*: Narita and Takahashi, 2008). Takahashi (1994) argued that the loss of petals in *T. apetalon* is caused by homeosis, with replacement of petals by stamens. Takahashi (1994) and Narita and Takahashi (2008) interpreted the petals of *Paris* and *Trillium* as 'andropetals', derived from stamens. This would imply that the androecium was originally three-whorled, although no evidence exists for this in relatives of Liliales. An alternative explanation is that vacant space has become occupied by stamens with a subsequent shift in position (Ronse De Craene, 2003). The distinction between sepals and petals in *Paris* is rather to be considered as derived from an original undifferentiated perianth and a consequence of the retardation in development of the petals (stamen-petal primordia are currently found in *Paris*). Stamens are in two whorls and in *Paris* the connective is apically extended. The gynoecium is superior, with either parietal or axile placentation and free stylodes. In *Trillium* nectar is secreted from the tepal bases. Dahlgren, Clifford and Yeo (1985) mentioned that septal nectaries are reported in *Trillium*, although this is unlikely. No nectaries were seen in *Paris*.

Smilacaceae

Fig. 6.14A,B. *Smilax aspera* L.

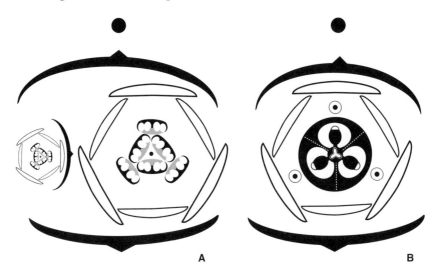

Fig. 6.14. *Smilax aspera* (Smilacaceae): A. staminate partial inflorescence; B. pistillate flower.

Staminate: ✳P3+3 A3+3 G0
Pistillate: ✳P3+3 A3°+0 G̲(3)

Smilax is dioecious, with flowers grouped in clusters along the stem. Staminate flowers tend to be grouped in pairs in the axil of a bract, with one adaxial bracteole each. Pistillate flowers tend to be solitary in the axil of the bract and with fewer flowers per cluster. In staminate flowers the basal section of the filament bears nectariferous trichomes. In pistillate flowers filiform staminodes occur opposite the outer tepals. The three-carpellate ovary bears one apical ovule per locule (rarely two). There is no trace of nectaries. The style is inexistent.

Tepals are sometimes fused into a short to long tube (e.g. *Heterosmilax*). Stamens are in two whorls of three, rarely a single whorl (*Heterosmilax*), or three whorls of three (up to 18: *Pseudosmilax*) (Dahlgren, Clifford and Yeo, 1985). Smilacaceae ressemble Dioscoreaceae in habit and flower structure, except that the ovary is inferior in the latter. Dahlgren, Clifford and Yeo (1985) reported septal nectaries in the family, but this is erroneous.

Pandanales

The order consists of five families (Triuridaceae, Stemonaceae, Cyclanthaceae, Pandanaceae, Velloziaceae) with highly diverging habits and floral morphologies. Triuridaceae are mycoheterotrophic herbs with small, specialized flowers (Maas and Rübsamen, 1986); Cyclanthaceae and Pandanaceae are small trees producing pseudanthial inflorescences (Rudall, 2003); Stemonaceae are herbs with variable merism; and Velloziaceae appears to be the only family with 'classical' monocot characteristics. The relationships of different families are still debatable and there was a highly divergent evolution associated with the exploration of different ecological niches. Stemonaceae are either two- or five-merous (*Pentastemona*). Staminate and pistillate flowers of Pandanaceae and Cyclanthaceae are reduced, with vestigial perianth and variable numbers of stamens or carpels.

Triuridaceae

Fig. 6.15A,B. *Lacandonia schismatica* E. Martínez & Ramos

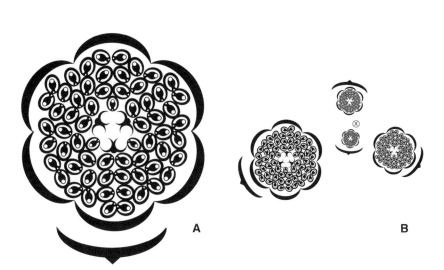

Fig. 6.15. *Lacandonia schismatica* (Triuridaceae): A. flower; B. inflorescence.

✶P3+3 G̲40–80 A3 (2–4)
General formula: ✶P3–10 A(2)3–6 G6–80

Triuridaceae are characterized by small, mostly unisexual, flowers, valvate tepals with pointed tips and many carpels (secondary increase?). Flowers are arranged in terminal racemes. The perianth consists of equal tepals connate at the base in one series; they are generally reflexed, resembling a star (Dahlgren, Clifford and Yeo, 1985). Appendages are frequently formed in the perianth and may function as osmophores (Rudall, 2003). In *Sciaphila* and *Triuris* staminate flowers have six tepals and three to six extrorse stamens on short filaments or fused into a column. *Sciaphila rubra* is dimerous, with hermaphrodite flowers and two stamens (Rudall, 2003). Stamens can be split in half-anthers, embedded in a staminal column (*Triuris hexophtalma*). In *Seychellaria*, three staminodes alternate with the stamens. Pistillate units have the same arrangement of tepals with a generally high number of carpels. Each carpel contains a single basal ovule and gynobasic stylode.

Lacandonia is unique among angiosperms because of its inward-out flowers with three distal stamens and several proximal carpels. Flowers are cleistogamous and are grouped in racemes; they are enclosed by six valvate tepals opening as a star. Because of its singular arrangement the species was placed in its own family Lacandoniaceae. Since its discovery in the late 1980s there was much speculation about the origin and homology of these flowers, interpreted as either the result of a homeotic mutation (e.g. Ronse De Craene, 2003; Vergara-Silva *et al.*, 2003) or a pseudanthium (e.g. Rudall, 2003). The pseudanthial hypothesis was recently questioned by Ambrose *et al.* (2006), mainly on the basis of floral developmental evidence and strong similarities with the development in pistillate *Triuris brevistylis*.

Lacandonia develops two sets of common primordia; the upper one initiates three apical stamens and two rows of carpels in centrifugal sequence, while the lower one develops only carpels. A similar development was shown for pistillate *Triuris*. The staminate flowers develop only three extrorse stamens on an androphore. Some flowers also produce central stamens (Vergara-Silva *et al.*, 2003). This indicates that *Lacandonia* belongs to Triuridaceae and that the upper stamens may not be the result of homeotic swapping of carpels and stamens (Ronse De Craene, 2003) because organs arise centrifugally on common primordia. Vergara-Silva *et al.* (2003) suggested a spatial displacement of expression patterns of B-function genes. Another plausible explanation is a mutational change of C-genes into B-activity on the distal part of the flower. The centrifugal development of several carpels in Triuridaceae is unique in monocots and is comparable to the increase of stamens on complex primordia. Multiplication of uniovulate carpel numbers to increase seed set is an alternative to increase of number of ovules within the ovary. More developmental studies in other genera of Triuridaceae would be helpful in understanding floral evolution in the group.

Velloziaceae

Fig. 6.16. *Xerophyta splendens* (Rendle) N. L. Menezes

✳ [P3+3 A3+3] Ğ(3)
General formula: ✳P3+3 A6-(18) G3

Flowers are showy and trimerous, inserted solitary or with few in terminal position. Tepals are undifferentiated and petaloid, and often bear ventral

Fig. 6.16. *Xerophyta splendens* (Velloziaceae). Note adaxial appendages on tepals and septal nectaries.

appendages (described as a corona) enclosing the short filaments and sometimes adherent to them. The appendages were interpreted as stipules of the stamens (e.g. Eichler, 1878; Arber, 1925; Dahlgren, Clifford and Yeo, 1985), although they are usually connected with the perianth and rarely with the filaments. The inferior ovary has three large septal nectaries ending in holes at the bottom of the flower. There is a short hypanthium connecting stamens and tepals. Although the basic number of stamens is six in two whorls, the stamens are often laterally increased by division, forming triplets and up to 18 stamens (*Vellozia*: de Menezes, 1980). There is variation between species with a hypanthium and fewer stamens (e.g. *Barbacenia, Xerophyta*) and those with higher stamen numbers and without hypanthium (*Vellozia*). Triplets may be free or fused into fascicles; the outer stamen whorl may be simple and the inner developed as triplets (*V. jolyi*). There is also a tendency for two lateral pollen sacs to become confluent by reduction of the middle partitions, leading to two-locular anthers.

6.3   The commelinids: Arecales, Commelinales, Poales, Zingiberales

Arecales

The Arecales consist of a single family Arecaceae, which contains five subfamilies (Stevens, 2001 onwards). The order has a basal position in the commelinids.

Arecaceae

Fig. 6.17A–C. *Ptychosperma gracile* Labill. based on Uhl (1976a,b).

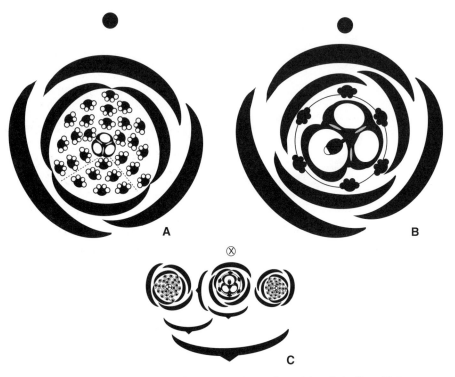

Fig. 6.17. *Ptychosperma gracile* (Arecaceae): staminate (A) and pistillate (B) flower; C. partial inflorescence. In A the broken line delimits groups of stamens opposite a petal; note septal nectary.

Staminate: ✳K3 C3 A3+3+6+3+6+3+3 G°(3)
Pistillate: ✳K3 C3 A°3+3 G̲(1:2°)
General formula: ✳K3 C3 A3–6–9–∞ G1–3(-4)

The basic floral pattern is trimerous with three imbricate sepals and petals, six stamens and three distinct uniovulate carpels (Moore and Uhl, 1982). Flowers are either bisexual or unisexual. Inflorescences range from solitary flowers to monopodial axes (Moore and Uhl, 1982). In the majority of palms, including

Ptychosperma, flowers are arranged in triads of two staminate flowers and an upper pistillate flower (Fig. 6.17C; Uhl, 1976a; Stauffer, Rutishauser and Endress, 2002). Flowers are mostly trimerous, but dimerous and tetramerous flowers occur in species of *Chelyocarpus*. The perianth is imbricate but petals often are valvate in bisexual and staminate flowers. Extreme dimorphism occurs in phytelephantoid palms (e.g. *Palandra*) with minute perianth segments in staminate flowers against very large petals in pistillate flowers (Moore and Uhl, 1982).

The majority of taxa have six stamens in two whorls. However, about 70 genera have more than six stamens, with polyandry superimposed on a basic trimery (Moore, 1973; Uhl and Moore, 1980). Nipoideae have an undifferentiated perianth; staminate flowers have only three outer (fused) stamens and no carpels; pistillate flowers lack staminodes.

Stamen development is either in centripetal (phytelephantoid palms) or centrifugal direction (caryotoid palms). In caryotoid palms variation in stamen number depends on the size of stamen primordia, of the floral sector and floral apex. The antepetalous sector is usually wider with space for more stamens, while the antesepalous sector generally has one row of stamens, or more (Uhl and Moore, 1980). Up to 200 stamens are formed in this way. In phytelephantoid palms flowers are tetramerous, with up to 1000 stamens in *Palandra*.

The gynoecium is tricarpellate, apocarpous to syncarpous; in several genera the ovary is pseudomonomerous (Fig. 6.17B). Carpels are increased to four and up to ten, especially in the phytelephantoid palms. Septal nectaries are variously developed, as palms can occasionally be wind-pollinated, but are mainly insect-pollinated.

In *Ptychosperma* (Fig. 6.17A) staminate flowers have a row of antesepalous stamens and groups of antepetalous stamens, with a pistillode exuding nectar. The pistillate flower has two trimerous whorls of staminodes (occasionally more) that are often fused. The ovary is pseudomonomerous with a single pendent ovule. Septal nectaries are formed above the locule (Uhl, 1976b). Stauffer, Rutishauser and Endress (2002) and Stauffer and Endress (2003) described a comparable pistillate flower for *Geonoma* and other Geonomeae. In *Caryota* no trace of a pistillode is visible (apparently homeotically replaced by a stamen: Uhl and Moore, 1980.

Commelinales

The order as circumscribed by molecular data contains five families and differs greatly from previous classifications (e.g. Dahlgren, Clifford and Yeo, 1985).

Oblique or transversal monosymmetry is widespread in Commelinaceae, Pontederiaceae, Phylidraceae and Haemodoraceae (Stevens, 2001 onwards;

Rudall and Bateman, 2004). Monosymmetry is often expressed in the androecium, as there is a strong tendency for heteranthy or the androecium is reduced to a single stamen in Phylidraceae or some Pontederiaceae.

The order shows much variation in floral structure and adaptations to various pollination syndromes.

Haemodoraceae

Fig. 6.18A. *Anigozanthos flavidus* DC

↙ [P3+3 A3+3] G̲(3)

Fig. 6.18B,C. *Xiphidium coeruleum* Aubl.

✳P3+3 A0+3 G̲(3)

The family is highly variable, with inflorescences ranging from monochasial cymes (a bifurcate cincinnus in *Anigozanthos*) to racemes. Flowers are bisexual and range from nearly polysymmetric to strongly monosymmetric. The perianth is imbricate in bud, but becomes valvate in *Anigozanthos* (Simpson, 1990). In monosymmetric flowers the outer median tepal is always in a posterior position (not so in polysymmetric *Xiphidium*) and tepals are imbricately arranged (Simpson, 1990, 1998a,b). In *Anigozanthos*, flowers have slighty oblique monosymmetry, which is linked with a slit formed on one side of the hypanthial tube connected to the tepals.

Several Haemodoraceae show a reduction to three stamens opposite the inner tepal whorl (e.g. *Xiphidium*, *Haemodora*: Simpson, 1998a,b). In *Pyrrorhiza* the androecium is reduced to a single abaxial stamen with two staminodes, as in *Hydrothryx* of Pontederiaceae. *Shiekia* has two supplementary staminodes opposite the abaxial outer tepals. In some genera with three stamens, including *Xiphidium*, the adaxial anther is larger (Simpson, 1990). The gynoecium of three carpels has axile placentation or ovules are basal in *Phlebocarya*. The ovary is pseudomonomerous in *Barberetta*, with abortion of the antero-lateral carpels (Simpson, 1998b).

Septal nectaries are usually present and sit on top of the ovary in epigynous flowers. *Xiphidium* is reported to lack septal nectaries but I observed three slits just below the insertion of the style. Simpson (1998a) demonstrated that ovaries have become secondarily superior in *Wachendorfia* and this is linked with a modification of the septal nectaries. *Wachendorfia* produces enantiostylous flowers, with left and right flowers differing in the orientation of the three stamens and style (Vogel, 1998b). The style of several Haemodoraceae, including *Xiphidium*, is also obliquely inserted.

Fig. 6.18. Haemodoraceae: A. *Anigozanthos flavidus*. *Xyphidium coeruleum*: B. flower; C. inflorescence. In A the thick broken line refers to the slit in the perianth tube.

Pontederiaceae

Fig. 6.19. *Eichhornia crassipes* (Mart.) Solms

↔P3+3 A3+3 G̲(3)

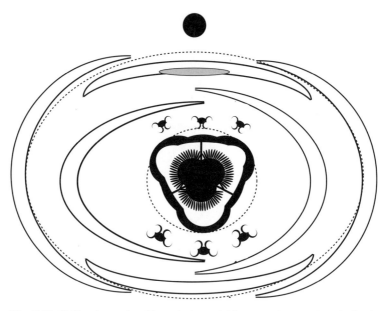

Fig. 6.19. *Eichhornia crassipes* (Pontederiaceae). The two trimerous tepal whorls are strongly disymmetrically flattened. Note the larger anterior stamens.

The family is sister to Haemadoracae and consists of freshwater, aquatic monocots. Flowers of *Eichhornia* are grouped in terminal racemes and lack bracts. Flowers are slightly monosymmetric but compressed between lateral outer tepals. All organs are basally connected by a hypanthium. Stamens in *Eichhornia* are arranged in three longer abaxial and three shorter adaxial stamens, which is also present in other Pontederiaceae (Strange, Rudall and Prychid, 2004). Nectaries were not seen by me, although they are reported to be present in *E. crassipes*, but not in all species (Strange, Rudall and Prychid, 2004). Septal nectaries were lost in relation to heteranthy in the family. The flower shows a tendency for disymmetry in *Eichhornia*, although the general tendency in the family is for median monosymmetry, which is expressed early in the floral development (Strange, Rudall and Prychid, 2004). Stamens are of unequal length or reduced to the three abaxial stamens in *Heteranthera*, or to one stamen and two staminodes in *Hydrothryx* (Strange, Rudall and Prychid, 2004). The ovary is tricarpellate, with axile or parietal placentation and a simple style with trilobate stigma. In *Pontederia* the ovary is pseudomonomerous by reduction of the two adaxial locules.

Commelinaceae

Fig. 6.20A,B. *Callisia warscewicziana* (Kunth & Bouché) Hunt

✳/↘K3 C3 A3+3 G̲(3)

Flowers are readily recognizable by the formation of scorpioid cymes (cincinni), subtended by a laterally inserted bract, and differentiation of sepals and

Fig. 6.20. *Callisia warscewicziana* (Commelinaceae): A. Flower, B. inflorescence. Anthers open sequentially.

petals. The number of stamens is basically six in two equal whorls but there is much variation in some genera with variable numbers of staminodia. Either the outer (e.g. *Palisota*: A3°+3) or inner whorl (e.g. *Aploleia, Tripogandra*: A3+3°) is staminodial, or staminodes are spread over two whorls (A1:2°+2:1°) (Faden in Dahgren, Clifford and Yeo, 1985).

In *Plowmanianthus* and *Cochliostema* flowers are obliquely monosymmetric by suppression of three stamens (remaining staminodial in *Cochliostema* and *P. perforans*), enhanced by the larger size of posterior sepal and occasionally a larger anterior petal and fusion of the fertile stamens (Hardy and Stevenson, 2000a; Hardy, Davis and Stevenson, 2004). As flowers are inserted sideways, orientation of flowers appears median, although it is oblique relative to the subtending bract. Other genera tend to be polysymmetric or slightly monosymmetric, with differences in maturation of the stamens (e.g. *Callisia*: pers. obs.). Staminodes can have variable forms, being stamen-like without anther, or with anthers and functioning as food stamens (e.g. *Tripogandra*). Flowers offer pollen as reward and lack nectaries, although mechanisms have developed in the family to restrict access to pollen and its availability (Faden, 2000). Floral deception is common by the development of yellow moniliform filaments luring insects looking for pollen (*Cochliostema*). Pollen dimorphism is usually related to heteranthy between different anther sets, which can be subtle to strongly developed (Faden, 2000). The ovary is generally trilocular, or bilocular, with simple terminal style.

Hardy and Stevenson (2000b) analysed the floral development of another species of *Callisia, C. navicularis*, and *Tradescantia*. They reported an unusual centrifugal initiation of the outer stamen whorl. This may be linked to the smaller petal primordia, providing more space for an earlier initiation of the antepetalous stamens, comparable to some obdiplostemonous eudicots.

Clear floral diagrams were shown in Hardy and Stevenson (2000a) and Hardy, Davis and Stevenson (2004).

Poales

The clade consists of about 15 families of mainly wind-pollinated plants lacking septal nectaries and with flowers variously reduced, or arranged in pseudanthia. Basal families to other Poales are Bromeliaceae and Rapateaceae, with insect-pollinated flowers possessing a differentiated perianth and septal nectaries. In other Poales nectaries are absent, except for Eriocaulaceae where they evolved secondarily on the inner perianth (Linder and Rudall, 2005). The order has undergone several flower reductions linked to a wind-pollination syndrome, such as lack of differentiation or loss of perianth, shift to unisexual flowers, reduction to a single ovule and loss of a stamen whorl (Linder and Rudall, 2005). In Cyperaceae and Typhaceae the perianth is reduced to bristles, while it is reduced to tiny lodicules in the Poaceae. The basic number of six stamens is also variously reduced by loss of the outer whorl (Restionaceae) or the inner whorl (Xyridaceae, Eriocaulaceae), or partial loss of adaxial or abaxial stamens (Xyridaceae, Poaceae, Cyperaceae). Carpels are variously reduced through pseudomonomery (e.g. Restionaceae), resulting in a single carpel (e.g. Poaceae, Cyperaceae). What is currently missing is a clear mapping of floral structural characters on a phylogeny to determine the reductive patterns in different floral whorls.

Restionaceae

Fig. 6.21A,B. *Hypodiscus aristatus* Nees

Staminate: ↔K3 C3 A0+3 G0
Pistillate: ↔K3 A0+3° G̲(1:1°)

Fig. 6.21C,D. *Elegia cuspidata* Mast.

Staminate: ↔K3 C3 A0+3 G0
Pistillate: ↔K3 A0+3° G̲(1:2°)

Restionaceae are dioecious plants, often with a strong dimorphism between staminate and pistillate inflorescences. Flowers are arranged in spikelets and are subtended by a single scarious bract. Staminate and pistillate reproductive structures are variously enclosed and protected by tepals, floral bracts and inflorescence bracts (Linder, 1991). The perianth is either undifferentiated or

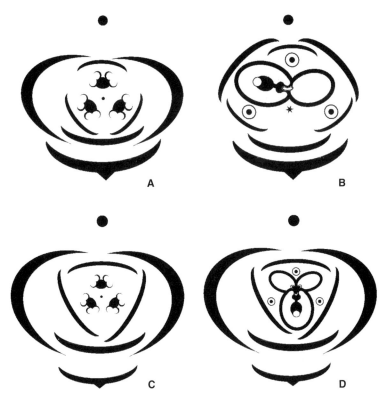

Fig. 6.21. Restionaceae: Staminate (A) and pistillate (B) flower of *Hypodiscus aristatus;* staminate (C) and pistillate (D) flower of *Elegia cuspidata*. Asterisk refers to lost carpel.

outer tepals differ strongly from inner tepals, with the outer lateral tepals strongly keeled (Linder, 1991, 1992a). The basic floral diagram of Restionaceae is trimerous, less often dimerous, and consists of two tepal whorls, a single whorl of stamens (inner) and three carpels. While this floral arrangement is found in some genera (e.g. *Dovea, Askidiosperma*), most genera have undergone various reductions of carpels (Linder, 1991, 1992a,b; Ronse De Craene, Linder and Smets, 2001, 2002). The ovary is often pseudomonomerous, with reductions affecting any of the three carpels (Fig. 6.21B,D; Ronse De Craene, Linder and Smets, 2002). This reduction is gradual, first affecting the fertility of locules, followed by loss of the sterile locules, which may still be represented by styles or vascular bundles (Linder, 1992a,b). Staminodes may be variously developed or are absent in pistillate flowers, as well as pistillodes in staminate flowers. A single pendulous, orthotropous ovule is present per locule, with one to three carinal styles.

Poaceae

Fig. 6.22A. *Oryza sativa* L.

✳P2 A3+3 G̲(1/2°)

Fig. 6.22B. *Anthoxanthum odoratum* L.

↔P2 A2 G̲(1/2°)*

* Arber (1934: Fig. 8.8) does not mention lodicules (overlooked?)

General formula:✳/↔/↓P0–3 A1–3(6-∞) G1/2°

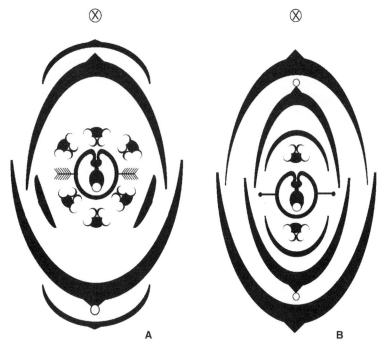

Fig. 6.22. Poaceae: A. *Oryza sativa*, B. *Anthoxanthum odoratum*. White dots represent awns; styles are shown in both flowers.

Flowers of Poaceae are highly reduced, with derived morphology, and there has been considerable speculation about their floral structure (see Rudall *et al.*, 2005; Sajo, Longhi-Wagner and Rudall, 2007, 2008). Grasses are adapted to wind pollination, although insects have been observed gathering pollen (Dahlgren, Clifford and Yeo, 1985).

Flowers are grouped in spikelets, short shoots with transformed bracts that make up the basic unit of the inflorescence. There is tremendous variation in the construction of spikelets. Two outer bracts (glumes) are usually present, enclosing a shoot with two rows of flowers in a distichous arrangement, or a single flower, and each individual flower (floret) is enclosed by two (lemma and palea) or more bracts. The palea is often a two-parted structure that is interpreted as the fusion product of two bracts and can be seen as developing from two primordia (e.g. Reinheimer, Pozner and Vegetti, 2005). The palea may be missing (e.g. *Alopecurus*). Glumes and inner bracts may bear appendages, such as hairs, bristles or awns. The flowers are mostly trimerous, rarely dimerous (e.g. *Anthoxanthum*: Fig. 6.22B). A single whorl of two to three green organs called lodicules, rarely more or none, is interpreted as representing the inner tepal whorl (Dahlgren, Clifford and Yeo, 1985; Cocucci and Anton, 1988; Sajo, Longhi-Wagner and Rudall, 2008). The majority of Poaceae have two lodicules and three stamens, as the adaxial lodicule is absent, leading to a monosymmetric flower. Rudall and Bateman (2004) illustrated the variation in stamen position within Poaceae, but more floral morphological investigations are required to understand the evolution of stamen positions in the family. Bambusoideae represent the basal group with three to six stamens and three lodicules. *Oryza sativa* (rice) belongs to this subfamily. It has six stamens but the perianth is reduced (Fig. 6.22A). The number of stamens is mostly reduced to three, two or one, rarely increased up to 170 stamens (*Ochlandra*). *Anthoxanthum* with two opposite stamens was interpreted as the result of a retention of one stamen of the outer whorl with one of the inner whorl (Rudall and Bateman, 2004). Suppression of either stamens or pistil leads to unisexual florets (Le Roux and Kellogg, 1999; Reinheimer, Pozner and Vegetti, 2005; Sajo, Longhi-Wagner and Rudall, 2007), in the same (e.g. *Panicum* group) or in distinct spikelets (e.g. *Zea mays*). The ovary is tricarpellate, but strongly pseudomonomerous with various degrees of reduction of two carpels comparable to Restionaceae (Philipson, 1985; Ronse De Craene, Linder and Smets, 2002; Sajo, Longhi-Wagner and Rudall, 2007, 2008). Two styles are formed on the sterile carpels, rarely three styles (*Streptochaeta*). The ovary with single ovule develops as a caryopsis (one-seeded, indehiscent fruit with fused testa and pericarp) and has a unique embryo development.

The interpretation of the grass flower remains controversial (Fig. 6.23). Some authors interpreted the palea as part of an outer perianth (the palea is occasionally two-keeled), consisting of two fused outer adaxial tepals with the abaxial outer tepal missing (Fig. 6.23A; e.g. Eichler, 1875; Stützel, 2006; Sajo, Longhi-Wagner and Rudall, 2008). The inner tepal whorl consists of two abaxial lodicules and the adaxial one is missing. The androecium of three stamens is usually interpreted as an outer stamen whorl alternating with the lodicules (e.g. Eichler, 1875; Stützel, 2006). Another interpretation regards lodicules and palea as bracts and the flower as naked (Fig. 6.23B; Dahlgren, Clifford and Yeo, 1985). Cocucci and Anton (1988) interpreted the androecium as basically two–whorled and the flower as monosymmetrics with adaxial stamens and lodicule reduced through the inhibitory influence of the palea (Fig. 6.23C). Rudall *et al.* (2005) favoured the first interpretation on the basis of the floral morphology of

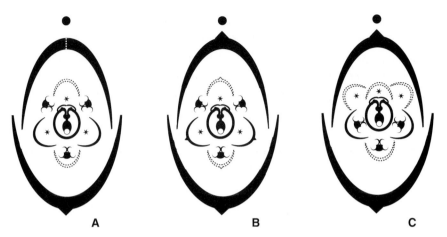

Fig. 6.23. Different interpretations of the floral evolution of Poaceae: A. Eichler (1875); B. Dahlgren, Clifford and Yeo (1985); C. Cocucci and Anton (1988). Broken lines refer to lost tepals or bracts; asterisks refer to lost stamens.

Ecdeiocolea in Ecdeiocoleaceae, which are considered to be the nearest sister group of Poaceae. Staminate flowers have six tepals and four stamens arranged in two whorls in a monosymmetric arrangement. The early diverging grass *Anomochloa* also has four stamens, arranged as three outer stamens and an adaxial inner stamen comparable to *Ecdeiocolea*. Among the sister groups of Poaceae, Flagellariaceae, Joinvilleaceae and *Georgeantha* of Ecdeiocoleaceae have a regular monocot floral formula (Rudall *et al.*, 2005). Sajo, Longhi-Wagner and Rudall (2008) recently demonstrated that the grass spikelet could be derived from an intermediate structure as in the basal grass genus *Streptochaeta*. Single flowers arise on top of (pseudo)spikelets surrounded by seven to eight basal bracts. Lower bracts (one to five) are interpreted as glumes; bract six (with an awn) bears a modified flower in its axis, or the flower arises in the continuation of the main axis. Bracts seven and eight represent outer tepals, one of which is lost, and the three inner tepals are comparable to the lodicules. Differentiation of glumes, palea and lemma arises later in the evolution of the grasses.

Zingiberales

Zingiberales consist of eight families with monosymmetric or asymmetric flowers. The floral diagrams of Zingiberales represent variations on the basic monocot diagram, with reductions and transformations affecting mainly the androecium.

In the order there is a remarkable variation in floral forms related with intricate pollination mechanisms. Two groups can be recognized: a more basal 'banana clade' including Musaceae, Lowiaceae and Strelitziaceae that often have an odd non-functional (?) staminode in the outer whorl, and the 'ginger group' including Heliconiaceae, where the pattern is reversed with a reduction of the abaxial stamen(s) leading to highly specialized constructs, including a further heterotopic transformation of stamens into petaloid staminodes and only a single (half-) stamen (Walker-Larsen and Harder, 2000; Rudall and Bateman, 2004). A hypanthial tube is usually developed connecting stamens and petals to various degrees. The labellum is a major morphological structure in Cannaceae, Lowiaceae, Costaceae and Zingiberaceae, but is not homologous between families (Kirchoff, 1992), consisting of variable parts of the androecium. Highly specialized asymmetric flowers have evolved in Marantaceae and Cannaceae, leading to the development of half-stamens. The ovary is always inferior and trimerous (or dimerous through the reduction of the abaxial carpel). Septal nectaries are present in all families of the order, except in Lowiaceae where they are aborted and in Zingiberaceae where they are highly transformed (cf. Rao, Karnik and Gupte, 1954; Pai and Tilak, 1965; Kirchoff, 1997). Flowers tend to be arranged in various inflorescences built on the same Bauplan, which is a cincinnus (Dahlgren, Clifford and Yeo, 1985). Floral diagrams of Zingiberales were presented by Kress (1990).

Musaceae

Fig. 6.24A,B. *Musa campestris* Becc.

↑P(5):1 A5 Ǧ(3)

The family contains two genera, *Musa* and *Ensete*. Flowers are arranged in tiers in the axil of large deciduous bracts. The number varies between a pair and up to 40 flowers (modified cincinni: Dahlgren, Clifford and Yeo, 1985). Individual flowers may be subtended by bracts, but these are absent in *M. campestris*. Flowers tend to be staminate in the distal portion of the inflorescence and pistillate in the proximal part. All floral organs are basally connected by hypanthial growth. The perianth is undifferentiated and completely fused, except for the adaxial side where the folds overlap and surround an odd petal. This petal could be mistaken for a staminode, but the presence of a sixth stamen or staminode in *Ensete* or some *Musa* contradicts this assumption. The median abaxial stamen is usually suppressed (Kirchoff, 1992). Stamens are spathulate with long anthers. The ovary is inferior with three axile placentae; locules are filled with trichomes and mucilage. The septal nectaries are merged into a triradiate structure emerging as an orifice at the base of the style on the adaxial side of the flower (cf. Pai and Tilak, 1965; Kirchoff, 1992).

Musaceae are closely related to Strelitziaceae and Lowiaceae, which have the perianth differentiated into sepals and petals, five stamens and one median adaxial staminode which is occasionally fertile or missing.

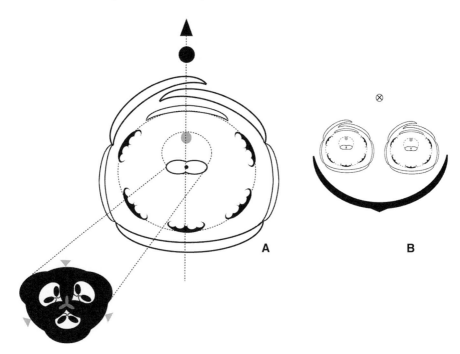

Fig. 6.24. *Musa campestris* (Musaceae). A. flower; B. partial inflorescence. Note the single aperture for the septal nectary.

Costaceae

Fig. 6.25. *Costus curvibracteatus* Maas

↓K(3) [C3 A(3°+2°):1] Ğ(3)

Flowers are arranged in terminal racemes. A coloured bract and bracteole are present, but are indistinguishable in colour from the perianth in *C. curvibracteatus*. The calyx is tubular. The large petaloid stamen is enclosed by a median abaxial petaloid labellum, which is similar in texture and shape. There are no lateral staminodes and the labellum is interpreted by Kirchoff (1988: Fig. 2) based on floral developmental evidence as consisting of five staminodes. The style fits between the two anther lobes as in Zingiberaceae. The inferior ovary has two or four rows of ovules on each placenta, with fewer ovules in the abaxial locule towards the bract. Contrary to Zingiberaceae, three glands are present in the locular zone of the ovary but do not have a slit-like shape characteristic of septal nectaries (Rao, Karnik and Gupte, 1954; Newman and Kirchoff, 1992; Endress, 1994). In some members of the family the anterior locule is reduced and the two antero-lateral septa are fused (Newman and Kirchoff, 1992).

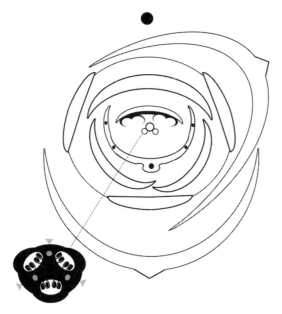

Fig. 6.25. *Costus curvibracteatus* (Costacaceae). Note the labellum consisting of five fused staminodes (black dots).

Zingiberaceae

Fig. 6.26A,B. *Roscoea cautleoides* Gagnep.

↓K(2) [C3 A3°+2°:1] Ǧ(3)

There is a single bract, occasionally with a second adaxial bracteole. Bracts can be strongly coloured, contrasting with the perianth (e.g. *Globba*). In *Roscoea* only two sepals are present; the adaxial median sepal is missing. Petals and stamens are fused into a tubular structure with apical appendages. Petals develop in a median-adaxial hood and two lateral appendages that connect with the androecium. There is only a single massive fertile stamen enclosing the single style. This stamen is flanked by two lateral staminodes in *Roscoea*, as in other members of tribe Zingibereae. In other tribes lateral staminodes are free, and in tribe Alpinieae they are very small to absent. A large labellum character-istic for Zingiberaceae is situated in a median abaxial position. The anther can develop lateral outgrowths (*Globba*: Box and Rudall, 2006). In other species of *Roscoea* (e.g. *R. purpurea*, *R. scillifolia*), the lower part of the anther is transformed in a lever mechanism resembling *Salvia* (Endress, 1994; pers. obs.). The style is flanked at the base by two glandular appendages. The inferior ovary has three parietal or axile placentae, depending on the tribe.

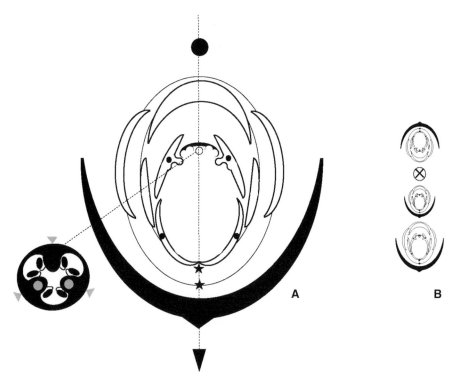

Fig. 6.26. *Roscoea cautleoides* (Zingiberaceae): A. Flower; B. Inflorescence. Lower asterisk refers to lost outer tepal; upper asterisk refers to lost staminode.

The interpretation of the labellum and glands has been controversial but both structures are closely interconnected. The two basal glands were interpreted as epidermal appendages, staminodes or stylodes (see Rao, 1963 and Newman and Kirchoff, 1992 for a discussion). If the labellum is a single organ it is part of the outer stamen whorl and the two glands could represent stamens. Some abnormal flowers of *Alpinia* have stamens instead of glands (Rao, Karnik and Gupte, 1954), supporting this interpretation. If the labellum is considered as made up of two staminodes it must belong to the inner whorl, and one anterior outer stamen is suppressed. A third interpretation considers the labellum as made up of three staminodes, and this is supported by the occasional presence of a third lobe.

Anatomical evidence suggests that the glands are extensions of septal nectaries, far removed from the androecium (Rao, Karnik and Gupte, 1954) and that the labellum is made up of two inner staminodes, occasionally with a third outer staminode. The glands can be variously developed as crescentic structures around the base of the style. In *Globba* it is likely that a third adaxial gland was suppressed (Box and Rudall, 2006). Floral development also supports the interpretation of two staminodial members forming the labellum (Kirchoff, 1997; Box and Rudall, 2006). Petals and inner stamens arise on common primordia and the anterior stamens fuse into a labellum. The outer stamen whorl arises after the inner and an outer adaxial stamen may be initiated but aborts later or is not initiated at all.

Cannaceae

Fig. 6.27. *Canna edulis* Ker Gawl

↯ K3 [C3 A1°–3°+ ½:2°] Ǧ(3)

Fig. 6.27. *Canna edulis* (Cannaceae): partial inflorescence.

Cannaceae and Marantaceae are closely related and share similar floral structures (Kirchoff, 1983). The flowers differ mainly in their pollination mechanism, which functions by an explosive movement of the style in Marantaceae (e.g. Classen-Bockhoff and Heller, 2006) and is passive in *Canna*. Both families share a secondary pollen presentation mechanism whereby the stamen enwraps the flattened style. In *Canna* pollen is deposited on the style by the presence of small hairs on one of its margins (pers. obs.).

Flowers are arranged in pairs, equal in Marantaceae, but one of which is smaller and subtended by a second bract in Cannaceae. The second flower is rarely fully developed (Kirchoff, 1983). Contrary to Zingiberaceae the median adaxial inner petal is smaller and enclosed by lateral petals. Only four to five members of the androecium are present (cf. Pai, 1965; Kirchoff, 1983) and the single fertile inner adaxial stamen is half-fertile, while the other half is petaloid. Pai (1965) interpreted half of the anther as reduced and the crest as a prolongation of the connective, but this is refuted by ontogenetic data showing an early division of two equal primordia, one developing into a theca and the other into the crest (Kirchoff, 1983). The other staminodes as well as the style are petaloid. One of the inner staminodes develops into a reflexed labellum while the third is erect and petaloid. Contrary to Zingiberaceae and Costaceae the labellum is a single organ (Pai, 1965; Kirchoff, 1983, 1997). The outer stamens are variously reduced or absent. In *Canna edulis* only one staminode is present (mostly the outer abaxial as in Zingiberaceae and Marantaceae). In *C. indica* two staminodes occur in the outer whorl.

The inferior ovary has axile placentation with ovules borne in two rows. Septal nectaries are present on the septal arms.

7

Early diverging eudicots: a transition between two worlds

The early diverging eudicots represent a transitional grade between basal angiosperms and core eudicots (Fig. 7.1). Ranunculales is the basal order of eudicots with the highest floral diversity (Ronse De Craene, Soltis and Soltis, 2003; Soltis *et al.*, 2005). Other intermediate orders (e.g. Proteales, Buxales, Trochodendrales) generally have much reduced, dimerous flowers and the link between Ranunculales and core eudicots remains unclear on a floral morphological basis (e.g. Ronse De Craene, 2004; Wanntorp and Ronse De Craene, 2005).

7.1   Ranunculales

The order is highly diverse in terms of floral structure. As such it occupies a transitional position between basal angiosperms and core eudicots (Ronse De Craene, Soltis and Soltis, 2003). Several derived characters tend to be concentrated in Menispermaceae, Ranunculaceae and Papaveraceae, such as median or transversal monosymmetry, sepal and petal differentiation and fusion, petal appendages in the shape of spurs, syncarpy and (pentamerous) cyclic flowers. Unisexual flowers with synandry have evolved in Menispermaceae and Lardizabalaceae (Endress, 1995b).

The androecium is highly variable, ranging from numerous spirally arranged stamens to a single stamen. All core Ranunculales share nectariferous petals that are of probably staminodial origin (Erbar, Kusma and Leins, 1998; Walker-Larsen and Harder, 2000; Ronse De Craene, Soltis and Soltis, 2003).

129

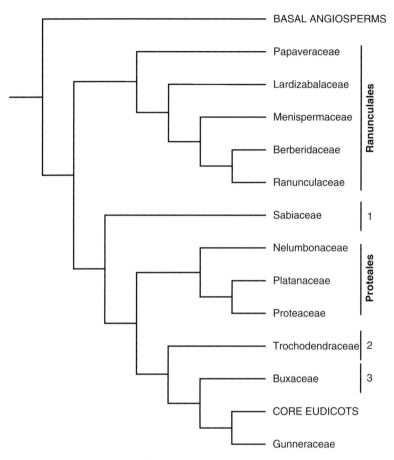

Fig. 7.1. Phylogenetic tree of the early diverging eudicots, based on Worberg *et al.* (2007). 1. Sabiales; 2. Trochodendrales; 3. Buxales.

Papaveraceae (including Fumariaceae)

Fig. 7.2A. *Eschscholzia californica* Cham.

↔K2 C2+2 A4+2+4+2+4+2+4 G̲(2)

Fig. 7.2B. *Macleaya microcarpa* Fedde

→K2 A2+2+4+(2+4) G̲(2)

Fig. 7.2C. *Corydalis lutea* (L.) DC., based on Ronse De Craene and Smets (1992b)

↔K2 C2+2 A4$^{1/2}$+2 G̲(2)
General formula: K2–3 (4) C(0)4–6-(8–12) A4–6-∞ G2-∞

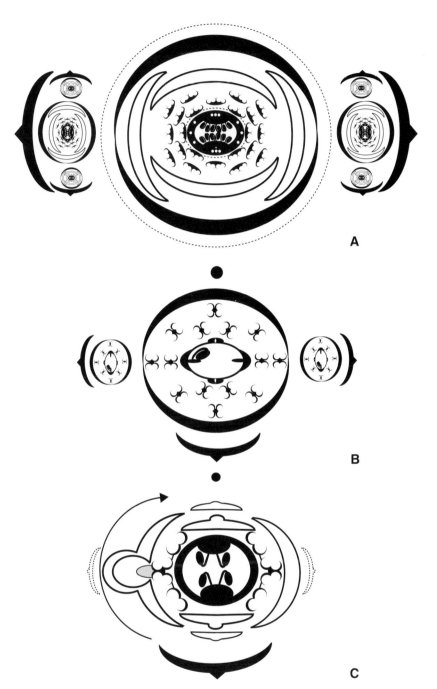

Fig. 7.2. Papaveraceae: A. *Eschscholzia californica*, partial inflorescence; B. *Macleaya microcarpa*, partial inflorescence; C. *Corydalis lutea*.

Flowers are basically dimerous and disymmetric; a few North American genera have trimerous flowers (e.g. *Platystemon, Meconella, Romneya*: Ernst, 1967). Sepals are early deciduous in all Papaveraceae; they tend to be large and saccate or calyptrate, or small and reduced in Fumarioideae. Petals are clearly arranged in two imbricate whorls and are often crumpled in bud. Stamens arise in regular whorls of four or two in dimerous flowers (forming pseudowhorls of six), and outer stamens are always inserted in two pairs or alternate with the petals (e.g. Murbeck, 1912; Ronse De Craene and Smets, 1990b; Karrer, 1991). The number of stamens can be highly fluctuating; in the specimens of *Eschscholzia* studied stamens ranged from four to 22, although the number can be higher. The regular arrangement of pairs and single stamens is interpreted as the result of the compression of an ancestral helical initiation, although others interpret the pairs as the result of doubling (e.g. Endress, 1987), with oscillations between dimerous and tetramerous whorls. In cases where stamen numbers are low, the four or six outer stamens are always present (e.g. Murbeck, 1912; Ernst, 1967). In the wind-pollinated *Macleaya* (and perhaps *Bocconia*), the four petals are replaced with stamens through a process of homeosis (Fig. 7.2B; Ronse De Craene and Smets, 1990b; Ronse De Craene, 2003) and the number of stamens varies between eight and 14. In *Sanguinaria canadensis*, the opposite of *Macleaya* occurs as floral attraction is enhanced by the transformation of the four outer stamens into petals leading to eight petals (Lehmann and Sattler, 1993). Some genera have very high numbers of stamens (e.g. *Romneya, Papaver*); stamens arise on a ring primordium from four alternipetalous forerunners and appear to be the result of a secondary increase (Ronse De Craene and Smets, 1990b; Karrer, 1991). The number of carpels is usually two (three), although a secondary increase within a whorl is frequent (e.g. *Papaver, Platystemon*). Ovules are formed in two rows on parietal placentae, although the placenta can become protruding-diffuse with increase in ovule number (Endress, 1995b). In *Macleaya* there is a single ovule inserted on a lateral placenta extending as an obturator. Endress (1995b) discussed the possible evolution of the monocarpellate gynoecium of Berberidaceae from a bicarpellate condition with parietal placentation as in Papaveraceae, suggesting two directions of evolution. The presence of one sterile placenta in *Macleaya* (Fig. 7.2B) is an indication for a derivation of the monocarpellate condition from the bicarpellate condition.

Fumarioideae are occasionally recognized as Fumariaceae and can be interpreted as reduced flowers of Papaveraceae. Contrary to Papaveroideae the sepal whorl is small and does not contribute to bud protection (Ronse De Craene and Smets, 1992b). Bracts and bracteoles are present, although the bracteoles may be lost. Outer petals are strongly different from inner petals, often with elaborations or spurs enclosing a nectary. While two transversal spurs develop in the regular disymmetric flower of *Dicentra*, only one spur develops in *Fumaria* and *Corydalis*, leading to transversal monosymmetry. Furthermore, resupination of 90° leads to median monosymmetry (Fig. 7.2C; Weberling, 1989). In *Hypecoum* the inner petals are elaborate and trilobed and contribute to a secondary pollen presentation mechanism (Endress and Matthews, 2006b). Stamens are arranged in two transversal triplets consisting of a central stamen with dithecal anther and two lateral stamens with half-anther by reduction of one half. The nectary is inserted abaxially at the base of the middle filament of a triplet; in monosymmetric flowers only the nectary opposite the spur is fully developed (Fig. 7.2C). The gynoecium is bicarpellate with two transversal carpels. Ronse De Craene and Smets (1992b) demonstrated that the androecium is derived from two whorls consisting of four outer half-stamens alternating with the petals and two transversal inner stamens. In *Hypecoum* with seemingly two median and two transversal stamens, the outer half-stamens fuse postgenitally into two larger median stamens (Ronse De Craene and Smets, 1992b). In *Pteridophyllum* the inner stamens are lost, resulting in four alternipetalous stamens.

Lardizabalaceae (incl. Sargentodoxaceae)

Fig. 7.3A–C. *Holboellia angustifolia* Wall.

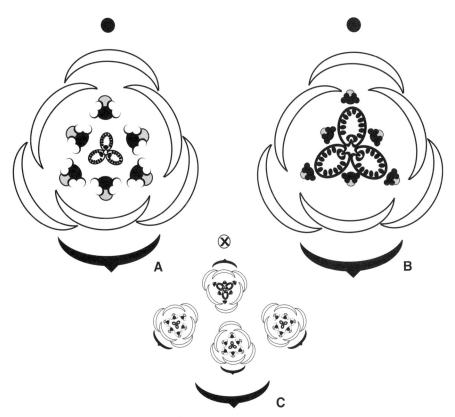

Fig. 7.3. *Holboellia angustifolia*. A. staminate flower; B. pistillate flower; C. partial inflorescence. Note the small nectar leaves attached to stamens and staminodes.

Staminate: ✳K3+3 [C3+3 A3+3] G3°
Pistillate: ✳K3+3 [C3+3 A3°+3°] G3

Lardizabalaceae share several floral characters with Menispermaceae, including trimerous, unisexual flowers, synandry, nectar leaves and a comparable floral formula. In *Holboellia* the sepal whorl is petaloid and the nectar leaves are small and attached as scales to the stamens; Fig. 7.3A,B). In *Decaisnea* no nectar leaves are present. Flowers of *Akebia* have only one whorl of sepals and show a strong floral dimorphism between staminate and pistillate flowers, but both are inserted on the same inflorescence. Stamens are either free or basally united in a synandrium and have extrorse anthers. Carpel number ranges from three to 12 (Eichler, 1878). The apocarpous gynoecium is plicate and has a laminar-diffuse placenta reminiscent of Nymphaeales (Endress, 1995b).

Sargentodoxa differs from other Lardizabalaceae in the higher carpel number in pistillate flowers (up to 90), linked with ascidiate, uniovulate carpels (Zhang and Ren, 2008). Staminate flowers are trimerous with a floral diagram resembling *Holboellia*, while bisexual and pistillate flowers have a highly variable merism with fluctuating perianth and carpel numbers. It is likely that the carpel number has been secondarily multiplied and that the flower has become increasingly chaotic as a result.

Menispermaceae

Fig. 7.4A,B. *Cocculus laurifolius* DC.

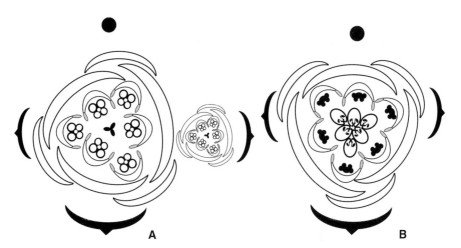

Fig. 7.4. *Cocculus laurifolius* (Menispermaceae): A. staminate partial inflorescence; B. pistillate flower.

Staminate: $*$K3+3 C3+3 A3+3 $\underline{G}$3°

Pistillate: $*$K3+3 C3+3 A3°+3° $\underline{G}$3+3

General formula (combined): $*$K(1)3–12 C(0)1–6 A(2)3–6(-∞) G(1)3–6(-∞)

Flowers of Menispermaceae are highly heterogenous, ranging from spiral, multistaminate (e.g. *Hypserpa*) to trimerous (e.g. *Cocculus*) or dimerous (e.g. *Cissampelos*) flowers with one to two stamen whorls (Schaeppi, 1976; Endress, 1995b). Compared with other Ranunculales there is high diversity in floral forms with occasional fusions of sepals or petals and even zygomorphy. Flowers are always unisexual (dioecious) with variable dimorphism between the genders, which is also expressed in the complexity of the inflorescences (Kessler, 1993; Endress, 1995b). The perianth is undifferentiated and spiral, or there is a distinction between sepals and petals. Sepals are inserted in several alternating whorls (e.g. *Sciadotenia*: diagram shown by Endress, 1995b) or more generally in two whorls. Petals or nectar leaves are much smaller and are homologized with staminodes as in Berberidaceae (Ronse De Craene and Smets, 2001a). They are occasionally absent (e.g. *Abuta*). Stamens range from many (e.g. *Hypserpa*: Endress, 1995b) to 12, 6, 3 or exceptionally 1 (Kessler, 1993). Carpels are free and range from 12 to 3 (rarely 1). Two ovules are present but only one develops as a seed.

Cocculus resembles Berberidaceae in its floral diagram, although flowers are unisexual with rudiments of carpels and stamens. Staminate flowers have three carpellodes rather than six, while pistillate flowers have six rudiments of stamens (cf. Wang *et al.*, 2006). The petals are sheathlike and enclose the filaments at maturity. Flowers are highly reduced in some dioecious genera (e.g. *Stephania*, *Cissampelos*), with strong dimorphism between staminate and pistillate flowers and no traces of lost organs (Eichler, 1878; Wang *et al.*, 2006). In staminate flowers with complete loss of the gynoecium, the androecium often develops as a synandrium consisting of three to four (e.g. *Stephania*: Wang *et al.*, 2006) or six stamens (*Dioscoreophyllum*: Schaeppi, 1976).

Berberidaceae

Fig. 7.5A. *Epimedium* × *versicolor* E. Morren

↔K2+2/2+2+2 A2+2 A2+2 $\underline{G}$1

Fig. 7.5B. *Podophyllum peltatum* L., based on Schmidt (1928) and DeMaggio and Wilson (1986)

$*$K3/3+3 C3+3/3+3+3 A3/3^2 +3^3 $\underline{G}$1

General formula: $*$K3–12 C4–6 A4–6(-∞) G1

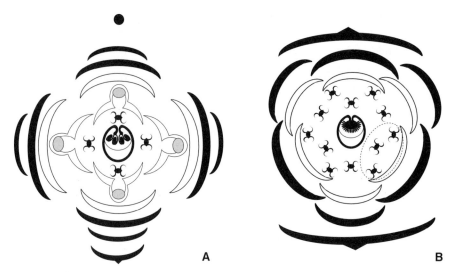

Fig. 7.5. Berberidaceae: A. *Epimedium* × *versicolor*, B. *Podophyllum peltatum*. Broken line refers to group of stamens.

Flowers are arranged in complex cymes, racemes or are solitary (*Podophyllum*). The flower of *Podophyllum peltatum* is terminal and inserted between the petioles of two leaves (DeMaggio and Wilson, 1986). Flowers are mostly trimerous, rarely variable or dimerous (e.g. *Epimedium*). The inflorescence of *Berberis* has one terminal pentamerous flower while lateral flowers are trimerous (Eichler, 1875, 1878; Endress, 1987). These have been extensively discussed as to whether pentamerous flowers are derived from trimerous flowers or the opposite.

The perianth is variable but differentiated in sepals and petals in a similar way as in Menispermaceae (e.g. *Berberis, Mahonia, Epimedium*) with two to three (occasionally more) whorls of sepals and generally two whorls of nectar leaves (petals) and two whorls of three stamens. *Vancouveria* and *Nandina* have several trimerous whorls of bracts below the six sepals. Petals of *Podophyllum* lack floral nectaries and are sometimes in double position. *Podophyllum peltatum* is unusual in having more than six stamens (ranging from 6 to 12–18), which were interpreted as a result of dédoublement (Payer, 1857; Schmidt, 1928; Terabayashi, 1983). DeMaggio and Wilson (1986) demonstrated that stamens arise in groups starting with three to six primordia arising opposite the petals, and that stamens may arise after initiation of the gynoecium, suggesting a secondary stamen increase. A secondary increase is mostly linked to the inner whorl. A perianth is secondarily lost in *Achlys*, leading to a breakdown of the regular merism of the flowers (Endress, 1989).

Nectar leaves develop from common primordia with the stamens (e.g. Schmidt, 1928; Brett and Posluszny, 1982) and they lag behind the stamens in development. They are best interpreted as stamen derived (cf. Brett and Posluszny, 1982; Endress, 1995b; Ronse De Craene, Soltis and Soltis, 2003). While most taxa have two marginal nectaries at the base of the nectar leaf, in *Epimedium* the nectar is concealed in a spur similar to *Aquilegia* (Ranunculaceae). The single carpel develops as a strictly ascidiate structure without ventral suture (Endress, 1995b). Placentation is marginal with two rows of ovules, sometimes protruding-diffuse with a high number of ovules (e.g. *Podophyllum*).

Ranunculaceae

Fig. 7.6A,B. *Ranunculus ficaria* L.

∗K3–5 C8–11 A15–31 G$\underline{9}$–10

Fig. 7.6C. *Aconitum lycoctonum* L.

↓P5 A°2:3 A∞* G3
*generally 26–27

Fig. 7.6D,E. *Aquilegia* sp.

∗K5 C5 A13 × 5 + 5°+5° G$\underline{5}$
General formula: ∗/↻/↓/↔P1-∞ or K3–8+ C0–13 A5-∞ G1-∞

The family is probably the most diverse in flower structural diversity among angiosperms. Inflorescences are variable and consist of terminal single flowers (e.g. *Ranunculus ficaria*), dichasia (e.g. *Clematis*) or racemes (e.g. *Actaea, Aconitum*). Bracts may hide lateral flowers enclosed by two bracteoles. In *Aquilegia*, terminal flowers are overtopped by the lateral branches. The flower of Ranunculaceae is in a flux, with several evolutionary novelties, such as the event of pentamery, staminodial petals, transitions between whorled and spiral flowers and mono-symmetry. Perianth evolution ranges from a tepalar perianth to a bipartite perianth consisting of outer tepals and inner nectar leaves. Monosymmetry is present in a few genera, superimposed on a regular, polymerous flower. Flowers are mostly insect pollinated, rarely wind pollinated (e.g. *Thalictrum*).

Ranunculaceae have been studied extensively for their phyllotaxis and the family was used as a model to interpret the evolution of petals in the angio-sperms (e.g. Eichler, 1878; Schöffel, 1932). The perianth is either spiral and undifferentiated, or whorled (dimerous, trimerous or pentamerous). Genera with petaloid tepals (e.g. *Anemone*) tend to have a variable number of tepals, and high numbers of stamens and carpels in a spiral. An involucre of three bracts is usually present and can be close to the perianth, mimicking a calyx (e.g. *Barneoudia, Hepatica*). Clear transitions from spirals to whorls are present in

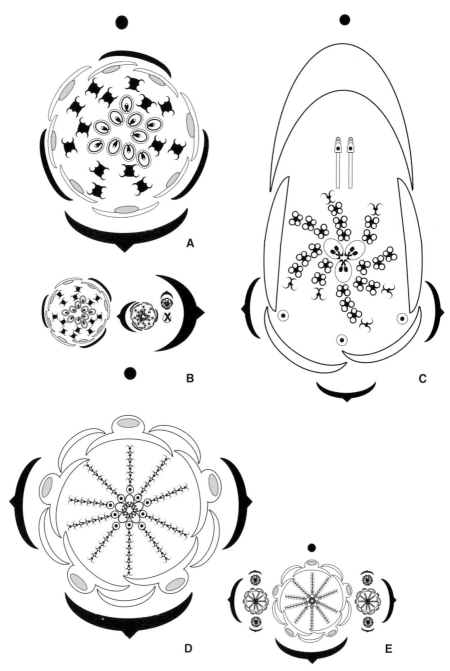

Fig. 7.6. Ranunculaceae: *Ranunculus ficaria*: A. flower; B. partial inflorescence; C. *Aconitum lycoctonum*; *Aquilegia* sp.: D. flower; E. partial inflorescence. Note the different types of nectar leaves.

Clematis with a dimerous perianth, a whorled arrangement of outer stamen whorls and a transition to a spiral arrangement in the gynoecium (Ronse De Craene and Smets, 1996a; Ronse De Craene, Soltis and Soltis, 2003). The flowers often have transitional series of Fibonacci numbers between different organ categories. Sepals are often five in number (2/5), with eight petals or nectar leaves (3/8) (e.g. *Adonis, Aconitum, Delphinium, Nigella*: Fig. 7.6B). Some pentamerous flowers with reduced stamen and carpel number can be confused with diplostemonous flowers (e.g. *Xanthorhiza, Ranunculus* sp.: Ronse De Craene and Smets, 1995c). The perianth is often whorled and trimerous (e.g. *Pulsatilla, Hepatica*); the outer stamens are arranged in three pairs following the trimerous perianth (Schöffel, 1932; Ronse De Craene and Smets, 1995c). Strictly dimerous flowers occur in *Clematis* (occasionally up to decamery), *Actaea* and *Thalictrum*. In dimerous flowers outer petals or stamens alternate with the sepals (Schöffel, 1932; Lehmann and Sattler, 1994). The petaloid tepals are arranged in two whorls. Petals are variously present (Eichler, 1878), and their presence or absence is linked with stamens through homeosis (Lehmann and Sattler, 1994).

Flowers have retained a spiral initiation throughout without fusion of parts. Stamens are variable in number, usually in a spiral, or arranged in trimerous or dimerous whorls. The only genus with a whorled arrangement throughout and syncarpy is *Aquilegia*; the regularity is caused by the development of spurred nectar leaves: Endress, 1987). *Aquilegia* is the only genus that possesses one or two inner whorls of staminodes.

Nectar is either concealed in large spurs formed by tepals or nectar leaves (Fig. 7.6B), or behind a basal adaxial flap on the petal. The evolution of the bipartite perianth is closely linked to the nectar leaves. Nectar leaves are interpreted as transformed staminodes, as in other Ranunculales (cf. Kosuge, 1994; Erbar, Kusma and Leins, 1998). In spiral flowers nectar leaves develop by differentiation of the outer stamens and provide only nectar (e.g. *Helleborus, Delphinium, Aconitum*: Fig. 7.6C) while tepals remain petaloid and provide attraction. An arrangement of nectar leaves in a whorl of five leads to a combination of rewards and attraction and a differentiation of a green calyx and coloured corolla (e.g. *Ranunculus, Aquilegia*: Fig. 7.6A,B). *Ranunculus* is highly variable in flower morphology, with generally pentamerous flowers with well-differentiated sepals and petals, a partial loss of nectar leaves (e.g. *R. auricomus*), or increase of nectar leaf numbers at the expense of stamens (e.g. *R. ficaria*: Fig. 7.6A, *R. chilensis*).

Ranunculaceae have a well-developed conical receptacle with apocarpous gynoecium. The number of carpels is variable and can be secondarily increased in some cases (e.g. *Myosurus*). Carpels are occasionally reduced to five, three to one carpel. Carpels have a single fertile ovule, or are developed as follicles with two rows of ovules.

Zygomorphy appears late in the development of flowers of *Delphinium* and *Aconitum* and includes a pentamerous undifferentiated perianth and two abaxial nectar leaves (initially eight: Eichler, 1878). The remaining nectar leaves (staminodes) are variously developed or reduced along a gradient running from the adaxial to the abaxial side of the flower. In some *Delphinium* one to four abaxial nectar leaves are developed, while the adaxial organs are barely visible at maturity.

The evolution of the flowers in Ranunculaceae appears remarkable, with several innovations that are reflected in core eudicots. However, the origins of pentamerous flowers and petals appear to be unique novelties not related to the evolution in the core eudicots.

7.2   The basal eudicot grade: Sabiales, Proteales, Trochodendrales, Buxales

Sabiales

Sabiaceae (Meliosmaceae)

Fig. 7.7. *Meliosma dilleniifolia* Wallp.

↘K4–5 [C3:2 A2:3°] G̲(2)

Sabiaceae is a small family of three genera, *Sabia*, *Ophiocaryon* and *Meliosma*, with an unresolved systematic position close to Proteales (e.g. Worberg *et al.*, 2007). Based on a series of micromorphological and embryological characters Sabiaceae appears close to Menispermaceae (data summarized in Ronse De Craene and Wanntorp, 2008). The flowers are arranged in terminal panicles and include numerous to solitary flowers that do not open in a regular succession. Flowers are enclosed by two bracteoles with variable position on the pedicel, occasionally intergrading with sepals. A single deciduous median bract subtending the flower is situated at the base of the pedicel. Bracteoles, sepals and petals are morphologically similar, especially visible early in their development. The pattern of insertion of the sepals appears more or less decussate.

Flowers of *Meliosma* are deceivingly complex (see Wanntorp and Ronse De Craene, 2007; Ronse De Craene and Wanntorp, 2008). They consist of four to five small sepals, five petals of two different sizes, three staminodes, two fertile anthers and a superior, bicarpellate ovary. Staminodes are inserted opposite the large petals and are basally adnate to them. Stamens are dorsally fused with a small petal. In buds stamens and staminodes are arranged in a closely coherent unit. Anthers consist of a broad basal platform, in some species extending into a crenulate rim, and bear two globular pollen sacs in an apical-adaxial position. The platform goes over into a narrow flattened filament. In young flower buds, staminodes surround the young styles in a coherent unit and the young anthers

Fig. 7.7. *Meliosma dilleniifolia* (Sabiaceae). Dots on nectary represent secretory appendages; numbers refer to order of initiation of the perianth.

fit in adaxial folds formed by the staminodes. At this stage filaments are abruptly bent inwards in the middle and the anthers are hidden from view. One stami-node is symmetrically developed and encapsulates one pollen sac of each anther; the two other staminodes have only one lobe developed, which enclose the other pollen sac. When the flower expands, filaments bend outwards and anthers become detached from the tight grip of the staminodes. The pollination mechanism might explain the heteromorphism of the petals; the broad petals enclose the bud and hold the staminodes erect, while the smaller often bifid petals do allow for the filament to curve outwards. Flowers of *Sabia* are regular with superposed sepals, petals and stamens, and without a dimorphic androe-cium. In all genera the gynoecium consists of two fused carpels bearing two parallel ovules, each with two connivent, occasionally twisted styles. The ovules become superposed at maturity due to space constrictions. The base of the ovary is surrounded by a conspicuous nectary, with five prominent appendages alter-nating with the stamens and staminodes. At maturity the nectary is either buoy-shaped with weak crenellations, or bears five prominent appendages topped with one to three stomata.

The arrangement of sepals, petals and stamens in opposite whorls is unusual for the angiosperms and is reflected in trimerous Berberidaceae and Menispermaceae. Wanntorp and Ronse De Craene (2007) derived the pentamerous flower of Sabiaceae from a spiral progenitor, but the opposite arrangement of stamens with petals fits well in the general Bauplan found in the early diverging eudicots.

Proteales

The association of Proteaceae, Nelumbonaceae and Platanaceae in a well-supported Proteales is one of the greatest surprises of molecular systematics, as the three families used to be placed in widely diverging groups (e.g. Cronquist, 1981: Rosidae, Magnoliidae, Hamamelidae). However, Proteales are probably end branches of a long evolution dating back about 120 million years, with a much higher fossil diversity as recognized for Platanaceae (Stevens, 2001 onwards; Von Balthazar and Schönenberger, 2009).

Proteaceae

Fig. 7.8. *Lomatia tinctoria* (Labill.) R.Br.

↓[P2+2 A2+2] G1

Most genera are bisexual, rarely dioecious (e.g. *Aulax*, *Leucadendron*). Flowers share a similar floral Bauplan throughout the family, although inflorescence structure can be highly variable (occasional formation of pseudanthia as in *Protea*). In Grevilloideae, flowers are always paired in the axil of a bract (Fig. 7.8), while in other subfamilies they are single. Douglas and Tucker (1996b) demonstrated that the two flowers are lateral branches of a short shoot. Individual flowers are generally subtended by a bract (not in *Lomatia*). The orientation of flowers relative to the main axis can be variable, ranging from straight to oblique (Douglas and Tucker, 1996b). The flowers are dimerous (apparently tetramerous), with two whorls of tepals (with valvate aestivation) and opposite adnate stamens. Stamens are rarely free from the tepals (e.g. *Symphyonema*). Interpretations of the perianth are controversial, ranging from petals to sepals (with loss of petals) or undifferentiated tepals. Tepals appear to be postgenitally fused and are apically reflexed at anthesis, exposing the attached anthers. Dehiscence of tepals is variable, either result-ing in a polysymmetric flower by equal partitioning, or monosymmetric by

Fig. 7.8. *Lomatia tinctoria* (Proteaceae): partial inflorescence. The upper level of the flower with stamen attachment is shown on the left flower.

a deeper slit on the adaxial side of the flower and the curving of all tepals towards the abaxial side (Fig. 7.8). A few Proteaceae are structurally monosymmetric, with a bilabiate perianth and abaxial (e.g. *Placospermum*) or adaxial (e.g. *Synaphea*) staminodes (Douglas and Tucker, 1996a, 1997). Half of the lateral anthers are sterile. In several *Protea* one stamen is sterile (Haber, 1966). The gynoecium is sessile or inserted on a gynophore and is always monocarpellate with two rows of ovules or a single basal or apical ovule. Orientation of the single carpel tends to be variable in the family with eight possible orientations in Grevilloideae, depending on the space available for initiation (Douglas and Tucker, 1996c). The stigma is club-shaped and functions as a receptor for pollen (secondary pollen presentation). Nectaries are present at the base of the ovary in alternation with the stamens (occasionally fewer: Fig. 7.8). They have been interpreted as petals (e.g. Eames, 1961; Haber, 1966), although there is no evidence to support this (Douglas and Tucker, 1996a; Ronse De Craene and Smets, 2001a).

Proteaceae appear to be more closely related with Platanaceae and share some floral characters, such as interstaminal scales and fruiting structures. *Platanus* has a whorl of scales alternating with stamens that were interpreted as staminodes and could represent a second stamen whorl, as would be the inner glands of Proteaceae (Von Balthazar and Schönenberger, 2009).

Nelumbonaceae

Fig. 7.9A,B. *Nelumbo nucifera* L., based on Hayes, Schneider and Carlquist (2000) and Eichler (1878)

✳/↔K2 C∞ A∞ G∞

Nelumbo was placed in Nymphaeaceae in premolecular classifications, based on strong convergent characteristics linked to a similar aquatic habitat (e.g. Ito, 1986b). However, a placement within Proteales comes as a surprise. The arrangement of solitary flowers on a submerged shoot is complex and little understood (Eichler, 1878; Stevens, 2001 onwards). Flowers are massive and have several characters in common with basal angiosperms (spiral petals, ascidiate carpels closed by secretion), although they are probably secondarily elaborated with an increase of petals, stamens and carpels (Hayes, Schneider and Carlquist, 2000). There is a hint of a dimerous flower in two outer sepals enclosing the flower. Petals arise in a spiral sequence and enclose the androecium of a very high number of stamens. The fact that stamens arise on a ring primordium (Hayes, Schneider and Carlquist, 2000) and a fascicled vascular supply (Ito, 1986b) is an indication that they were secondarily increased. Carpels are scattered and embedded in a flattened receptacle. Each carpel encloses a single apical ovule. There is no nectary tissue.

Trochodendrales

Trochodendraceae

Fig. 7.10. *Tetracentron sinense* Oliv.

✳P2+2 A2+2 G̲(4)

The family Trochodendraceae consists of two monospecific genera, *Tetracentron* and *Trochodendron*, with highly diverging morphologies. However, both share several less conspicuous characters (anther morphology, pollen, nectaries, carpels: Endress, 1986).

Trochodendraceae share a dimerous flower structure and absence of differentiation of the perianth with other early diverging eudicots.

Tetracentron is dimerous with flowers arranged spirally in catkin-like spikes (Chen *et al.*, 2007). Young flowers look conspicuously like some Araceae, although the carpels are diagonally inserted, a condition found also in Potamogetonaceae (Alismatales). In *Trochodendron*, flowers have indefinite stamen and carpel numbers (variably spiral or whorled). The number of stamens in *Trochodendron* ranges from 39 to around 70, and stamens are initiated in a helical sequence (Endress, 1986). Tepals are inconspicuous in *Tetracentron*, but reduced

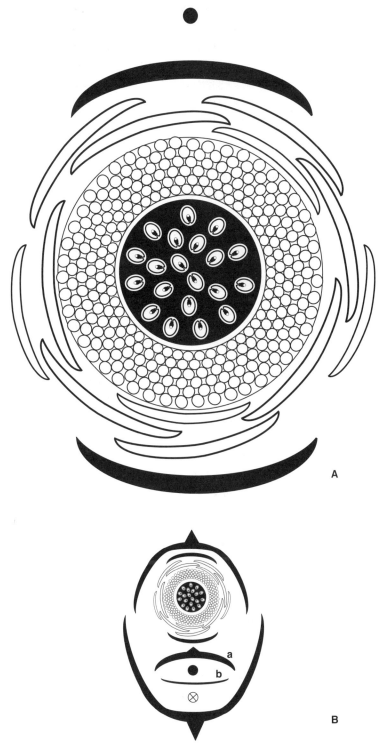

Fig. 7.9. *Nelumbo nucifera* (Nelumbonaceae): A. flower; B. flowering shoot. White dots represent stamens arising centripetally on a ring primordium. a, leaf; b, stipule of leaf.

Fig 7.10. *Tetracentron sinense* (Trochodendraceae). Nectary shown on the abaxial side of the carpels.

to small scales in *Trochodendron*, which has occasionally been interpreted as lacking tepals (Endress, 1986). Wu, Su and Hu (2007) recently confirmed the presence of tepals and suggested that the perianth was secondarily lost based on the presence of micromorphological characters of petals on several floral parts of *Trochodendron*. As in other early diverging eudicots (e.g. Buxaceae) or basal core eudicots (e.g. Gunneraceae), distinction between bracts and tepals is unclear (see Von Balthazar and Endress, 2002a,b; Wanntorp and Ronse De Craene, 2005; Ronse De Craene, 2008). *Trochodendron* has two tiny lateral bracteoles that could be homologized with the lateral tepals of *Tetracentron*. In both genera the anthers open by two lateral valves.

The ovaries of *Tetracentron* and *Trochodendron* are half-inferior, with a small residual area and basally fused plicate carpels (Chen *et al.*, 2007). In *Trochodendron* the carpel number is increased. Two rows of ovules are found on marginal placentae.

The dorsal side of the ovary develops as a nectary in both genera (Endress, 1986), and this is comparable to nectaries found in pistillate *Buxus*.

Buxales

The order contains two families, Buxaceae and Didymelaceae. Flowers are simple, dimerous (or spiral) and apetalous, as in several other core eudicots.

Buxaceae

Fig. 7.11A–C. *Buxus sempervirens* L., based on Von Balthazar and Endress (2002a,b)

Fig. 7.11. *Buxus sempervirens* L. (Buxaceae): A. staminate flower; B. pistillate flower; C. inflorescence.

Staminate: ↔P2+2 A2+2 G0

Pistillate: *P0 A0 G̲(3)

General formula: staminate: */↔P(0)2–4 A4-∞ G0; pistillate: *P0 A0 G(2)3(4)

Floral morphology of the family has been extensively documented by Von Balthazar and Endress (2002a,b). Flowers are unisexual and basically dimerous or spiral. Inflorescences are varied but are mostly short compact spikes composed of lateral staminate flowers and a terminal pistillate flower in *Buxus* (Fig. 7.11C). Within inflorescences the arrangement of flowers can fluctuate between a decussate and spiral phyllotaxis (Von Balthazar and Endress, 2002b).

Staminate flowers invariably have a dimerous arrangement with four tepals and four stamens. In *Styloceras* a perianth is missing. In *Notobuxus* median stamens are arranged in a series of six or eight stamens, but it remains uncertain whether there are one, two or three whorls of stamens (Von Balthazar and Endress, 2002b: Fig. 10). In staminate flowers the gynoecium is replaced by a flat nectariferous pistillode or is absent (*Styloceras*).

Pistillate flowers invariably arise in a terminal position. The two to three carpels are surrounded by a variable number of bract-like phyllomes arising in a spiral sequence. In *Buxus*, pistillate flowers bear nectaries between the stylar lobes that appear convergent with the septal nectaries of monocots (Smets, 1988), or are probably homologous to the nectaries of Trochodendraceae. Each carpel is topped by a massive style with ventral stigma and contains two parallel pendent ovules on an axile placentation. In *Styloceras* and *Pachysandra* a false septum divides the ovary in four locules (Von Balthazar and Endress, 2002a). In *Pachysandra* and *Sarcoccoca* only staminate flowers are nectariferous and nectarless pistillate flowers take advantage of this (Vogel, 1998b).

Especially for pistillate flowers it is unclear whether a perianth is present or not. Bracts and tepals cannot be distinguished morphologically as they arise in a continuous spiral sequence, except for a slightly longer plastochron between presumed sepals and bracts, and occasional fusions of the upper bracts (Von Balthazar and Endress, 2002b). On this basis Von Balthazar and Endress (2002b) interpreted the bracts immediately preceding the reproductive organs as a weakly differentiated perianth and an experimental phase in the elaboration of a perianth, as found in core eudicots. However, it is more likely that a perianth is missing in pistillate flowers and perhaps in staminate flowers, as part of an evolutionary process of reduction affecting the early diverging eudicots and culminating in Gunnerales (see p. 149).

8

Basal core eudicots: the event of pentamerous flowers

Core eudicots are a strongly supported clade, which represents 70% of all Angiosperms (Fig. 8.1; Soltis *et al.*, 2005). Pentamerous flowers with a differentiation of a calyx and corolla are generalized, and there are good indications that a bipartite perianth, two stamen whorls and isomerous carpel whorl represent a condition acquired very early in the clade. However, lack of resolution among major lineages of core eudicots does not allow for a clear understanding of floral evolution, although progress is being made in solving this issue.

8.1 Gunnerales and Berberidopsidales

Gunnerales

The small order consists of two monogeneric families, Gunneraceae and Myrothamnaceae. Recent molecular analyses (e.g. Soltis *et al.*, 2003) have placed Gunnerales as the sister group of all core eudicots. Gunneraceae have traditionally been associated with Haloragaceae (e.g. Schindler, 1905), mainly because their reduced flower morphologies look similar. The dimerous floral Bauplan of Gunnerales tends to be more similar to the basal eudicot grade, such as Buxales or Trochodendrales, than to the core eudicots and this has led to a questioning of the origin of pentamery from dimerous ancestors at the base of core eudicots (Ronse De Craene, 2004; Wanntorp and Ronse De Craene, 2005). Within Gunneraceae there is a general pattern to floral reduction and unisexual flowers, and it appears structurally difficult to derive pentamerous flowers from a prototype such as *Gunnera* (Wanntorp and Ronse De Craene, 2005; Ronse De Craene and Wanntorp, 2006). Absence of clear intermediates, both extant and fossil, makes the understanding of floral evolution difficult (Ronse De Craene,

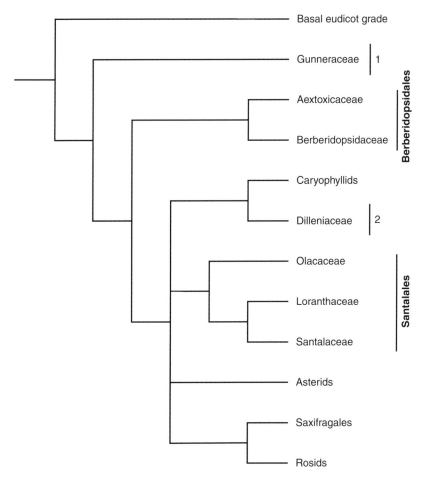

Fig. 8.1. Phylogenetic tree of core eudicots, based on Soltis *et al.* (2003). 1. Gunnerales; 2. Dilleniales.

Soltis and Soltis, 2003; Ronse De Craene, 2008) and the floral evolution of Gunnerales may well be a unique specialization linked to wind pollination, as suggested by Wanntorp and Ronse De Craene (2005). Flowers of Myrothamnaceae are even more reduced, without clear differentiation of a perianth and a distinction between staminate and pistillate plants. (e.g. Jäger-Zürn, 1966). Flowers tend to be labile, with variable number of stamens and carpels and a perianth of uncertain homology.

Gunneraceae

Fig. 8.2. *Gunnera manicata* Linden ex André

↔K2 C2 A2 Ǧ(2)

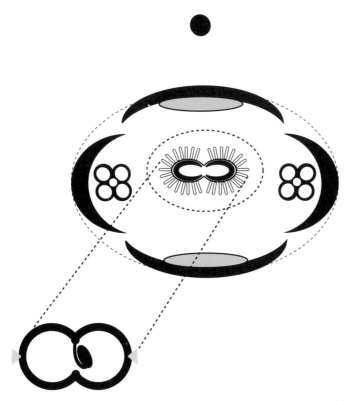

Fig. 8.2. *Gunnera manicata* (Gunneraceae). Glandular tissue consists of hydathodes.

Flowers are grouped in dense compound inflorescences, with each partial inflorescence ending with a fully developed terminal flower. The basic floral condition is generally found in subgenus *Panke* to which *G. manicata* belongs (Ronse De Craene and Wanntorp, 2006; González and Bello, 2009). There is a progressive reduction from the distal side of the inflorescence to the proximal side with loss of petals and stamens. Stamens and petals tend to be genetically connected, and arise from common primordia (González and Bello, 2009). Loss of stamens tends to be correlated with reduction of petals and corresponds with several independent derivations of unisexual flowers in the different subgenera of *Gunnera*. Ronse De Craene and Wanntorp (2006) provided a diagram linking different flower types. They demonstrated a clear tendency for unisexuality to be linked with the loss of perianth parts. Flowers are wind-pollinated and petals tend to play a role in anther protection and pollen release (González and Bello, 2009).

Flowers are unusual in the dimerous, almost distichous arrangement of petals, stamens and carpels. Sepals are median and persistent and were compared to bracteoles (e.g. Soltis *et al.*, 2003); they have an inflated base by the presence of a hydathode, while the apical part shrivels at maturity (Ronse De Craene and Wanntorp, 2006; González and Bello, 2009). Flowers have a half-inferior to inferior ovary with a single apical ovule; two strongly developed styles are usually present, especially in unisexual flowers, where they take up the greatest area of the flower (so called style-flowers: Ronse De Craene and Wanntorp, 2006).

A possible interpretation for the unusual distichous arrangement of petals, stamens and carpels was proposed by Wanntorp and Ronse De Craene (2005) as the result of partial loss of organs from a dimerous, decussate ancestor similar to Buxaceae or *Tetracentron*.

Berberidopsidales

The order was created by Soltis *et al.* (2003) to accommodate two families, Berberidopsidaceae and Aextoxicaceae, at the base of the core eudicots. Berberidopsidaceae used to be placed in the heterogeneous Flacourtiaceae, and Aextoxicaceae used to have an uncertain position somewhere close to Euphorbiaceae. Morphological evidence for a close affinity between the two families was given by Ronse De Craene (2004, 2007), who also demonstrated the basal position of *Berberidopsis* in the core eudicots on the basis of floral developmental evidence. Both families represent a transitional stage in the evolution of core eudicots, as they possess spiral flowers with an undifferentiated perianth or with petals of a transitional nature. Berberidopsidales have been variously placed in core eudicots, at the base of the clade (Soltis *et al.*, 2003) or more recently as sister to a clade consisting of asterids and caryophyllids (Wang *et al.*, 2009).

Berberidopsidaceae

Fig. 8.3. *Berberidopsis corallina* Hook. f.

✳P12 A8 G̲(3)
General formula: ✳P12–17/K5 C5 A6–13-∞ G3–5

Two species of *Berberidopsis* and the monotypic *Streptothamnus* make up Berberidopsidaceae. The flower of *Berberidopsis* has an undifferentiated coloured perianth, although the outer tepals remain persistent in fruit. Floral development is spiral throughout, with an average of 12 tepals and 8 stamens. There are three carpels with parietal placentation. Ronse De Craene (2004) showed that the spiral flowers have a predisposal for a pentamerous arrangement as found in core eudicots and hypothesized that the pentamerous bipartite flower of core eudicots was derived by a progressive differentiation of an outer and inner whorl of tepals into sepals and petals. Kubitzki (2007a) criticized the assumption that *Berberidopsis* is a basal link in floral evolution, on the basis

Fig. 8.3. *Berberidopsis corallina* (Berberidopsidaceae); numbers give order of initiation of the perianth.

that the family has several derived characteristics, including the parietal ovary, differentiation of calyx and corolla and multiple androecium in *Streptothamnus*. However, it needs to be emphasized that *Berberidopsis* is to be interpreted as a model for the evolution in core eudicots, and not necessarily as a progenitor, as it is a well-known fact that characters can evolve in different degrees in flowers.

Berberidopsis corallina has eight stamens in a single whorl, while the number is much higher in other species but is currently unknown. Anthers are broad with protruding connective and the filament is short. A nectary develops as an extrastaminal ring.

Aextoxicaceae

Fig. 8.4A,B. *Aextoxicon punctatum* Ruíz & Pav.

∗K5–6 C5–6 A5–6 G̲1

The monotypic genus *Aextoxicon* is morphologically highly different from *Berberidopsis*, although several characters tend to be common to both genera,

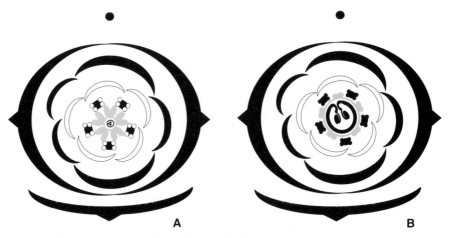

Fig. 8.4. Staminate (A) and pistillate (B) flower of *Aextoxicon punctatum* (Aextoxicaceae). Note the difference in size between nectaries of staminate and pistillate flowers.

including wood anatomy and micromorphological evidence (Carlquist, 2003; Ronse De Craene, 2004, unpubl. data). Flowers are functionally staminate or pistillate, with a differentiation of genders arising late in the development. Two bracteoles enclose the flower bud as a calyptra that is shed at anthesis. The same decussate arrangement as the bracteoles is found in the outer sepals, where a transition to a spiral initiation occurs (Ronse De Craene, unpubl. data). The ovary is monocarpellate, with variable position and with two apical pendent ovules. The style is very short and compressed against the globular ovary. The nectary consists of broad wing-like appendages inserted between the stamens; glandular tissue is more extensively developed in staminate flowers due to space restrictions in the pistillate flowers.

Aextoxicon compares well with *Berberidopsis* in the spiral flowers, with clear homology of perianth parts. Both families represent a clear transitional stage in the evolution of the core eudicots. Character mapping has shown that there is a sudden transition from simple dimerous flowers, as in the early diverging eudicots immediately preceding the core eudicots, to the pentamerous core eudicots with bipartite perianth (Ronse De Craene, 2008). Berberidopsidales appear to fit as a morphological transition between Ranunculales and core eudicots although this is not clear from character reconstructions.

8.2   Santalales

The order contains seven families, with a basal paraphyletic Olacaceae (Nickrent and Malécot, 2001; Malécot and Nickrent, 2008). While Santalales have been considered a natural group for a long time, affinities of the order are still little supported. Santalales have several synapomorphies or shared tendencies as families become increasingly parasitic along the lineage, and this is reflected in the reduction of flowers. The basal Olacaceae fit in the core eudicot Bauplan as they have the most regular flowers with a differentiated perianth, two stamen whorls and three to five carpels (Wanntorp and Ronse De Craene, 2009).

All Santalales share broad valvate petals, which develop at the expense of the calyx. While a calyx is present and accrescent in some Olacaceae, it is lost or replaced by a rim (calyculus) in other families (Loranthaceae, Opiliaceae). No trace of sepals is present in Santalaceae, Viscaceae and Misodendraceae. The calyculus has been variously interpreted in the past as a receptacular emergence (e.g. Eichler, 1878; Engler and Krause, 1935), as a reduced calyx (e.g. Sleumer, 1935; Endress, 1994) or as derived from bracteoles (Venkata Rao, 1963). On the basis of floral developmental evidence, Wanntorp and Ronse De Craene (2009) supported the interpretation of the calyculus as a structure of bracteole origin. Obhaplostemony is present in most taxa by loss of antesepalous stamens. A few genera of Olacaceae are diplostemonous (e.g. *Heisteria*, *Strombosia*). The gynoecium is rarely isomerous, mostly tricarpellate and superior to inferior with a single style. The placentation is free-central, usually with three ovule primordia that grow into appendages without integuments. A nectary develops as a disc around or on top of the ovary. Several Santalales have tufts of hair between the petals and the stamens. Their function is currently unknown but they may play a role in the retention of nectar (Endress and Matthews, 2006b).

The floral structure and development of several of the former Olacaceae is still unknown but can give important clues about the evolution of flower and perianth in the order (cf. Wanntorp and Ronse De Craene, 2009 and ongoing research).

Olacaceae

Fig. 8.5. *Coula edulis* Baill.

$\ast$K?C5A5^3+5 $\underline{G}$(3)
General formula: $\ast$ K(0)3–6 C3–6 A3–12(-18) G(2-)3(-5)

Olacaceae are recognized as paraphyletic and several genera, including *Coula*, belong to different clades. *Olax* is more closely related to Loranthaceae and shares a similar calyculus (Wanntorp and Ronse De Craene, 2009). More basal Olacaceae (Olacaceae *sensu stricto*) have a well-differentiated calyx and

Fig. 8.5. *Coula edulis* (Olacaceae). Note alternipetalous triplets of stamens and calyculus.

corolla with diplostemonous flowers (e.g. *Strombosia*, *Heisteria*) and a tendency for loss of antesepalous stamens (e.g. *Diogoa*), very rarely antepetalous stamens (e.g. *Heisteria pentandra*: Sleumer, 1935). The calyx, while being narrow and basally fused, is accrescent in some genera (e.g. *Heisteria*) and plays an important role in seed dispersal by a difference in colour with the ovary. In other genera a small rim is found which is occasionally difficult to differentiate from a calyculus, but in *Schoepfia* (now placed close to Loranthaceae) a calyculus consisting of the fused bract and bracteoles is found below the rim (Sleumer, 1935). Petals are usually thick and valvate, compressing the stamens against the ovary in bud. *Olax* has hexamerous flowers with petals fused into pairs and a variable number of stamens alternating with staminodes. The ovary is (two-)three- to five–carpellate and septa are often formed leading to axile placentation. However, there is a tendency for septa to become reduced (e.g. *Cathedra*). Integuments are normally developed.

   Coula is exceptional in Santalales because it has 12–20 stamens in a diplostemonous arrangement, with antesepalous stamens arising as triplets. The antepetalous stamens are occasionally missing. The calyx is represented by a shallow rim with narrow teeth, comparable to a calyculus. Ontogenetic studies should confirm whether this is derived from the fusion of two bracteoles or whether it represents a reduced calyx (Wanntorp and Ronse De Craene, 2009). Previous authors interpreted the smaller lateral stamens of *Coula* as pairs. Sleumer (1935: 7) showed floral diagrams of several Olacaceae, including *Coula* with 15 stamens.

Loranthaceae

Fig. 8.6. *Phthirusa pyrifolia* Eichler

Fig. 8.6. *Phthirusa pyrifolia* (Loranthaceae): partial inflorescence.

✳ K0 [C6/3+3 A6/3+3] Ğ(3)

Viscaceae have often been associated with Loranthaceae (e.g. Eichler, 1878; Engler and Krause, 1935). However, their affinities are closer to Santalaceae with whom they share simplified flowers without calyculus (Malécot and Nickrent, 2008).

Several former Olacaceae appear to be closely related to Loranthaceae and Opiliaceae (including *Olax* and *Schoepfia*). All share the presence of a calyculus enwrapping the flower. The calyculus is of obvious bracteole origin and in *Phthirusa* it results from a further reduction of complex dichasial inflorescences as the calyculus is surrounded by an external tier of two bracteoles and a bract (Fig. 8.6). In *Struthanthus* there are no extra bracteoles (Wanntorp and Ronse De Craene, 2009).

In most Loranthaceae flowers are arranged as triplets (basically a dichasium), and bracts and bracteoles (in cases where they are present) are basally fused

(Engler and Krause, 1935). Flowers are bisexual or unisexual, with the laterals of the triplet staminate (*Nuytsia*: Narayana, 1958b). Flowers are tetramerous, pentamerous or hexamerous (Fig. 8.6), or even occasionally heptamerous (e.g. *Phthirusa, Nuytsia*). Hexamerous flowers appear trimerous (cf. Polygonaceae) because petals and stamens are arranged in two whorls of different size (e.g. *Phthirusa, Loranthus*). The petals have valvate aestivation and can have a long basal tube (e.g. *Psittacanthus*) and stamens appear to be fused with the petals, as in Proteaceae (Venkata Rao, 1963). The flower may appear monosymmetric by unequal division of the corolla lobes. This is probably caused by space restrictions. The ovary is inferior and is often covered by a broad disc surrounding a central erect style. The ovary is rarely septate (*Lysania*: Narayana 1958a), usually unilocular with central placenta or is solid (e.g. *Phthirusa*). Ovules lack a nucellus and integuments. Only one ovule develops as a single seed. Narayana (1958b) presents a progressive reduction series in the family, from a well-developed central placenta with ovules to a complete suppression of the central column.

Santalaceae

Fig. 8.7. *Thesium strictum* Berg

Fig. 8.7. *Thesium strictum* (Santalaceae): partial inflorescence.

✳ K0 [C5 A5] Ğ(3)
General formula: ✳ K0 C3–5 A3–5 G (2–5)

Santalaceae closely resemble Loranthaceae morphologically but they lack the typical calyculus of the latter. Flowers are bisexual or unisexual with a residue of the other gender (e.g. *Osyris*). Inflorescences are varied but a dichasial arrangement of flowers is most common. Flowers are pentamerous, tetramerous or trimerous, mostly subtended by a bract and two bracteoles, which may be rarely absent (e.g. *Thesium ebracteatum*: Pilger, 1935). In *Quinchamalium* the bract and bracteoles fuse into a cup-like involucre that resembles a calyx and was described as a calyculus by Pilger (1935). Santalaceae have relatively simple flowers with a valvate corolla and antepetalous anthers. Petals have occasionally been described as tepals, but a comparison with other Santalales supports the interpretation of a corolla. The odd petal is abaxial in pentamerous flowers, as is to be expected for petals. Filaments are usually very short and anthers are broadly flattened. A tuft of hairs usually develops between petal and stamen. Stamens and petals are often lifted by a hypanthium enclosing the inferior ovary. The gynoecium is often inferior, rarely half-inferior or superior, and the style is embedded in a broad dish-like nectary that often extends as lobes between the stamens (e.g. *Santalum, Osyris*). The number of carpels is usually three, rarely less (two) or more (four to five). No septa are formed and the central axis develops as a placental column bearing three ovular protuberances without integuments (see Sattler, 1973 for *Comandra*). The placental column remains short or can grow into a coiled snake-like protuberance filling the ovarian cavity.

8.3   Dilleniales

Dilleniaceae

Fig. 8.8. *Hibbertia cuneiformis* (Labill.) Smith

$*$ K5 C5 A5^4 G5
General diagram: $*$/$\downarrow$ K(3)4–5(-18) C(2)3–5(-7) A(1–3)5–400(-900) rarely 5 +5 G1–10(20)

Dilleniaceae were viewed as an archaic family at the base of Dilleniidae in the past (e.g. Cronquist, 1981). The position of Dilleniaceae is debatable, as the family has been variously associated with caryophyllids and asterids (e.g. Stevens, 2001 onwards; Soltis *et al.*, 2003). However, there is little morphological evidence to link Dilleniaceae with Caryophyllales, except for the centrifugally developing multistaminate androecium, persistent calyces, campylotrous ovules and successive cambia in the wood. A better understanding of floral structure and development of basal Caryophyllales, such as Rhabdodendraceae, might give clues about the relationships of Dilleniaceae (Horn, 2007).

The family is highly variable in number and arrangement of parts, although the greatest variation is found in the androecium. Dilleniaceae are nectarless pollen flowers, principally bee pollinated and usually with many stamens

Fig. 8.8. *Hibbertia cuneiformis* (Dilleniaceae). Note the protruding styles between the stamen fascicles.

(Endress, 1997). Characteristic features of the family are: imbricate perianth aestivation (quincuncial in pentamerous flowers), accrescent calyces, showy, (usually) polysymmetric petals crumpled in bud, secondary polyandry with common primordia or a ring primordium, and basifixed anthers (Judd and Olmstead, 2004; Horn, 2007).

Flowers can be terminal and solitary, or are arranged in various cymose inflorescences.

Dilleniaceae have a spiral perianth with imbricate sepals and petals, but the flower is clearly pentamerous with several evolutionary patterns, especially in the androecium. This correlates well with other core eudicots, especially rosids. Merism can be highly variable, although pentamery is predominant and is probably the ancestral condition. The highest variation in merism is found in the calyx of *Tetracera*, ranging from two smaller and three larger sepals to 7–15 helically arranged sepals (Horn, 2007). The number of petals can be reduced to three in some species of *Hibbertia*. *Davilla* has three smaller outer and two larger inner sepals. The latter enclose the bud as two lids before and after anthesis (pers. obs.).

Tucker and Bernhardt (2000) showed a remarkable variation in stamen number and development in *Hibbertia*. The number of stamens is increased by the development from common primordia, but several patterns have led to further increases or reductions. The total number of stamens depends on the number of common primordia and the duration of meristematic activity. Most Dilleniaceae are polyandric and there is a remarkable correlation between the number of carpels and the extent of development of the androecium. The most common development is five common stamen primordia alternating with five carpels. With three carpels, two stamens remain simple and the three alternating with the carpels are common primordia dividing into more stamens (e.g. *H. stellaris*: Ronse De Craene, unpubl. data). Many more stamens (up to 900 in some *Dillenia*) arise by the development of a ring primordium and this is often correlated with a lateral increase of carpels up to ten (20) and much larger flowers (Endress, 1997). The outermost stamens often remain sterile.

Heteranthy occurs in some genera, with different sets of inner and outer stamens.

Some species of *Dillenia*, *Hibbertia* and *Didesmandra* are monosymmetric by the reduction of one side of the flower. Petal initiation is unidirectional instead of helical and the androecium is restricted to the adaxial side of the flower, occasionally with the presence of staminodes on the abaxial side (*H. empetrifolia*, *H. hypericoides*: Tucker and Bernhardt, 2000). Zygomorphy is correlated with a reduction of carpels to two and can result in a single stamen only.

Hibbertia (subg. *Adrastea*) *salicifolia* was described as an obdiplostemonous flower with ten stamens (Eichler, 1878). The floral development is highly unusual in the centrifugal initiation of the antepetalous stamens on a ring primordium and heteranthy between the two whorls of stamens. In my opinion the development appears to be an interrupted centrifugal multiplication of the androecium, similar to an analogous development in *Triumfetta* (Malvaceae: van Heel, 1966).

Several authors have suggested that polyandry is ancestral in the family (e.g. Dickison, 1970; Horn, 2007), although this seems of little consequence in the light of the high variation in stamen development that can be linked to a single stamen whorl in most cases.

The gynoecium is normally apocarpous. In some *Dillenia* with a specific pollination mechanism carpels can be fused by basal concrescence to the receptacle (Endress 1997). Higher carpel numbers are clearly derived, and arrangement is in one whorl, rarely two whorls (*H. grossulariifolia*: Tucker and Bernhardt, 2000). The latter condition is questionable, as carpels may have become displaced in the flower. Carpels are always opposite the petals in cases where they are isomerous; with fewer carpels the position can vary. The orientation of monosymmetric flowers of *Hibbertia* is transverse and so are the two carpels, in contrast to *Hibbertia* subg. *Adrastea* with median orientation (see Eichler, 1878; Horn, 2007). Ovules are inserted in two, up to six rows per carpel, or reduced to one to two with basal placentation.

A diagram of *H. cuneiformis* was shown by Baillon (1868a); diagrams of other *Hibbertia* were shown in Tucker and Bernhardt (2000: 1919).

9

Caryophyllids: how to reinvent lost petals

The Caryophyllid clade or Caryophyllales *sensu lato* contains about 29 families grouped in two major clades (Fig. 9.1; Cuénoud *et al.*, 2002; Brockington *et al.*, 2009). A natural Caryophyllales ('Centrospermae') has been recognized for a long time, mainly on the basis of embryological and phytochemical characters (e.g. Mabry, 1977). Inclusion of molecular characters has increased the size of the order dramatically by adding carnivorous plant families (including Droseraceae and Nepenthaceae), knotweeds (Polygonaceae) and some halophytic groups (Frankeniaceae, Tamaricaceae and Plumbaginaceae). Except for carnivorous families, these taxa were often associated with Caryophyllales in the past. Whether these taxa should be grouped with core Caryophyllales or be separated within the Polygonales depends on a better resolution of phylogenetic relationships (e.g. Cuénoud *et al.*, 2002; Brockington *et al.*, 2009).

Core Caryophyllales have a number of floral features in common that clearly separate the clade from remaining core eudicots. Flowers are basically pentamerous and apetalous, with a derived insect-pollination syndrome. The ovary regularly has a free-central placentation by break-up of septa. Styles are typically separate or there are well-developed style branches. Ovules and seeds are characteristically curved (campylotrous) and the reserve tissue of seeds often contains perisperm. Nectaries are typically situated on the inner side of the stamens or stamen tube.

However, the addition of families previously thought to be unrelated alters the context in which to approach characters of the group. Members of Polygonales contain families where petals are present and others without (e.g. Polygonaceae, Nepenthaceae). Another characteristic, which appears synapomorphic, is the absence of bracteoles and the lateral position of the outer sepals (Ronse De Craene, 2008). Placentation is mostly parietal, spreading on the ovary floor, or basal.

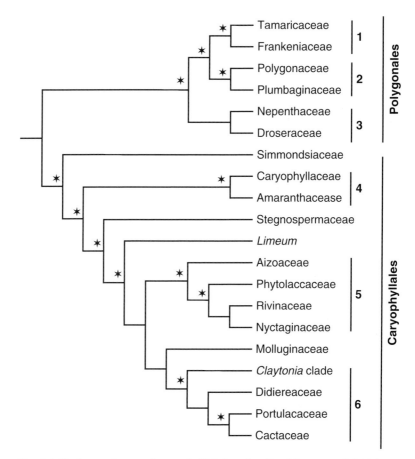

Fig. 9.1. Phylogenetic tree of caryophyllids based on Brockington *et al.* (2009).
1. Tamaricaceae-clade; 2. Polygonaceae-clade; 3. Droseraceae-clade;
4. Caryophyllaceae-clade; 5. Aizoaceae-clade; 6. Portulacaceae-clade. Asterisk
denotes branches with more than 90% support.

The affinities of caryophyllids are unclear (e.g. Soltis *et al.*, 2005). They emerge
from the core eudicot polytomy in a weak sister group relationship with
Dilleniaceae or with asterids. Floral and vegetative morphologies indicate that
caryophyllids have become widely adapted to extreme environments, with
unique floral mechanisms.

9.1   Polygonales

'Droseraceae-clade'

Molecular data have demonstrated that five carnivorous families
(Droseraceae, Nepenthaceae, Drosophyllaceae, Dioncophyllaceae, Ancistrocladaceae)

are in fact closely related (see e.g. Williams, Albert and Chase, 1994). They share a syndrome of glandular hairs becoming specialized in trapping insects. However, floral morphology is highly variable, with loss of petals in Nepenthaceae. Ancistrocladaceae and Dioncophyllaceae are basically diplostemonous.

Nepenthaceae

Fig. 9.2A,B. *Nepenthes alata* Blanco (staminate) and *N. hirsuta* Hook. f. (pistillate)

Staminate: ✳ K4 A4+4+1* G0
 *apical stamen not present in all flowers
Pistillate: ✳ k4 A0 G̲(4)
General formula (combined): ✳ K (3)4–(5–6) A4–24 G(3)4

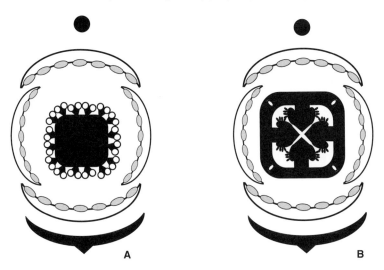

Fig. 9.2. Nepenthaceae. A. *Nepenthes alata*: staminate flower, B. *N. hirsuta*: pistillate flower.

Inflorescences are racemose, although flowers can be paired and all flowers mature synchronously suggesting a derivation from a cymose inflorescence (Macfarlane, 1908). Flowers are dioecious, subtended by a single bract and are mostly tetramerous. The perianth is similar in pistillate and staminate flowers and is interpreted as a calyx with an initiation of two alternating pairs; there are rarely two whorls of three sepals (Eichler, 1878). Sepals bear large glands on the adaxial side (Macfarlane, 1908; Endress, 1994). Stamens range from four to 24, often with a terminal anther and no trace of the gynoecium. Stern (1917) described the development of the stamens as alternating whorls of four. The extrorse stamens are grouped on a common synandrium with expanding andro-phore. In cases where only four stamens are present they alternate with the sepals. Preliminary observations of the development of staminate flowers (Ronse De Craene, unpubl. data) show the initiation of alternating whorls. A pair or a single stamen may be apically present or a stub is visible in young stages. In older buds all thecae (with two pollen sacs) are arranged equidistantly as anther tissue is embedded in the common column. This arrangement closely resembles staminate flowers of Myristicaceae. Pistillate flowers have four carpels opposite the sepals and broad, mostly sessile, stigmatic lobes. The placentation is generally described as axile, but is in reality a deeply intruding parietal placentation (Stern, 1917; Ronse De Craene, unpubl. data). Ovules are small and arise in two to four rows.

Droseraceae

Fig. 9.3A,B. *Drosera capensis* L.

✳K5 C5 A5 G̲(3)

General formula: ✳K5 C5 A5-∞ G3–5

Fig. 9.3. *Drosera capensis* (Droseraceae): A. flower, numbers refer to initiation sequence of stamens; B. inflorescence.

Flowers arise in an erect or curved cincinnus. No bracteoles are present (erroneously presented by Eichler, 1878) and the outer sepals are consequently inserted in a lateral position. The corolla is imbricate (2/5), contorted or cochleate ascending. Stamens are haplostemonous, or secondarily increased to up to 20 stamens by an irregular lateral division of antesepalous and antepetalous stamens (*Dionaea*: Ronse De Craene, unpubl. data). Initiation of stamens in

Drosera is inversed in a similar fashion to many Caryophyllales, starting with a stamen opposite sepal four and ending with the one opposite sepal one (Ronse De Craene, unpubl. data). Anthers are extrorse as in *Nepenthes* (not in *Dionaea*). Carpels are antepetalous in cases where isomerous, or most often three with parietal-basal placentation. In *Dionaea*, ovules spread over the ovary floor in five indistinct groups. Styles are short, carinal and variously branched, comparable to Polygonaceae. Nectaries appear to be absent in the family.

'Tamaricaceae-clade'

Tamaricaceae and Frankeniaceae share several floral characters besides their halophytic adaptations, including fused sepals (long tubes in *Frankenia*), free contorted petals (often with basal adaxial appendages overlapping each other in a direction opposite to the petal margins), a basically diplostemonous androecium with stamen tube and a superior ovary with parietal placenta.

Tamaricaceae

Fig. 9.4. *Reaumuria vermiculata* L., based on Ronse De Craene (1990)

∗K5 C5 A5+5$^{\infty}$ G̲(5)

General formula: ∗K(4–5)C(4–5)A(5–10–∞) G̲(3–5)

Fig. 9.4. *Reaumuria vermiculata* (Tamaricaceae); numbers refer to first initiated stamens.

Inflorescences are racemose or flowers are terminal surrounded by several bracts (*Reaumuria*). The latter was interpreted as a derivation from a cymose inflorescence by loss of lateral flowers (Ronse De Craene, 1990). Bracteoles are typically absent and the outer sepals are in lateral position (Eichler, 1878). Petals have a variable aestivation (contorted to quincuncial) and broad lateral appendages are found in *Reaumuria*, but not in the other genera (Ronse De Craene, 1990). Stamens are clearly of different length with an obdiplostemonous arrangement (*Myricaria*) or the androecium is often haplostemonous in *Tamarix*. Stamens are extrorse to introrse and basally fused. The inside of the stamen tube of *Tamarix* develops into a nectary that can develop into various shapes (Zohary and Baum, 1965). Carpels are antepetalous in cases where they are isomerous; with three carpels, two are oriented in an adaxial position (Eichler, 1878). The ovary contains two to several ovules grouped at the base of parietal placentae, which are occasionally intruding but never connected in the middle. The styles are carinal (*Tamarix*) or commissural (*Myricaria*).

The androecium of *Reaumuria* is polyandrous, with a diplostemonous groundplan and a centrifugal proliferation of antepetalous stamens. The androecium develops as broad wings covered with stamens (Ronse De Craene, 1990). Tamaricaceae superficially resemble some Malvales (Cistaceae) in the similar gynoecium and contorted aestivation. The ovary shows the same development as in Droseraceae.

Frankeniaceae

Fig. 9.5A,B. *Frankenia laevis* L.

✳K(5) C5 A5+1 G(3)
General formula: ✳K5 C5 A(3)-6(-24) G3

Inflorescences are terminal dichasia with lateral flowers enclosed by transversal bracteoles. A second set of bracteoles arises within each set and the four bracteoles are united at the base. The calyx is tubular and valvate with an unusual initiation of two pairs followed by a single sepal between the first pair (Ronse De Craene, unpubl. data), contrary to the 2/5 sequence reported by Payer (1857) for *F. pulverulenta*. This corresponds with a transition of a decussate arrangement of bracts to a pentamerous flower. Petals have a contorted aestivation and a broad basal appendage interlocking neighbouring petals. The androecium consists of six stamens (rarely three or more) apparently arranged in two trimerous whorls of different length. However, in reality five stamens are situated opposite the sepals, and a sixth opposite one of the petals (Payer, 1857; Ronse De Craene, unpubl. data). The androecium can be interpreted as basically diplostemonous with loss of the four antepetalous stamens. This may be caused by pressure of the gynoecium in the limited space of small flowers; *F. boissieri* has five stamens only, and in *F. triandra* the number varies from

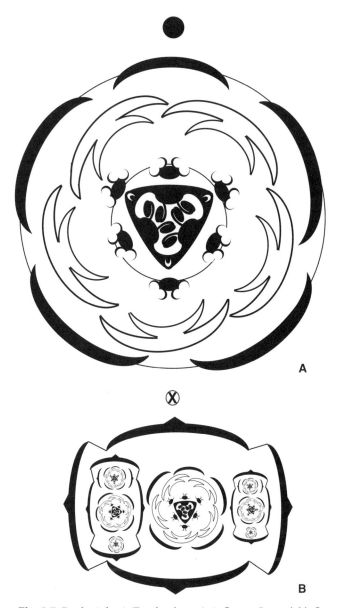

Fig. 9.5. *Frankenia laevis* (Frankeniaceae): A. flower; B. partial inflorescence.

three to six, occasionally with the shorter ones staminodial (Niedenzu, 1925). *Hypericopsis* has a stamen increase to 20–24 stamens. All stamens have inflated filaments with extrorse anthers and are fused into a tube. The gynoecium is trimerous with parietal placentation and an inversed orientation with the odd carpel in adaxial position, as in Tamaricaceae (Eichler, 1878).

The ventral scales of *Frankenia* and *Reaumuria* are comparable to ventral petal scales in Caryophyllaceae (e.g. Endress and Matthews, 2006b) and could indicate a closer affinity with Caryophyllales, as indicated by earlier authors (e.g. Friedrich, 1956). However, petals in Caryophyllales are interpreted as stamen-derived (see Ronse De Craene, Smets and Vanvinckenroye, 1998), while this is less clear in Polygonales.

'Polygonaceae-clade'

The clade shares a lateral insertion of the outer sepals with the other members of Polygonales. It differs in the basal position of the single ovule and presence of bracteoles.

Polygonaceae

Fig. 9.6A. *Rheum webbiana* Royle

✲K3+3 C0 A6+3 G(3)

Fig. 9.6B,C. *Persicaria lapathifolia* (L.) Gray

✲/↔K5 C0 A5+1 G(2)
General diagram: ✲K2–6C0 A2–9-∞ G2–3(-4)

Inflorescences are variable, with partial inflorescences consisting of cluster-like dichasial or monochasial cymes subtended by a saccate bract (Fig. 9.6C; Brandbyge, 1993). The bracteoles are considered to be fused into an ocreola and shield one flower or a group of flowers; free bracteoles are present in *Triplaris* and *Coccoloba* (Eichler, 1878). Larger sheathing bracts may surround groups of flowers, which appear axillary at the nodes. Flowers are basically apetalous with a mostly petaloid calyx. All flowers have a hypanthium and superior ovary. Some clades are wind-pollinated with inconspicuous perianth (e.g. *Rumex*), while insect pollination has developed in all major subfamilies. Flowers are hermaphrodite, and occasionally dioecious with aborted anthers and ovules (e.g. *Fallopia japonica*, *Rumex*). Although the perianth of Polygonaceae is always spirally initiated, there is a tendency for a cyclic arrangement. The outer perianth parts are often well differentiated from the inner, especially in hexamerous (trimerous) flowers at fruiting stage (e.g. *Fallopia*, *Rumex*, *Triplaris*), or the differentiation is gradual with a difference in petaloidy (e.g. *Polygonum*, *Persicaria*). Stamen number ranges between nine and three in trimerous flowers and eight and three in pentamerous flowers. A secondary increase of stamens is found in *Symmeria*, but also *Calligonum* with a lateral increase of stamens in each whorl (Galle, 1977). A well-developed hypanthium

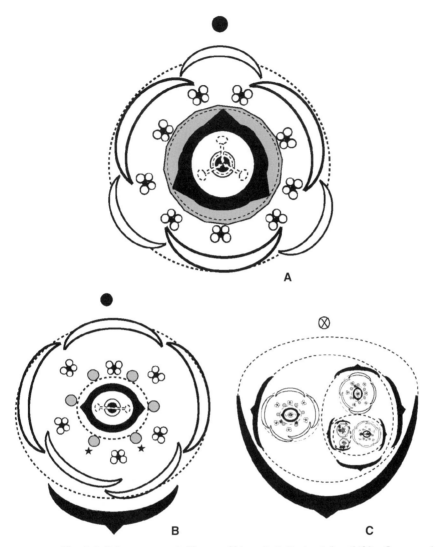

Fig. 9.6. Polygonaceae: A. *Rheum webbiana*, B–C. *Persicaria lapathifolia*, flower and partial inflorescence.

is present, placing the stamens and tepal lobes at different levels relative to the gynoecium. The floral diagram of Polygonaceae is highly variable, although mostly stable in major clades. In pentamerous flowers, outer sepals are in lateral position (Fig. 9.6B). As a result the orientation of carpels appears inversed (with three carpels) or transversal (with two carpels). The illustration in Eichler (1878) gives the wrong orientation. The ovary is either trimerous or dimerous and is highly similar throughout the family. Styles are separate and carinal with globular stigma. There is a single basal orthotropous ovule filling the ovarian cavity. The nectary is highly variable in the family, from inconspicuous and embedded in the hypanthium (e.g. *Polygonum*, *Fallopia*) to globular protuberances (e.g. *Fagopyrum*, *Persicaria*) or a broad disc surrounding the ovary (some *Persicaria*, *Rheum*) (Ronse De Craene and Smets, 1991b).

Galle (1977: 467) gave a schematic series comparing the different flower diagrams of the Polygonaceae. Several phylogenetically important patterns were summarized in Ronse De Craene and Akeroyd (1988). Another diagram with a potential evolution between different floral types in the family is shown in Fig. 9.7.

The Bauplan of the flower of Polygonaceae was (and still is) controversial, in that the family has members with trimerous or dimerous flowers (Rumiceae) and members with pentamerous flowers. The high resemblance between the floral diagram of Polygonaceae and monocots led earlier authors to believe that trimery is ancestral in Polygonaceae with a basic floral formula of P3+3 A3+3 G3, as is found in the genus *Pterostegia* (e.g. Payer, 1857; Eichler, 1878). Although ancestral trimery has been questioned by some authors (e.g. Bauer, 1922), it was generally accepted by others (e.g. Geitler, 1929; Laubengayer, 1937; Ronse De Craene and Akeroyd, 1988), because of seemingly convincing evidence in the perianth and androecium for a transition of trimerous flowers to pentamerous flowers. The third perianth part is half outer – half inner and often bears morphological characters of an outer and inner whorl, even in the anatomy (Ronse De Craene and Akeroyd, 1988). The presence of outer stamen pairs is another argument in favour of basic trimery, linking Polygonaceae with outer stamen pairs found in basal angiosperms. However, accepting a basically trimerous Bauplan for Polygonaceae does not fit with phylogeny: Polygonaceae is nested within an essentially pentamerous clade (Cuénoud *et al.*, 2002) and molecular studies within Polygonaceae indicate that the trimerous taxa are nested within pentamerous clades (Lamb-Frye and Kron, 2003; Sanchez and Kron, 2008).

The presence of stamen pairs in Polygonaceae can be explained either by dédoublement, induced by pressure of the carpels opposite the outer stamens, or as the result of a displacement of alternisepalous stamens. The first interpretation implies that the ancestral androecium was haplostemonous. However, no evidence exists for a splitting (e.g. Bauer, 1922; Laubengayer, 1937), or observations are inconclusive (Galle, 1977) and the sister group Plumbaginaceae has stamens alternating with the sepals (obhaplostemony). The second interpretation of a the shift of alternisepalous stamens in pairs, as suggested elsewhere for Caryophyllales (p. 176), does not require dédoublement and implies that the androecium consists of two whorls, with a loss of two stamens opposite sepals one and two (Fig. 9.7A,B). Inner stamens often have a different orientation from the outer in Polygonaceae and are occasionally extrorse while the outer are

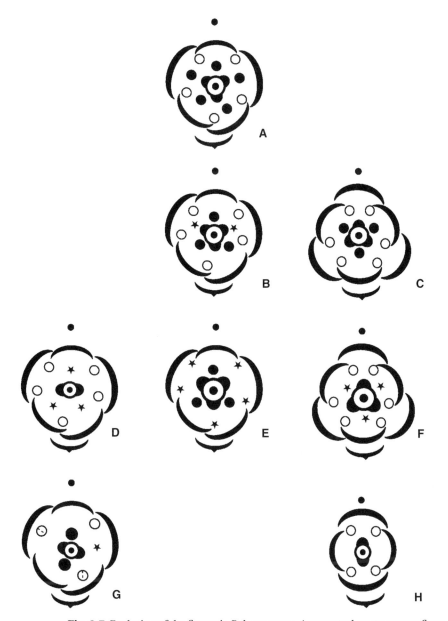

Fig. 9.7. Evolution of the flower in Polygonaceae. A. ancestral pentamerous flower of Polygonaceae; B. pentamerous flower with eight stamens; C. hexamerous flower of *Rheum* with nine stamens; D. *Persicaria sp.* with five stamens and two carpels; E. *Polygonum sp.* with three stamens; F. *Rumex sp.* with six stamens; G. *Polygonum molliaeforme* with two stamens derived by a pairwise fusion of two pairs; H. tetramerous flower of *Oxyria digyna*. White dots: outer alternisepalous stamens; black dots: inner antesepalous stamens; asterisks: lost stamens.

introrse (e.g. *Fagopyrum*). Again this is remarkably concordant with the evolution in Caryophyllales and supports the view of a closer affinity between Polygonaceae and core Caryophyllales (cf. Cronquist, 1981). Polygonaceae are basically apetalous and a transition from five to six in the calyx can be accompanied by an arrangement in two whorls without the stabilizing influence of the petals (cf. *Manilkara* in Sapotaceae, Loranthaceae).

Pentamerous Polygonaceae (e.g. *Coccoloba*, *Persicaria*) basically have eight stamens: four outer in two pairs, one transitional (opposite the third perianth part) and three inner (Fig. 9.7B). Trimerous Polygonaceae have nine stamens, three pairs opposite the outer whorl of perianth and three inner stamens (e.g. *Rheum*: Fig. 9.7C). Trimerous (hexamerous) Rumiceae such as *Rheum* and *Rumex* are linked with dimerous *Oxygonum* by loss of a sector in each whorl (Fig. 9.7C,F, H). Comparative floral morphology has shown that within pentamerous Polygonaceae there are several parallel pathways for flower reduction, leading to dimerous (tetramerous) flowers (Fig. 9.7D; e.g. *Persicaria virginiana* with K2+2 A4+1 G2: Ronse De Craene and Akeroyd, 1988), or trimerous flowers (e.g. *Koenigia islandica* with K3 A3 G3: Ronse De Craene, 1989a). In the genus *Polygonum* and allies (e.g. *Polygonella*) there is a general pattern leading to sterilization or loss of outer stamens (Fig. 9.7E). Reduction of stamens is often correlated with a reduction of the ovary to two carpels. (Fig. 9.7D; cf. Bauer, 1922; Ronse De Craene and Akeroyd, 1988; Ronse De Craene, Hong and Smets, 2004). The outer alternisepalous stamens may be replaced by single stamens, as in *Polygonum molliaeforme* (Fig. 9.7G; Ronse De Craene, Hong and Smets, 2004).

In summary, changes in merism are confusing in the Polygonaceae because of the small, reduced flowers, with easy transitions to an increase (e.g. *Calligonum*) or a reduction (*Oxyria*, *Koenigia*), mixing up trimery with hexamery, and dimery with tetramery.

Plumbaginaceae

Fig. 9.8. *Ceratostigma minus* Stapf ex Prain

✱K(5) [C(5) A(5)] G̲(5)

Inflorescences develop as monochasial units. Two bracteoles are usually present, except in *Armeria*.

De Laet *et al.* (1995) studied the floral development of a selection of species. They found that five common stamen-petal primordia alternate with the sepals in all species. Common zonal growth leads to a stamen-petal tube, except in

Fig. 9.8. *Ceratostigma minus* (Plumbaginaceae): partial inflorescence.

Plumbago where petals and stamens are separate. The ovary bears a single basal anatropous ovule and five stigmatic lobes facing the sepals. At anthesis the ovule becomes inversely orientated toward the top of the ovary by coiling of the funiculus and becomes connected with a stylar obturator. Nectary tissue is closely associated with the androecium and is confined to the inner area of the stamen tube.

Plumbaginaceae share more morphological characters with some core Caryophyllales than with its sister group Polygonaceae, except for the single basal ovule (also present in several Caryophyllales). Friedrich (1956) points to the similarity between *Aegilitis* and *Stegnosperma* (Stegnospermaceae) in the stamen tube and attachment of petal lobes on the abaxial side of it. It is possible that petals of Plumbaginaceae are derived from stamens in the same way as petals of Caryophyllaceae or Stegnospermaceae.

9.2   Core Caryophyllales

Floral diagrams of the (core) Caryophyllales have following aspects in common: they are basically pentamerous without petals, and nectaries are associated with the stamen bases. Tetramerous flowers are occasionally found and could be described as dimerous (e.g. subfamily Rivinoideae of Phytolaccaceae). The basal sister group of core Caryophyllales is Simmondsiaceae, which has apetalous, unisexual flowers (Brockington *et al.*, 2009). The perianth is basically homologous to sepals and different degrees of petaloidy have evolved independently in different clades. The five sepals arise in a 2/5 sequence and have an imbricate aestivation (except for Nyctaginaceae).

Sepals (especially the outer) are characteristically hooded or bear a dorsal crest, similar to some Polygonaceae. Green sepals characterise basal clades, such as Caryophyllaceae and Amaranthaceae; in more derived clades sepals become petaloid ('Portulacaceae-clade', Phytolaccaceae), or staminodia evolve into petaloid appendages on several occasions (e.g. Aizoaceae, Molluginaceae *pro parte*; for a discussion see Ronse De Craene, Smets and Vanvinckenroye, 1998; Ronse De Craene, 2007).

The androecium is the most variable and complex floral organ, ranging from a single stamen to very high numbers arising in a centrifugal sequence. Trying to understand the ancestral androecium of Caryophyllales remains difficult despite attempts to understand its evolution (see Hofmann, 1993; Ronse De Craene and Smets, 1993, 1994; Ronse De Craene, Vanvinckenroye and Smets, 1997; Ronse De Craene, Smets and Vanvinckenroye, 1998; Ronse De Craene, Volgin and Smets, 1999). The variation of the androecium was explained as a derivation of a classical diplostemonous stamen arrangement in the Caryophyllales, and floral diagrams were misused in abundance to support this view (e.g. Walter, 1906; Lüders, 1907; Franz, 1908; Friedrich, 1956). However, the pattern of stamen initiation throughout the Caryophyllales is totally different from diplostemony by the centrifugal or inversed spiral stamen initiation. Therefore this stamen initiation was described as pseudodiplostemony (Ronse De Craene, Smets and Vanvinckenroye, 1998). In all cases with more than ten stamens, initiation is centrifugal and the outer stamens may be sterile or petaloid.

There is strong evidence that staminodial petals have arisen more than once in the core Caryophyllales (at least five to seven times: Ronse De Craene, 2008; Brockington *et al.*, 2009). This is linked with the strong association of petals with the androecium (arising from the same primordium, often as part of the same common primordium, or fusion with a staminal tube). Petals in Caryophyllaceae, Aizoaceae, Stegnospermaceae and Molluginaceae *sensu lato* are staminodes inserted on the outside of the stamen tube.

The basic androecial configuration is whorled with stamens in antesepalous and antepetalous sectors. Some taxa have a unique combination of three whorls of stamens arising in a centrifugal sequence (5+5+5). This arrangement is found in Molluginaceae *sensu lato* (e.g. *Hypertelis salsoloides*) and some Phytolaccaceae, but also in Caryophyllaceae where the outer stamens develop as petaloid appendages. Alternisepalous stamens become frequently shifted in pairs opposite sepals one, two and three, and more rarely opposite sepals four and five, similar to Polygonaceae (e.g. Phytolaccaceae). Ronse De Craene, Vanvinckenroye and Smets (1997) and Ronse De Craene, Smets and Vanvinckenroye (1998) interpreted these pairs as an ancestral condition originating with the derivation of pentamerous flowers from trimerous flowers. However, in the context of the relationships of Caryophyllales this seems unlikely. Alternatively a more likely suggestion for the occurrence of pairs is a shift of the alternisepalous stamens towards the middle of the sepal.

The androecium has an unusual development with an inversion in the direction of initiation of the antesepalous stamens, different from a regular development of diplostemonous flowers (described as pseudodiplostemony by Ronse De Craene, Smets and Vanvinckenroye, 1998). This pattern of initiation is found all over core Caryophyllales, but also in several Polygonales (Ronse De Craene, unpubl. data). The first antesepalous stamens to arise are those opposite sepals four, five and three, followed by those opposite sepals two and one. This sequence is inversely correlated with reductions and losses of stamens, as the first initiated stamens are the last to be lost (Caryophyllaceae: Ronse De Craene, Smets and Vanvinckenroye, 1998).

Two processes appear to be superimposed in the evolution of the androecium of Caryophyllales:

- a clear tendency for centrifugal development and stamen increase affecting most clades.
- a tendency for reduction of the antesepalous stamens linked with an inversed initiation and a displacement of alternisepalous stamens opposite the sepals where they behave as antesepalous stamens.

A centrifugal initiation of stamens starts with an upper whorl of ten and is superimposed on these two stamen whorls. It remains controversial whether the centrifugal stamen initiation is a reversal or a secondary increase. Both possibilities are present in Caryophyllales. The arrangement and number of carpels highly influence the number and position of antesepalous stamens, which tend to alternate with the carpels. Stamen losses in the antesepalous whorl are always those opposite the outer sepals, alternating with carpels, often leading to eight stamens (e.g. Caryophyllaceae, *Limeum*, *Macarthuria*: Hofmann, 1993; Ronse De Craene, Smets and Vanvinckenroye, 1998).

The outer stamens occupy the alternisepalous sectors but are often shifted into pairs opposite the outer sepals. Pairs may converge and fuse into single stamens (e.g. *Mollugo*).

Several clades can be defined by the position of the stamens. Aizoaceae and Nyctaginaceae have stamens in alternisepalous position, with a potential increase of stamens in centrifugal direction. Amaranthaceae and Caryophyllaceae have stamens preferentially in antesepalous position. Flowers with eight stamens are often found in Phytolaccaceae, Nyctaginaceae (e.g. *Bougainvillea*), Polygonaceae

(p. 172), Molluginaceae *sensu lato* and occasionally in Caryophyllaceae. Two different evolutionary pathways are linked with the partial reduction of the antesepalous stamens and the fusion of the alternisepalous stamens opposite the sepals. In Nyctaginaceae, Aizoaceae and part of Molluginaceae the antesepalous stamens are lost and the remaining stamens are in alternisepalous position. In Aizoaceae a secondary stamen increase leads to higher stamen numbers and petaloid staminodes. In the portulacaceae-clade stamens are opposite the sepels by partial loss of stamens and fusion of pairs.

The gynoecium consists of five carpels, always in antesepalous position. The number is occasionally reduced to four (in tetramerous flowers), three or two carpels.

The ovary is basically septate with axile or basal placentation but the septa become reduced or dissolve in several clades, leaving a free-central placentation (e.g. Caryophyllaceae: Hofmann, 1993). Placentation is basically axile but there are diverging patterns, either to a free-central placentation in Caryophyllales or a basal placentation in Polygonales, becoming secondarily parietal in some families. A secondary carpel increase occurred in some *Phytolacca* (Phytolaccaceae), *Pleuropetalum* (Amaranthaceae) and Cactaceae, with several carpels arising in a single whorl. This increase is associated with an increase in the number of stamens. A reduction to a single carpel is characteristic of Rivinoideae, Phytolaccaceae and Nyctaginaceae.

Hofmann (1993) provided a series of floral diagrams to illustrate the evolution in the Caryophyllales.

Simmondsiaceae

Fig. 9.9A–C. *Simmondsia chinensis* C. K. Schneid.

Staminate: $*$K5 C0 A5+3* G0
Pistillate: $*$K5 C0 A0 G̲(3)
*The number of inner stamens is variable, ranging from one to five.
Köhler (2003) gives a range of 8–16.

Simmondsiaceae occupy a basal position in core Caryophyllales, immediately behind Rhabdodendraceae with which they share wind-pollinated flowers (Brockington *et al.*, 2009). Flowers are (tetra-) penta- to (hexa-)merous. *Simmondsia* is dioecious, with staminate inflorescences grouped in axillary clusters. Flowers lack bracteoles and outer sepals are lateral in staminate flowers. The extrorse stamens are arranged in two distinct girdles, the outer is complete and alternisepalous, while the inner is generally reduced in number (Ronse De Craene, unpubl. data). Pistillate flowers are solitary and pendent. Two bracteoles are closely connected with the strongly imbricate calyx and outer sepals are arranged in median position. The three carpels bear an apical-axile placenta with one (two) ovules and three reflexed styles (Köhler, 2003). A nectary is absent.

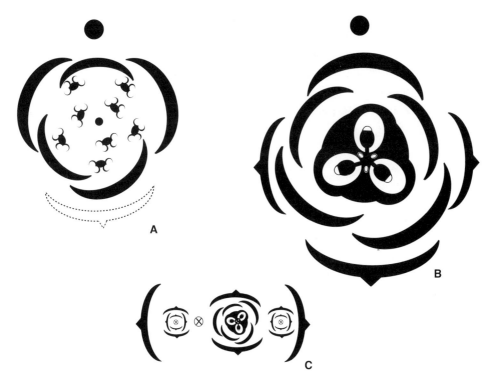

Fig. 9.9. *Simmondsia chinensis* (Simmondsiacea)e: staminate (A) and pistillate (B) flower; C. arrangement of pistillate flower on the main stem.

'Caryophyllaceae-clade'

Caryophyllaceae

Fig. 9.10A,B. *Arenaria tmolea* Boiss. (Alsinoideae)

✳K5 A(5°+5+5) G̲(3)

Fig. 9.10C. *Chaetonychia cymosa* Sweet (Paronychioideae)

✳K5 A(5+5°) G̲(2)
General formula: ✳K(3)4–5 C0–5 A1–10 G̲2–5(-10)

Flowers of Caryophyllaceae appear highly diverse, with different evolutionary patterns affecting the flower of three subfamilies Caryophylloideae, Alsinoideae and Paronychioideae. The basic inflorescence arrangement is a dichasium, occasionally a monochasium (Bittrich, 1993). Bracteoles are well developed and become occasionally associated with the flower in an epicalyx. Sepals are imbricate, mostly free or occasionally fused into a long tube (e.g. *Dianthus*). In Paronychoideae sepals generally have a dorsal appendage (Fig. 9.10C). A hypanthium is mostly well developed, especially in apetalous species. In Caryophylloideae (e.g. *Dianthus*) petals, stamens and carpels are lifted on an anthophore separate from the calyx (Fig. 1.4D). Petals are strongly associated

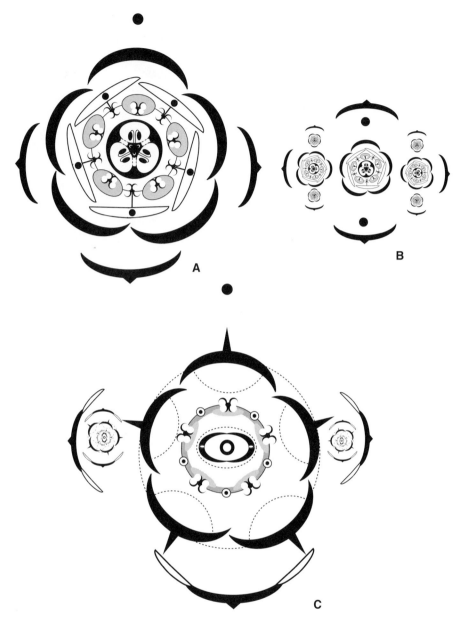

Fig. 9.10. Caryophyllaceae: *Arenaria tmolea*, A. Flower; B. partial inflorescence; C. *Chaetonychia cymosa*: partial inflorescence. Note large stipules on bracts and depressions opposite the sepals.

with the androecium and were interpreted as staminodes by Ronse De Craene, Smets and Vanvinckenroye (1998). Petals vary much in shape; they are generally clawed with a bifid apex, and there is often an appendage at the top of the claw. Petals are occasionally reduced, replaced by stamens (*Scleranthus perennis*), or absent (Fig. 9.10C). In most Caryophyllaceae the antesepalous stamens are inflated at the base and connected with the antepetalous stamens; the petaloid staminodes are inserted outside the stamen tube. Nectaries are situated inside the fused stamens or develop externally at the base of the antesepalous stamens and may extend laterally towards the antepetalous stamens (Alsinoideae: Fig. 9.10A). When fully developed, the androecium is (ob)diplostemonous (Ronse De Craene and Smets, 1995b). Initiation of the antesepalous stamens follows an inverted spiral starting with stamens opposite sepals four and five, while antepetalous stamens more often arise simultaneously (Ronse De Craene, Smets and Vanvinckenroye, 1998). When fewer in number, stamens are in antesepalous position (except for *Colobanthus*). In cases with a single whorl of staminodes (Fig. 9.10C) it is not clear whether this represents the petaloid staminodes or the antepetalous stamens. Ronse De Craene, Smets and Vanvinckenroye (1998) demonstrated that a further reduction of antesepalous stamens is inversed to their sequence of initiation and this is closely correlated with the carpel numbers. The number of stamens can be further reduced to two or even a single stamen, situated opposite sepals four and five (e.g. *Scleranthus annuus*: Ronse De Craene, Smets and Vanvinckenroye, 1998).

The ovary is syncarpous with a free-central placenta arising through the breakdown of septa and generally with numerous ovules, or basal with a single ovule (Bittrich, 1993). Styles are in a commissural position (cf. Eichler, 1878), but occasionally carinal (Fig. 9.10A).

Amaranthaceae (incl. Chenopodiaceae)

Fig. 9.11. *Blutaparon vermiculare* (L.) Mears

✳ K5 C0 A(5) G̲(2)
General diagram: ✳ K(0–1) 3–5 C0 A(1-)5 (-8–9) G2–3 (-6)

Inflorescences are mostly spicate or capitate but are basically cymose. Flowers are bisexual to unisexual. Each flower is subtended by a bract and two bracteoles. Sepals (tepals) are mostly three to five (rarely absent or single in some *Amaranthus* and *Salicornia*), with imbricate aestivation, free or basally connate, with the two inner occasionally shorter or absent. Sepals are often persistent and hardened in fruit. Sepals can be petaloid or are greenish. Stamens are equal in number as the sepals and opposite to them (rarely fewer). *Pleuropetalum* is exceptional in having a higher number of stamens and carpels (Ronse De Craene, Volgin and Smets, 1999). Whether this represents a secondary increase or a plesiomorphic condition is difficult to verify, although stamens arise in a similar pattern as in Phytolaccaceae (see p. 189). Stamens are free (in Chenopodiaceae) or more commonly fused into a tube. Interstaminal

Fig. 9.11. *Blutaparon vermiculare* (Amaranthaceae).

appendages (pseudostaminodes: Eliasson, 1988) are frequently present on the tube and have various forms. These were interpreted as stipular by Eichler (1878) or as staminodes (see Ronse De Craene and Smets, 2001 for a review). Presence of these appendages is inversely correlated with broad filaments (Eliasson, 1988). The ovary consists of two to three carpels and is unilocular with basal placentation, highly similar to some Caryophyllaceae or Plumbaginaceae (Figs. 9.8, 9.10C, 9.11). A single to several ascending ovules can be found. In *Pleuropetalum* several ovules develop centrifugally on a gynoecial dome and are interpreted as a secondary multiplication (Ronse De Craene, Volgin and Smets, 1999). Floral nectaries are formed as a ring at the inner base of the filaments or extending between the filaments. Nectaries are absent in unisexual flowers and there is a tendency for flower reduction linked with wind pollination (*Chenopodium*).

'Aizoaceae-clade'

Molluginaceae *sensu lato*

Fig. 9.12A. *Corbichonia decumbens* (Forsk.) Exell

✳ K5 C0 A(5°+5°+10°+10+5+5+5) G̲(5)

Fig. 9.12B. *Glinus lotoides* L.

✳ K5 C0 A(3°:1+5) G̲(3)
General formula: ✳ K(4)5 A°0–5–∞ A(3-)4–5–∞ G(1)2–5

Fig. 9.12. Molluginaceae *sensu lato*: A. *Corbichonia decumbens*, partial inflorescence.
B. *Glinus lotoides*, partial inflorescence. Note incomplete partly staminodial
antesepalous stamen whorl.

Recent molecular phylogenies have shown Molluginaceae to be paraphyletic,
containing at least three to five clades (*Macarthuria*, *Limeum*, Lophiocarpaceae
(*Lophiocarpus* and *Corbichonia*), *Hypertelis* and Molluginaceae *sensu stricto*, including

Glinus (Cunéoud *et al.*, 2002; Brockington *et al.*, 2009). Four genera (*Macarthuria, Limeum, Corbichonia, Glinus* p.p.) share a comparable floral diagram and development and possess petaloid organs in variable degrees (e.g. Hofmann, 1973, 1993). The five sepals have quincuncial aestivation. Stamens arise centrifugally in pentamerous or decamerous whorls with the alternisepalous stamens largest and in upper position (Fig. 9.12A). In cases where they are present, antesepalous stamens range from two to three to five, and arise sequentially after the alternisepalous stamens (Hofmann, 1993; Ronse De Craene, unpubl. data). The two upper whorls are fused in a tube, which is nectariferous on the inside and more whorls of stamens or petaloid staminodes may arise externally in regular whorls (*Corbichonia*: Fig. 9.12A; Ronse De Craene, 2007). In cases where the antesepalous stamen whorl is incomplete, as in *Macarthuria* or *Limeum*, missing stamens are those opposite the outer sepals and the remaining stamens alternate with the carpels. The androecium of *Glinus lotoides* is extremely variable, occasionally comparable to some *Corbichonia* (Hofmann, 1973). In my material only the alternisepalous whorl was complete, with staminodes and odd stamen opposite petals (Fig. 9.12B).

In Molluginaceae *sensu stricto* (including *Mollugo, Adenogramma, Pharnaceum*, among others) the androecium consists generally of five stamens alternating with the sepals. In *Mollugo* the number of stamens ranges from five (*M. cerviana*) to three (*M. nudicaulis*), the latter arising by a pairwise fusion of four stamens (Batenburg and Moeliono, 1982; cf. Cucurbitaceae).

The ovary is isomerous with antesepalous carpels or is reduced to three (two). Placentation is axile with narrow partitions and styles are carinal.

Aizoaceae

Fig. 9.13. *Mesembryanthemum nodiflorum* L.

✳ K 5 C0 A5°+10°+10°+ … + 5°+10°+10°+10+5 -G- (5)
General floral diagram: ✳ K (3-)5(-8) C0 A(1-)5- 5$^\infty$ G(1-)2–5

Aizoaceae is a large family with about five subfamilies. Flowers are often tetramerous with sepals in a decussate arrangement, or pentamerous with a 2/5 arrangement (rarely with six or up to eight sepals). In most subfamilies, outer sepals continue the sequence of the leaves and are different from the inner sepals, being thicker and hood-like. The complex androecium of Aizoaceae has been known for a long time (e.g. Payer, 1857; Ihlenfeldt, 1960). The androecium consists basically of five alternisepalous (complex) stamens, which remain simple (e.g. *Plinthus*), or divide in pairs (*Galenia*), or centrifugally in triplets (e.g. Aizooideae), or in large groups of stamens (Mesembryanthemoideae, Ruschioideae: Ihlenfeldt, 1960; Hofmann, 1973, 1993). Exceptionally there are

Fig. 9.13. *Mesembryanthemum nodiflorum* (Aizoaceae). Note false septa in locules; broken line delimits five common stamen primordia.

two whorls of five stamens in some species of *Sesuvium*. Taxa with higher number of stamens tend to develop a ring primordium (e.g. *Mesembryanthemum*, *Aptenia*). Outer primordia may develop in coloured staminodes, effectively replacing a corolla. The number of stamens and staminodes developing centrifugally can be variable or is consistent (Fig. 9.13). Ihlenfeldt (1960) and Haas (1976) listed various possibilities for differentiation into fertile inner stamens and outer staminodes. The transition can be progressive or abrupt. This is strongly taxon-bound, but is highly similar to the development in some Molluginaceae (e.g. *Corbichonia*). Species without stami-nodes usually have an adaxially coloured calyx (e.g. *Sesuvium*, *Tetragonia*), while it is green in taxa with staminodes. The ovary is often half-inferior or inferior by the development of a hypanthium lifting androecium and sepals. A nectary develops on the inner side of the stamen-petal tube. The ovary is generally isomerous with axile placentation, and a sterile central apical part can sometimes be strongly developed (columella: Haas, 1976).

Nyctaginaceae

Fig. 9.14. *Bougainvillea spectabilis* Willd.

✳ K(5) C0 A(5+3) G̲1
General formula: ✳ K(3-)4–5(-7) C0 A(1-)5(-40) G̲1

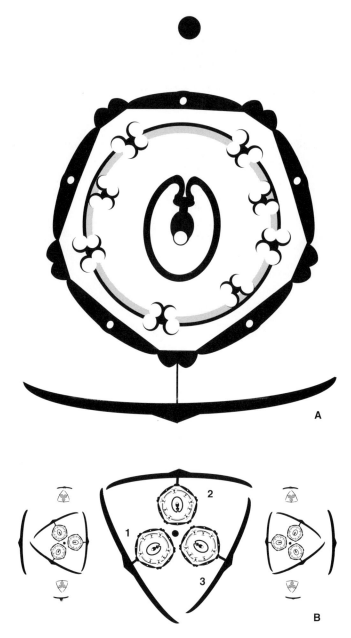

Fig. 9.14. *Bougainvillea glabra* (Nyctaginaceae): A. flower; B. inflorescence. Numbers refer to sequence of maturation of flowers in a triplet.

The family is distinctive by its fused perianth tube consisting of an upper petaloid section that shrivels at fruiting and a lower persistent section enclosing the ovary (anthocarp).

Flowers of *Bougainvillea* are typically arranged in three-flowered dichasia. Flowers are enclosed by large, often petaloid, bracts (involucre). In some cases bracts may be confused with sepals, as in *Mirabilis* where five fused bracts enclose the flower in the manner of sepals. This has occasionally been regarded as an incipient acquisition of a bipartite perianth, as bracts cannot be distinguished from sepals (Brockington *et al.*, 2009). The single flower in *Mirabilis* is the central remainder of an originally five-membered inflorescence surrounded by an involucre (Sattler and Perlin, 1982). The perianth is pentamerous, rarely tetramerous with valvate or plicate (folded) aestivation (appearing contorted). The anthocarp has been variously interpreted as a perianth tube (e.g. Rohweder and Huber, 1974) or a hypanthium (Buxbaum, 1961: 'Achsenbecher'). The fact that stamens are variously fused to the tube and the dual nature of the perianth points to a nature of a hypanthium.

The number of stamens in Nyctaginaceae varies from (one) two to three (*Boerhaavia*) to up to 40 (some *Pisonia*). Eight stamens occur in some *Pisonia* and in *Bougainvillea* (Fig. 9.14; Vanvinckenroye *et al.*, 1993). Stamens in *Bougainvillea* are initiated in a 3/8 sequence, following the 2/5 initiation of the perianth (Sattler and Perlin, 1982; Vanvinckenroye *et al.*, 1993). Five stamens alternate more or less with the sepals while three are positioned more or less opposite the inner sepals. At maturity stamens have different lengths. Most other Nyctaginaceae have five stamens alternating with the sepals (e.g. *Mirabilis*). When lower than five, stamen numbers can be variably in alternation or opposite sepals (*Boerhaavia*, *Oxybaphus*). Stamens are free or monadelphous. A nectary is situated on the adaxial side of stamens or below the gynoecium (Zandonella, 1977). The ovary is always superior and monocarpellate with single basal ovule.

Fiedler (1910) provided several diagrams of the flower of Nyctaginaceae, concentrating on the androecium that he interpreted as basically diplostemonous; higher numbers as in *Pisonia* were seen as the result of dédoublement. However, his diagrams have little value as he analysed flowers with a preconceived approach that does not correspond to the real development and structural arrangement of the flower. It is possible that higher numbers of stamens result from a centrifugal increase, as in Phytolaccaceae.

Phytolaccaceae (incl. Rivinoideae)

Fig. 9.15A. *Phytolacca dodecandra* L'hérit., based on Ronse De Craene, Vanvinckenroye and Smets (1997)

$\ast$ K5 C0 A5+5 G̲(5)

Fig. 9.15B. *Trichostigma peruvianum* (Moq.) H.Walt.

$\ast$ K2+2 C0 A4+2+2+4 G̲1

Fig. 9.15C. *Ercilla volubilis* A. Juss.

$\ast$ K5 C0 A5+3–4 G̲(5)

General formula: $\ast$ K5 C0 A (5-)8(-30) G̲1–5(-17)

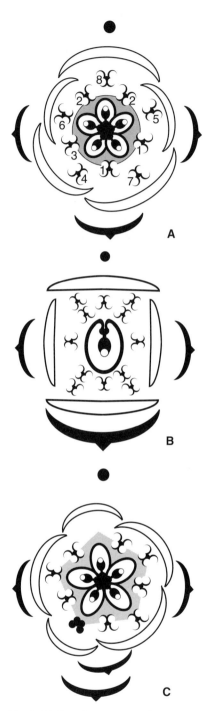

Fig. 9.15. Phytolaccaceae: A. *Phytolacca dodecandra*, B. *Trichostigma peruvianum*, C. *Ercilla volubilis*. Numbers refer to order of initiation of stamens.

Phytolaccaceae are paraphyletic, with Rivinoideae more closely related to Nyctaginaceae (Brockington *et al.*, 2009). Flowers are constantly pentamerous or tetramerous (apparently dimerous) in Rivinoideae, and arranged in racemose inflorescences (Rohwer, 1993b). Flowers are generally bisexual, rarely dioecious (e.g. *P. dioica*, *Monococcus*).

Sepals are inconspicuous or petaloid, mostly with quincuncial aestivation. *Phytolacca dodecandra* with 10–12 stamens occupies a pivotal position in the family (Fig. 9.15A; Ronse De Craene, Vanvinckenroye and Smets, 1997): outer stamens arise simultaneously in two pairs opposite the outer sepals; a single stamen arises sideways of sepal three, before three other stamens initiate in sequence opposite sepals four, three and five. Two more stamens arise opposite sepals one and two. *Ercilla* and some *Phytolacca* species (e.g. *P. esculenta*) with eight stamens have no stamens opposite sepals one and two (except for alternisepalous stamens converging in pairs) or one is staminodial.

Carpel numbers are closely linked with the number of stamens and tend to alternate with them. In some species of *Phytolacca* the number of carpels can be secondarily increased and this affects the number and arrangement of inner stamens, which are occasionally paired as to alternate with the carpels (*P. americana*: ten stamens and ten carpels; *P. esculenta*: eight stamens and eight carpels). A secondary centrifugal stamen increase can also be superimposed on this basic arrangement (e.g. *P. dioica*, *P. acinosa*: Hofmann, 1993; Ronse De Craene, Vanvinckenroye and Smets, 1997).

The ovary is basically syncarpous, with a single basal ovule per carpel, but can become secondarily apocarpous (Rohweder, 1965). A short gynophore may be present. The placenta is deeply embedded in the apical meristem, appearing to arise separately from the carpel wall (Ronse De Craene, Vanvinckenroye and Smets, 1997). A nectary is usually situated as a rim between the base of the gynoecium and the inner stamens.

In Rivinoideae flowers are either pentamerous (e.g. *Seguieria*) or tetramerous (e.g. *Trichostigma*, *Rivina*). The androecium of *Trichostigma* is comparable to the three-whorled pentamerous arrangement of some Molluginaceae. Initiation is centrifugal and this was interpreted as a tendency for reduction (Ronse De Craene and Smets, 1991c), as some Rivinoideae have only four alternisepalous stamens (e.g. *Hilleria*, *Rivina*). The flower of *Petiveria* is monosymmetric, with four diagonally inserted sepals, six stamens and an oblique ovary position. Ronse De Craene and Smets (1991c) interpreted the distorted position as a result of a reduction of the size of the bracteoles, influencing the position of four sepals and the rest of the flower. *Monococcus echinophorus*, an Australian dioecious endemic, shares the same floral arrangement as *Petiveria*; staminate flowers have a multistaminate androecium arising in four alternisepalous groups and without trace of a carpel (Vanvinckenroye, Ronse De Craene and Smets, 1997). The gynoecium is generally monocarpellate with a flattened crest-like stigma. No nectaries are present in Rivinoideae (Rohwer, 1993b).

It is best to interpret the androecium of *P. dodecandra* and all Phytolaccaceae as derived from two initial whorls, with an alternisepalous whorl with stamens converging in pairs opposite sepals one and two (contrary to the interpretation of Ronse De Craene, Vanvinckenroye and Smets, 1997), and an antesepalous whorl arising in an inversed sequence and centrifugally relative to the first whorl.

'Portulacaceae-clade'

Portulacaceae, Didiereaceae, Basellaceae and Cactaceae are associated in a clade. While the three first families share a bipartite involucre, the numerous spirally inserted tepals of Cactaceae could be derived by a division of initially fewer perianth parts (as in *Lewisia* of Portulacaceae) or by displacement of bracts in the confines of the flower (Ronse De Craene, 2008).

Cactaceae differ from other Caryophyllales in that they are not pentamerous, but have a spiral, indefinite perianth.

Portulacaceae

Fig. 9.16A,B. *Lewisia columbiana* (J. T. Howell) Rob.

✴ K7* C0 A(5) G̲(3)

*K up to 12 and G up to 8 in some *Lewisia*.

General formula: ✴ K(2-)5(12) A(1-)5-∞ G̲2–5(-8)

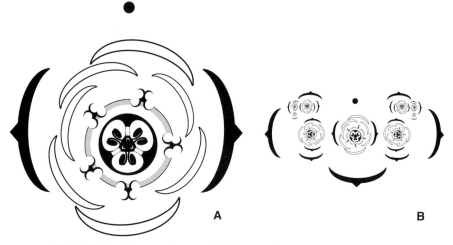

Fig. 9.16. Portulacaceae: *Lewisia columbiana*. A. flower; B. partial inflorescence.

Inflorescences are mostly cymose with dichasia deviating into monochasia (Carolin, 1993). Most Portulacaceae are pentamerous with five petaloid sepals enclosed by two (three) large median involucral bracts. Bracts are often interpreted as sepals and the sepals as petals (e.g. Eichler, 1878; Franz, 1908; Milby, 1980; Carolin, 1993), although morphological evidence suggests that the petals are comparable to petaloid sepals as found in other Caryophyllales. Sepals always arise in a 2/5 sequence with the outer in a lateral position. The median involucral bracts alternate with two outer transversal bracts. *Lewisia* has a higher and variable number of tepals, possibly caused by doubling of inner tepals (tepals are occasionally bilobed; see Eichler, 1878). The number of involucral bracts can also be much higher in the genus (e.g. *L. rediviva*: Franz, 1908).

The androecium of Portulacaceae is highly variable, ranging from one to many stamens. In several species of *Talinum* the first stamens arise in pairs opposite the outer tepals. The androecium is restricted to these five stamens in *T. napiforme*, while ten stamens arise in other species, with an initiation comparable to *Phytolacca dodecandra* (Fig. 9.16A; Vanvinckenroye and Smets, 1996). More stamens may arise centrifugally in alternating whorls (e.g. *T. portulacifolium*). In other Portulacaceae the initiation of the ten upper primordia is rendered less clear by the development of a ring primordium (e.g. *Calandrinia, Anacampseros, Portulaca*: Hofmann, 1993; Vanvinckenroye and Smets, 1999). This increase is superimposed on the initial whorl of 10–13 stamens as found in *Talinum*. As in other Caryophyllales the inner side of the stamen tube develops as a nectary.

Portulacaria afra has five alternisepalous stamens, while in *Claytonia* (*Montia*) the five stamens are antesepalous but can be reduced to three opposite sepals three, four and five (*M. fontana*: Payer, 1857). The question is what makes these positions different: is it a polyphyletic origin, or are the petals in *Montia*, as well as Basellaceae and Didiereaceae, of staminal origin? An evo-devo investigation should clarify this.

Sepals of *Montia* arise basipetally from common primordia with the stamens (Milby, 1980; Hofmann, 1993). This is caused by a shift of the protective function towards the involucre, linked with a retardation of the sepal development. In some species of *Talinum*, tepals also have a retarded development (Vanvinckenroye and Smets, 1996). In Portulacaceae the gynoecium is often trimerous (rarely dimerous or pentamerous) and develops septa. The ovary is superior in most genera, to (half-)inferior in *Portulaca*. The ovules arise on a central column by the later disintegration of the septa. Portulacaceae share a single style with long terminal stigmatic lobes with Cactaceae.

Vanvinckenroye and Smets (1999: 191) presented a number of floral diagrams of *Anacampseros* and *Talinum*.

Cactaceae

Fig. 9.17. *Echinopsis formosa* (Pfeiff.) Jacobi

Fig. 9.17. *Echinopsis formosa* (Cactaceae).

✳ P∞ A∞ G̲(7)

Flowers are formed solitary on short shoots and bracts are well demarcated from the petaloid tepals, which are indefinite in number. The androecium is multistaminate and develops centrifugally on a ring primordium (see also Boke, 1963, 1966; Ross, 1982; Leins and Schwitalla, 1985). Through the development of a deep hypanthium the gynoecium occupies a (half-)inferior position and the stamens are placed in several tiers on the slopes of the hypanthium. The number of carpels ranges from (three) four to five, up to 18, centred around a central residual floral apex (Boke, 1963, 1966). Placentation is basal-laminar because septa fail to extend beyond small outgrowths (Leins and Erbar, 2007). The ovules arise on the carpel wall at the base of the septa and, depending on the genus, extend beyond the septa in a waveline, or become restricted to one to two ovules (*Pereskia aculeata*). The inner side of the androecium produces a narrow nectary.

Pereskia is considered to be the most primitive genus in the family on the basis of vegetative and floral characteristics (Boke, 1963). In *Pereskia* five to eight androecial areas are demarcated on the ring primordium alternating with the inner tepals and carpels (Ross, 1982; Leins and Schwitalla, 1985). Although this could represent the basic condition in Cactaceae, it does not correspond with the position of stamens in the nearest sister groups where stamens have an antesepalous position.

10

Rosids: the diplostemonous alliance

The rosid clade is a well supported, but the least resolved major clade of core eudicots, containing more than a quarter of all angiosperm species (Schönenberger and von Balthazar, 2006). Saxifragales are generally linked with rosids, although support is not high and the order has been associated with caryophyllids (e.g. Soltis *et al.*, 2003). It is clear that Saxifragales represent an ancient early diverging lineage in the core eudicots (Soltis *et al.*, 2005; Magallón, 2007). A recent analysis incorporating a high number of genes has clarified the internal relationships of rosids with the recognition of two main clades, fabids (or Fabidae) and malvids (or Malvidae), and the inclusion of Saxifragales as a basal order and Vitaceae as sister to the rosid clade (Wang *et al.*, 2009). Figure 10.1 represents a phylogenetic tree of the rosids based on Wang *et al.* (2009).

10.1   Saxifragales

Saxifragales *sensu* APG (2003) comprises an assemblage of highly diverse families (about ten), including core saxifrages and allies and part of the former Hamamelidae of Cronquist (1981) with strongly reduced flowers.

The general flower morphology fits well with the syndrome found in rosids. Flowers tend to be generally pentamerous or tetramerous and share a hypanthium, often with half-inferior ovary. Obdiplostemony is a common feature in families with two whorls of stamens (e.g. Saxifragaceae, Haloragaceae, Crassulaceae: Ronse De Craene and Smets, 1995b). Diplostemony is also reconstructed as the ancestral character state in the Saxifragales by Soltis *et al.* (2005). There is a general pattern to haplostemony (e.g. Iteaceae, Grossulariaceae), and this is often linked with a reduction of the size of the petals and petaloidy of the

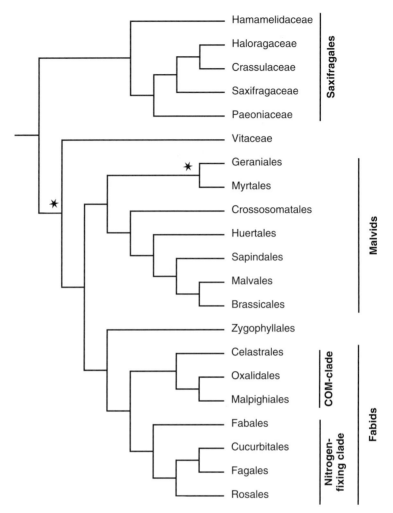

Fig. 10.1. Phylogenetic tree of the rosids, based on Wang *et al.* (2009). Asterisks point to branches with low support.

hypanthium and sepal lobes. Obdiplostemony may be linked to the general weakening of the petals. Within Saxifragales there is an entire range of transitions between flowers with well-developed petals (e.g. *Ribes* p.p., *Bergenia*, *Saxifraga*) to apetalous genera (e.g. *Ribes* p.p., *Rodgersia*, *Chrysosplenium*). Sympetaly is rare and has only been found in some Crassulaceae where it is linked with hypanthial growth. Hamamelidaceae show high variation in floral diagrams, linked with wind pollination and the development of pseudanthia (Endress 1967, 1976).

Paeoniaceae are anomalous in having massive flowers without clear boundaries between bracts, sepals and petals, a multistaminate centrifugal androecium, and a variable number of basally united carpels. It is postulated that the increase in stamens triggered a disturbance of the genetic boundaries of different organ categories and this has led to a proliferation of perianth parts (Ronse De Craene, 2007). Two carpels are very frequently found in Saxifragales, and Magallón (2007) argued that they may be a plesiomorphic condition for core eudicots.

Frequent recurrent morphological characters in the order are: two carpels, a hypanthium, obdiplostemony and a reduction or retardation of the petals.

Hamamelidaceae

Fig. 10.2A,B. *Hamamelis japonica* Sieb. & Zucc.

✳ K(4) C4 A4+4° -G-(2)

General formula: ✳ K(0-)4–5(-7) C0–4–5 A(1-)4–5 (-24) G2

Fig. 10.2. *Hamamelis japonica* (Hamamelidaceae): A. Flower; B. partial inflorescence with three flowers.

The family is highly diverse and the variability is often linked with the evolution of a wind pollination syndrome. The ovary is usually bicarpellate in the family, partly fused and with free styles. Flowers of *Hamamelis* are grouped in triplets or pairs on a short shoot bearing a terminal aborted apex. The genus

Hamamelis is closest to other Saxifragales in having narrow petals, a hypanthium, a bicarpellate gynoecium and weak obdiplostemony. Petals, when present, are narrow and coiled, and superposed to filament-shaped nectariferous staminodes in *Hamamelis*. In other genera such as *Corylopsis*, ten intrastaminal protuberances produce nectar besides the staminodes, or nectaries are present on the petal bases (*Disanthus*). The fertile antesepalous whorl bears dithecal monosporangiate anthers dehiscing by adaxially curved flaps (also in *Exbucklandia*, but disporangiate in other genera). The ovary is superior to (half-)inferior and has two apical ovules (rarely more), of which one aborts (Eichler, 1878). Diagrams of inflorescences and flowers of *H. virginiana* were shown in Mione and Bogle (1990).

In some genera the number and arrangement of floral parts become irregular, with the reduction and loss of petals or the entire perianth and the aggregation of several reduced flowers in elaborate pseudanthial structures with bracts functioning as attractive organs (e.g. *Parrotia*, *Distylium*, *Rhodoleia*, *Parrotiopsis*: Endress, 1978; Bogle, 1989). In apetalous genera flowers are often unisexual (Endress, 1978). A few taxa have undergone a secondary increase in the androecium (*Fothergilla*, *Matudea*: Endress, 1976, 1978). Some taxa have an additional whorl of 'sterile phyllomes' or scales within the staminodial whorl (e.g. *Maingaya*, *Loropetalum*, *Corylopsis*) that were interpreted together with the petals as staminodial in origin (Mione and Bogle, 1990). This would make the androecium ancestrally multiwhorled. However, this interpretation is uncertain as scales arise very late in ontogeny (Endress, 1967) and the nearest sister groups of Hamamelidaceae are diplostemonous. The inner scales are best interpreted as receptacular emergences. Magallón (2007) analysed character evolution in the Hamamelidoideae relative to Saxifragales; she concluded that pentamerous and tetramerous flowers were derived three times, and that a bicarpellate gynoecium is plesiomorphic in the family.

Paeoniaceae

Fig. 10.3. *Paeonia potaninii* Kom.

↺ K2–5 C7–9 A5$^\infty$ G3–4
General floral formula: ↺ K3–5 C5–13 A5$^\infty$ G2–15

Flowers are usually solitary at the end of branches, often with a smaller flower just below. Flowers are large and have a variable number of bracts, sepals and petals, with a spiral initiation throughout. The flower bears a hypanthium that can be more or less deeply developed. Bracts are leaf-like structures closely pressed to the flower and progressively transgressing into the sepals, followed by the petals. Sepals and petals intergrade and transitional organs are frequently found. Stamens are always numerous, arising centrifugally from five common primordia alternating with the petals (Hiepko, 1964; Leins and Erbar, 1991). The

Fig. 10.3. *Paeonia potaninii* (Paeoniaceae). Note stamens arising on five common primordia.

outermost stamens may be occasionally staminodial. Between androecium and gynoecium a crenelated receptacular disc (nectary) develops. This was interpreted as of staminodial origin on the basis of vasculature (Melville, 1984), although not supported by floral ontogeny. According to Hiepko (1964, 1966) the disc should be interpreted as a receptacular emergence. Hiepko (1966) mentioned that the disc is not secretory, although secretion from crushed inner stamens occurs in some species. It is possible that a nectary was present ancestrally, but that it has become defunct due to a transfer to pollen rewards through the many stamens. The gynoecium is apocarpous with a variable number of follicular carpels. Ovules develop in two rows on the folded margins. Paeoniaceae were associated with Ranunculaceae or Dilleniaceae in the past because of superficial similarities in the flower construction. Its placement in Saxifragales is supported by the presence of a hypanthium and apocarpous gynoecium, although the multistaminate androecium and apocarpy are clearly derived (see Soltis *et al.*, 2005). The spiral flower with variable number of parts is aberrant within core eudicots and could be interpreted as a reversal linked with a disruption of the boundaries between different whorls (Ronse De Craene, 2007).

A simplified floral diagram of *Paeonia officinalis* was presented by Leins and Erbar (1991).

Haloragaceae

Fig. 10.4. *Haloragis erecta* (Murr.) Oken

∗ K4 C4 A4+4 Ǧ(4)

General formula: ∗ K(0)2–4 C(0)2–4 A4–8 G2–4

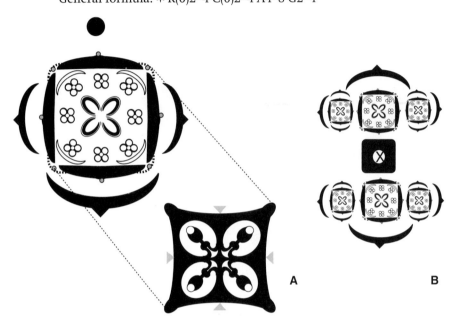

Fig. 10.4. *Haloragis erecta* (Haloragaceae): A. Flower; B. partial inflorescence.

Inflorescences are variable, containing dichasial partial inflorescences. Flowers are bisexual or occasionally unisexual, arranged in a dichasium with staminate or bisexual flowers distal to the basal pistillate flowers. Flowers are generally tetramerous (rarely dimerous) with persistent, valvate sepals. Petals are contorted (occasionally absent or rudimentary: pistillate *Myriophyllum*), and drop off with the anthers, leaving filiform filaments behind. The androecium is diplostemonous with all stamens inserted at the same level. Antepetalous stamens are lost in some genera. In *Proserpinaca* the flowers are trimerous and haplostemonous with reduced petals (Eichler, 1878). The ovary is inferior, two to four carpellate with free stylodes on top of the inferior ovary. In *Haloragis* the ovary has four wings and is partitioned by four septa with one apical ovule per locule. There are occasionally two present but one aborts (Schindler, 1905). APG II (2003) included *Penthorum* and *Tetracarpaea* in Haloragaceae, but this reduces the morphological circumscription of the family (cf. Kubitzki, 2007b).

Fig. 10.5. Crassulaceae: Flower (A) and inflorescence (B) of *Kalanchoe fedtschenkoi*; flower (C) and inflorescence (D) of *Sempervivum* × *fauconnettii*.

Crassulaceae

Fig. 10.5A,B. *Kalanchoe fedtschenkoi* Raym.-Hamet and H.Perrier

∗ K(4) [C(4) A4+4] G̲(4)

Fig. 10.5C,D. *Sempervivum* × *fauconnettii* Reut.

∗ K18 C18 A18+18 G̲(18)

Flowers are solitary or grouped in branched cymes. Flowers are hypogynous or slightly perigynous. Pentamery is the basic condition in the family (Mort *et al.*, 2001), but occasionally the merism of the flower can become very high (up to 32) by a simultaneous increase in each whorl (e.g. *Aeonium, Sempervivum*: Fig. 10.5C). Petals are well developed with an imbricate aestivation. Fused petals are common in the family (e.g. *Kalanchoe*). A petal tube is linked with hypanthial growth as the basal third of the tube was found not to be petaloid and the antepetalous stamens are inserted higher than the antesepalous stamens (Berger, 1930). The androecium is obdiplostemonous or haplostemonous. Obdiplostemony is linked with a tendency for reduction of antepetalous stamens (occasionally sterile in *Sempervivum*: Berger, 1930). *Crassula* is haplostemonous. The gynoecium is apocarpous, usually with four to five antepetalous carpels. Placentation is axile and styles are free. Nectaries consist of a scale at the base of each carpel, with its vascular supply derived from carpellary traces (Tillson, 1940). Mort *et al.*, (2001) demonstrated that fused petals and a higher merism have arisen more than once in the family.

Grossulariaceae

Fig. 10.6. *Ribes speciosum* Pursh

∗ [K5 C5 A5] Ğ(2)

Grossulariaceae have racemose inflorescences, and flowers are generally subtended by a bract and two smaller bracteoles (occasionally absent). When bracteoles are missing, the outer sepals can be positioned transversally (e.g. *Ribes alpinum*, *R. sanguineum*: Eichler, 1878). Flowers are bisexual, rarely unisexual with staminodes in pistillate flowers, and (tetra-) pentamerous. A short to deep petaloid hypanthium is present, topped by calyx lobes of the same or a different colour. Calyx lobes are rarely erect (e.g. in *R. speciosum*), usually spreading or reflexed. Petals are narrow with apert aestivation, rarely absent. The androecium is haplostemonous. Stamens are included, when the hypanthium is long (e.g. *R. malvaceum*), or strongly exserted (e.g. *R. speciosum*). In some species the connective bears a distal nectary. A well-developed nectary is present as a five–lobed disc, with lobes alternating with stamens (e.g. Weigend, 2007) or an

Fig. 10.6. Grossulariaceae: flower of *Ribes speciosum*.

inconspicuous annular zone on the hypanthium. The ovary is inferior to half-inferior with two parietal placentae covered with numerous ovules. The ovary is topped with two carinal styles in median position.

Saxifragaceae

Fig. 10.7A,B. *Rodgersia aesculifolia* Batalin

✳ K5 C0 A5+5 -G-(2)

Fig. 10.7C. *Saxifraga fortunei* Hook. f.

↘ K5 C5 A5+5 G̲ (2)
General formula: ✳ (↓) K5C0-5A5+5 or 5G2

Molecular phylogenies have greatly improved the delimitation of Saxifragaceae by removing several genera that were previously included as subfamilies (e.g. *Bauera*, *Francoa*, *Parnassia*, *Lepuropetalon* to rosids and Hydrangeoideae, *Vahlia*, Montinioideae and Escallonioideae to asterids: Morgan and Soltis, 1993). Closest relatives of Saxifragaceae are Grossulariaceae, Iteaceae and Pterostemonaceae, which share similarities in floral morphology and anatomy (e.g. Bensel and

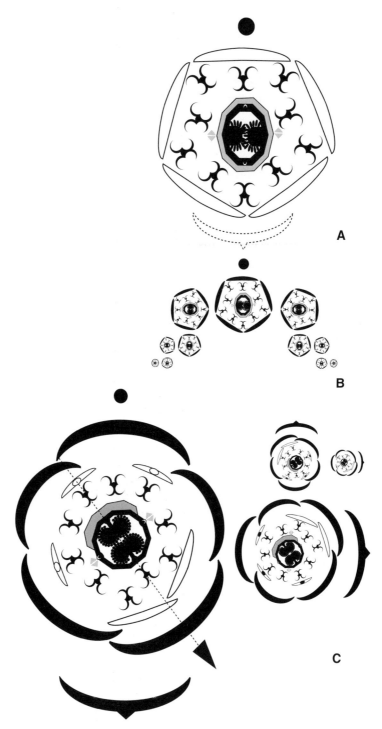

Fig. 10.7. Saxifragaceae: Flower (A) and partial inflorescence (B) of *Rodgersia aesculifolia*; (C) partial inflorescence of *Saxifraga fortunei*.

Palser, 1975b). Inflorescences are variable and built on a monotelic or polytelic pattern. Sepal lobes are large and are linked to the hypanthium, thus appearing basally fused. There is much variation in the extent of development of petals, ranging from well-developed imbricate petals (e.g. *Saxifraga*, *Lithophragma*) to small, weakly developed appendages (e.g. *Mitella*, *Tiarella*) or their complete loss (e.g. *Rodgersia*). As in Grossulariaceae, hypanthium, bracts and sepals may be petaloid in case petals are lost (*Chrysosplenium*, *Rodgersia*: Fig. 10.7A). Petals have a small insertion base and are often narrow and fimbriate (e.g. Endress and Matthews, 2006b). Obdiplostemony is commonly found and tends to be correlated with a retardation of the petals and the development of the hypanthium (Gelius, 1967; Ronse De Craene and Smets, 1995b). In the genus *Mitella* the androecium can exceptionally be more variable, ranging from obdiplostemonous to (ob)haplostemonous arrangements (Cronquist, 1981).

A hypanthium is usually present, weakly to strongly developed, and there is a nectary at its base or on the ovary. Ovary position can fluctuate from superior to inferior, even within a genus (e.g. *Lithophragma*: Kuzoff, Hufford and Soltis, 2001) and is generally restricted to two carpels in a median position. The ovary is usually partially fused with transitions between axile and parietal placentation and free carinal styles (Bensel and Palser, 1975b).

Zygomorphy is rare and is found in genera with predominantly polysymmetric flowers (e.g. *Saxifraga fortunei*: Fig. 10.7C, *Heuchera richardsonii*); it is more rarely characteristic for a genus (e.g. *Tolmiea* with loss of one abaxial petal: Klopfer, 1973).

10.2   Malvids

Figure 10.8 shows the phylogenetic tree of malvids based on Wang *et al.* (2009). Malvids contains three basal orders, Geraniales, Myrtales and Crossosomatales, and an upper clade of Malvales, Sapindales and Brassicales. Although the basal rooting of the latter three orders is not fully understood (with a basal order Huertales, sister to Malvales–Brassicales: Worberg *et al.*, 2008), their circumscription is well supported on several morphological and phytochemical characters (Ronse De Craene and Haston, 2006).

10.2.1   Early diverging malvids: Geraniales, Myrtales

Geraniales

Five families are recognized in this morphologically heterogenous order.

All Geraniales share obdiplostemonous flowers, with isomerous antepetalous carpels and a persistent calyx. Nectaries are extra- or intrastaminal and occur as separate entities.

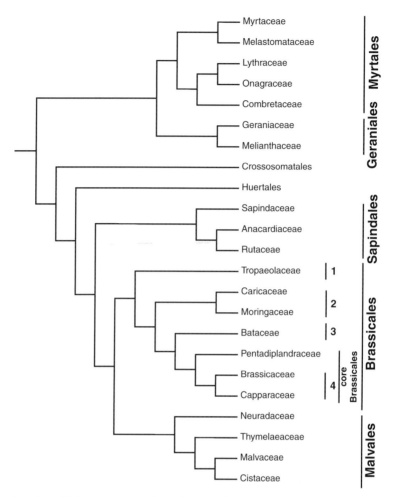

Fig. 10.8. Phylogenetic tree of Malvids, based on Wang *et al.* (2009) and Ronse De Craene and Haston (2006). 1. Tropaeolaceae-clade; 2. Moringaceae-clade; 3. Bataceae-clade; 4. Brassicaceae-clade (core Brassicales).

Ronse De Craene *et al.* (2001) discussed several morphological characters linking Francoaceae and Melianthaceae. A number of characters is shared by other Geraniales, supporting molecular evidence: a strong tendency to evolve median monosymmetric flowers with displacement of the extrastaminal receptacular nectary to the adaxial side, basically axile placentation with clear separation of the style from the ovary, and obdiplostemony with a clear trend towards haplostemony by sterilization or loss of the antepetalous stamens.

Geraniaceae

Fig. 10.9A,B. *Erodium leucanthemum* Boiss.

✳/↓ K5 C5 A(5+5°) G̲(5)

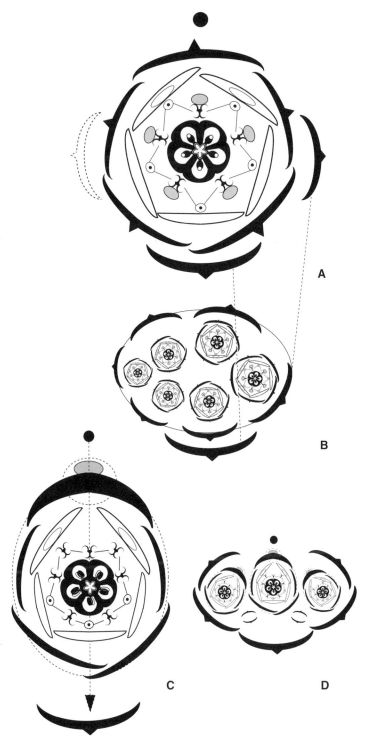

Fig. 10.9. Geraniaceae: Flower (A) and corresponding partial inflorescence (B) of *Erodium leucanthemum; C. Pelargonium abrotanifolium.* The dotted bracteole in A is lost.

Fig. 10.9C,D. *Pelargonium abrotanifolium* (L.f.) Jacq.

↓ K5 C5 A(5+2:3°) G̲(5)

General formula: ✳/↓K5 C5 A5–15 G5

Inflorescences are cymose, with flowers seemingly umbellate through contraction of internodes (Fig. 10.9B,D). Flowers of Geraniaceae are basically polysymmetric and pentamerous, with isomery in all floral whorls. While *Geranium* or *Monsonia* are strictly polysymmetric, *Erodium* has a tendency for the adaxial petals to differ from the abaxial in size and texture. *Pelargonium* (Fig. 10.9C) represents truly monosymmetric flowers with a reduced number of stamens. The two upper petals differ in shape, colour and size from the adaxial petals, with variably descending aestivation. In *P. mutans* the median adaxial petal is absent (pers. obs.). This pattern is also expressed in the nectaries which are found as five equal dorsal bumps at the base of the filaments in *Geranium*, as unequally sized nectaries in *Erodium* and as a single adaxial nectary in *Pelargonium* located at the base of a spur (Vogel, 1998c). The spur is hypanthial in nature and produces nectar at the base of a tube extending below the adaxial sepal. Nectar is withheld in the flower through the erect adaxial sepal and posterior petals, while other sepals and petals may be reflexed in some species. Petals are retarded in their development and occasionally arise from common primordia with the stamens (e.g. Payer, 1857; Ronse De Craene, Clinckemaillie and Smets, 1993; Erbar, 1998). The androecium of Geraniaceae is obdiplostemonous, but this is not caused by a shift in position of the antepetalous stamens; they arise outside the antesepalous stamen whorl possibly on common primordia with the petals (Ronse De Craene and Smets, 1995b; Erbar, 1998). In *Erodium* antepetalous stamens develop into filamentous staminodes. A secondary stamen increase is found in *Monsonia* and *Hypseocharis* by lateral division of antepetalous stamens. In *Pelargonium* three adaxial staminodes are formed through unequal development (Sattler, 1973); they are opposite the anterior petals. Except for *Hypseocharis* the gynoecium has a stylar beak separated from the ovary.

Carpels are superior, isomerous and antepetalous. In *Pelargonium* ovules are superposed by constricted space.

Melianthaceae

Fig. 10.10A. *Francoa sonchifolia* Cav., based on living material and Ronse De Craene and Smets (1999b).

✳ K4 C4 A4+4 -G̲(4)-

Fig. 10.10B. *Melianthus major* L., based on Ronse De Craene *et al.* (2001)

↓ K5 C4 A4 G̲(4)

General formula: ✳/↓K4–5 C4–5 A4–5(+4–5) G4–5

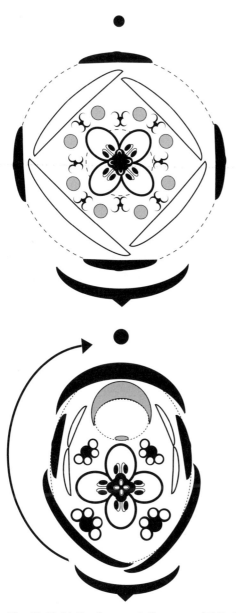

Fig. 10.10. Melianthaceae: A. *Francoa sonchifolia*; B. *Melianthus major*.

There is high floral diversity in the small family Melianthaceae with five genera *Bersama*, *Melianthus*, *Francoa*, *Greyia* and *Tetilla*. *Francoa* and *Greyia* are occasionally treated as separate families, or *Greyia* is treated as part of the solely southern African Melianthaceae (e.g. Stevens, 2001 onwards), although *Greyia* appears to be closer to *Francoa* morphologically (Ronse De Craene and Smets, 1999; Ronse De Craene *et al.*, 2001). Both *Greyia* and *Francoa* share the presence of conspicuous receptacular nectaries, which were interpreted as staminodes by several authors. Flowers are pentamerous or tetramerous (*Francoa* is occasionally pentamerous), with obdiplostemonous androecium, half-inferior ovary and axile placentation. Ovules are numerous and arranged in two rows per locule. Petals are characteristically contorted to cochleate-ascending.

Bersama and *Melianthus* are closely related but are remarkable in their divergent evolution. *Melianthus* has monosymmetric flowers, which are exacerbated by unequal growth of a shallow hypanthium, comprising a large cup-like adaxial nectary, and shift of the petals to the adaxial side. Bilateral symmetry is initiated early in the development of the flower, disrupting the 2/5 sequence of initiation of the sepals. The abaxial petal is missing, although initiated but suppressed later in the development (asterisk in Fig. 10.10B; Payer, 1857; Ronse De Craene *et al.*, 2001). The adaxial sepal encloses the nectary as a hood-like appendage. The nectary develops as a rim partially surrounding a depression in which black nectar accumulates. On the side of the gynoecium a protrusion of the receptacle develops into a beak. The beak is sometimes interpreted as a posterior staminode (Eichler, 1878), although there is no ontogenetic evidence in support of this. As the flower resupinates at maturity, the petals and adaxial sepal withhold the nectar. Petals are marginally connected by trichomes. Principal pollinators of *Melianthus* are sunbirds attracted by abundant nectar. In *Bersama* flowers are nearly polysymmetric by the equal development of the five petals and the smaller adaxial nectary, which is sometimes circular. Early initiation of flowers of *Melianthus* and *Bersama* is monosymmetric, an unusual pattern in the rosids. The androecium and gynoecium are tetramerous. Two rows of two to four ovules are initiated within each carpel of *Melianthus*, while only a single basal ovule develops in *Bersama* by reduction of the other.

Melianthaceae share several similarities (Ronse De Craene and Smets, 1999; Ronse De Craene *et al.*, 2001), including racemose inflorescences, the absence of bracteoles and the presence of a sterile bract on top of the inflorescence, a tendency for developing tetramerous flowers, resupination of flowers (not in *Francoa*), a shallow hypanthium with sunken ovary, and stomata on the anthers. Some species of *Bersama* have five stamens, indicating a derivation from pentamery and polysymmetric flowers. In Francoaceae, the genus *Tetilla* is monosymmetric (slightly so in *Greyia*), indicating a fluctuating level of symmetry in the Geraniales.

Myrtales

Myrtales is a homogenous order of 12–13 families, which has not chan-ged much since the introduction of molecular systematics, except for inclusion of Vochysiaceae as sister to Myrtaceae. The order shares a number of striking floral morphological characters that enable an easy recognition. Flowers are generally bisexual and tetra- or pentamerous (occasionally hexamerous). A common characteristic is the presence of a deep hypanthium lifting perianth and androecium high above the gynoecium. The ovary consists of two to five carpels, is inferior or less often superior, with axile placentation, often with numerous ovules on protruding placentae. Petals are always free and the androecium is basically diplostemonous (Dahlgren and Thorne, 1984), with a secondary stamen increase in some families. Stamens are incurved in bud and there is a single terminal style (Judd and Olmstead, 2004). Petals are reduced or lost in many taxa (e.g. Combretaceae). Monosymmetry is rare and weakly developed, except in Vochysiaceae where it results from reduction (Litt and Stevenson, 2003a,b).

There is a strong tendency for building obhaplostemonous flowers in Myrtales, which are present in eight families. Some small families are exclu-sively obhaplostemonous (e.g. Rhynchocalycaceae, Oliniaceae, Penaeaceae, Alzateaceae), with a tendency for reduction or loss of the petals (Schönenberger and Conti, 2003). The single stamen of Vochysiaceae arises opposite a petal but is occasionally displaced towards a sepal in some species (Litt and Stevenson, 2003b). Obhaplostemonous flowers of Myrtaceae have undergone a secondary stamen increase (Ronse De Craene and Smets, 1991a). Obdiplostemony, which is associated with a reduction of the antepetalous stamens, is not common (some Combretaceae: Ronse De Craene and Smets, 1995b). In cases where a nectary is present, it is usually associated with the hypanthium as an inconspicuous disc or is epigynous.

Onagraceae

Fig. 10.11. *Epilobium angustifolium* L.

$*/\downarrow$ K4 C4 A4+4 or 4+0 Ğ(4)
General formula: $*/\downarrow$K2–7 C2–7 A(1–2)4–8 G (2-)4–7

Onagraceae and Lythraceae are sister groups, sharing a valvate calyx (Judd and Olmstead, 2004).

Flower merism is mostly four, rarely two (*Circaea*) or a higher number (five to seven in some *Ludwigia*: Eyde, 1977). Flowers have a strongly developed hypanthium linking perianth and androecium. The gynoecium is inferior and the style protrudes through the narrow hypanthium. There is an epigynous

Fig. 10.11. *Epilobium angustifolium* (Onagraceae). Note the different maturation of the stamens.

nectary. The flower is well adapted to long-tongued insects (hawkmoths) or birds. Flowers are regular to weakly monosymmetric. In *Epilobium*, monosymmetry is caused by a larger abaxial sepal and wider space between anterior petals; stamens and style are curved to the anterior side (Fig. 10.11). Petals are often clawed and usually present; they are lost in some species of *Fuchsia*. The androecium is diplostemonous (not obdiplostemonous as stated by Eichler, 1878: pers. obs.; Sattler, 1973) or haplostemonous (some *Ludwigia*, *Circaea*); there is no secondary increase of stamens.

Only *Lopezia* is strongly monosymmetric with two median stamens, of which the adaxial one is a petaloid staminode (Eyde and Morgan, 1973). The gynoecium is isomerous and inserted opposite the petals. The ovary has axile placentation with protruding placentae covered by small ovules or fewer ovules are uniseriate (Eyde, 1977).

Lythraceae (incl. Punicaceae and Sonneratiaceae)

Fig. 10.12. *Cuphea micropetala Kunth.*

↓ [K6 C6/0 A5+6] G(2)

General formula: ✱/↓ K4–6 C(0-)4–6 A4+4/6+6 G2–4–6(-12)*

*flowers are rarely 8- to 16-merous

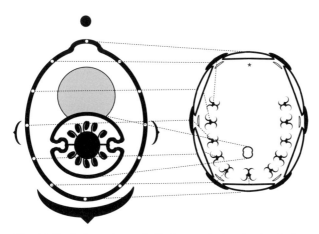

Fig. 10.12. *Cuphea micropetala* (Lythraceae), showing level of ovary and level of insertion of the stamens and perianth.

Tobe, Graham and Raven (1998) described the floral morphology and presented a number of floral diagrams but did not include *Cuphea*. Floral morphology is highly diverse with several morphologies, but all taxa share a long tubular hypanthium and superior ovary. Sepal and petal lobes, as well as stamens, are inserted at different levels on the tube. Hexamery is very common in the family (70%), while the remainder is tetramerous or has a variable and higher merism (e.g. *Lafoensia*). Petals are well developed (crumpled in bud), sometimes reduced or missing in some genera. In *Cuphea* they are barely visible and are often overlooked. An epicalyx is present at the top of the hypanthial tube in *Cuphea* and was interpreted as emergences of the congenitally fused sepals (Mayr, 1969), which is not satisfactory as explanation. The androecium is basically diplostemonous with occasional reductions to haplostemony or more rarely obhaplostemony. Paired stamens occur occasionally opposite the sepals. When stamens are situated at the base of the hypanthium, a secondary increase is centrifugal from antepetalous primary primordia, as in *Lagerstroemia* (Ronse De Craene and Smets, 1991a). In *Punica* a secondary stamen increase runs centripetally from antepetalous sectors (Leins, 1988).

The ovary can be occasionally (half-)inferior but is mostly superior. Apart from the highly aberrant ovary of *Punica* with one to more extra carpel whorls within the ovary (see e.g. Ronse De Craene and Smets, 1998b), the gynoecium is mostly bicarpellate with axile placentation (with septa interrupted at the top of the ovary). The number of carpels is occasionally isomerous (six) to higher (*Sonneratia*: 12).

A nectary is variously present on the inner slopes of the hypanthium, at the base of the ovary, or can be absent (Tobe, Graham and Raven, 1998). In *Cuphea* it arises on the lower part of the ovary, as in *Decodon*, but shifts towards the junction between ovary and hypanthium. Zygomorphy is rare and occurs late in the development of the flower. In *Cuphea* this is emphasized by the loss of a stamen and the adaxial development of the nectary at the base of the gynoecium (Fig. 10.12).

Myrtaceae

Fig. 10.13A. *Callistemon citrinus* (Curtis) Skeels

✶ K5 C5 A0+5$^\infty$ Ǧ(3)

Fig. 10.13B,C. *Syzygium australe* (Link) B. Hyland

✶ K4 C4 A∞ Ǧ(2)
General formula: ✶ K4–5 C4–5 A4-∞ G2–3

Myrtaceae is a distinctive family, characterized by an inferior ovary and a broad hypanthial cup lined with nectary tissue. Sepals, petals and stamens are inserted on the upper margins of the cup. Bracteoles are variously developed in *Callistemon*; they determine the position of the first sepals. In *C. citrinus* the outer sepals are lateral in the absence of bracteoles (Fig. 10.13A), contrary to species with bracteoles (e.g. Orlovich, Drinnan and Ladiges, 1999). Sepals and petals usually have an imbricate quincuncial or cochleate (in pentamerous flowers), or decussate aestivation (in tetramerous flowers). Petals are early caducous or occasionally calyptrate (*Eucalyptus*). Petals arise in a 2/5 sequence following sepal initiation and cannot be distinguished from the calyx in early stages. In *Callistemon citrinus* petals are green as the sepals. Stamen number in Myrtaceae is very variable (4–1000+ per flower: Schmid, 1980), but can most often be described as complex obhaplostemonous. Primary primordia arise opposite the petals (occasionally from common stamen-petal primordia: *Eucalyptus*: Drinnan and Ladiges, 1989) and stamens have a centripetal or lateral development, sometimes covering the upper inner margins of the cup (Fig. 10.13B). The

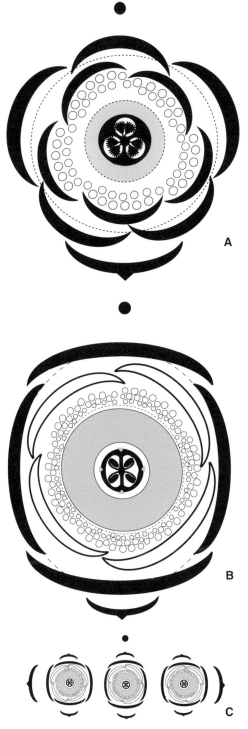

Fig. 10.13. Myrtaceae: A. *Callistemon citrinus*; *Syzygium australe*, B. flower, C. partial inflorescence.

stamens remain in clear fascicles (e.g. *Melaleuca*, *Lophostemon*) or expand into a girdle (e.g. *Callistemon*, *Syzygium*) (Orlovich, Drinnan and Ladiges, 1996, 1999). The expansion and growth of the hypanthium is responsible for a lateral and centripetal stamen increase (Ronse De Craene and Smets, 1991a; Carrucan and Drinnan, 2000). Stamens are typically incurved in bud.

Some Myrtaceae have less developed hypanthia with an obhaplostemonous androecium (e.g. *Backea*, *Micromyrtus*) or with stamens grouped in pairs or triplets (e.g. *Thryptomene*, *Chamaelaucium*) (Carrucan and Drinnan 2000). The reports of (ob)diplostemony for *Psiloxylon*, *Heteropyxis* and other genera enumerated by Schmid (1980) are suspicious without the backing of floral developmental evidence.

In *Syzygium* the first stamen primordia arise opposite the petals but extend in the intermediate areas finally covering a narrow margin of two to three rows around a broad sloping nectary (Fig. 10.13B; Ronse De Craene and Smets, 1991a; Belsham and Orlovich, 2003). Addition of more stamens is made possible by expansion of the floral bud. The inferior ovary has a single terminal style; placentation is axile and two rows of ovules are arranged on a U-shaped placenta in each locule. The number of ovules ranges from few to numerous (Fig. 10.13A).

Belsham and Orlovich (2003) provided an (incomplete) floral diagram of *Syzygium*. Leins (1965) gave the diagram of *Melaleuca nesophila* with one whorl of antepetalous stamen fascicles.

Myrtaceae can be confused with some Rosaceae in their general flower construction. The main differences are the arrangement of the stamens and single terminal style.

Melastomataceae

Fig. 10.14. *Medinilla magnifica* Lindl.

✳ [K5 C5 A5+5] Ğ(5)
General formula: ✳/↓ K(3)4–5(-8) C(3)4–5(-8) A5–10(-∞) G3–6

Flowers are mosty tetra- or pentamerous (occasionally from three- up to ten-merous). Flowers have a deep hypanthium with superior to inferior ovary. Sepals are free lobes, or fused into a calyptra (*Conostegia*). Petals are generally strongly contorted and inserted with the stamens on top of the hypanthial rim. The androecium is mostly diplostemonous and the filaments are characteristically incurved in bud, with the anthers fitting in holes formed in the hypanthial tissue surrounding the ovary (Fig. 10.14). Appendages of the connective, which are often elaborate and different in colour from the anthers, are a main feature of the family. Anthers are curved

Fig. 10.14. *Medinilla magnifica* (Melastomataceae): partial inflorescence. Note the depressions around the ovary with fitting anthers.

and poricidal, hanging over on the abaxial side of the flower. Antepetalous stamens are rarely staminodial or suppressed (*Poteranthera*, *Dissochaeta*: Eichler, 1878). Polyandry is found in a few genera (e.g. *Conostegia*, *Clidemia*, *Plethiandra*). In *Conostegia* the stamen number is increased in a single whorl, often in correlation with a lateral carpel increase (Puglisi, 2007). The ovary is mostly isomerous and antepetalous (not antesepalous as suggested by Eichler, 1878), or reduced to two. Placentation is mostly axile on protruding placental columns.

Most Melastomataceae are nectarless as they produce pollen or secretions on the connective as award. However, in some genera that reverted to nectar production, nectary is exuded from a cleft at the site where the filaments are sharply bent in bud (e.g. *Myconia*: Vogel, 1997). *Medinilla* exudes nectar from its petal tips in bud.

Combretaceae

Fig. 10.15. *Guiera senegalensis* J. F. Gmel.

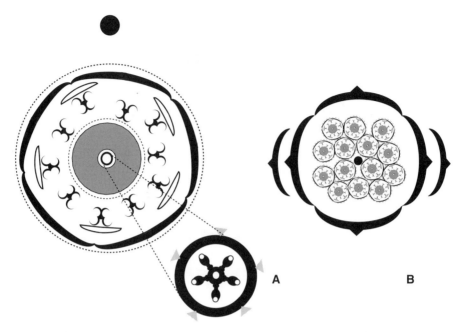

Fig. 10.15. *Guiera senegalensis* (Combretaceae): A. flower, B. partial inflorescence.

✳ K5 C5 A5+5 Ǧ(5)

General formula: ✳ K4–5 C4–5/0 A4–5(+4–5) G2–3–5?

Inflorescences are capitate or spicate, rarely racemose. Flowers are penta-merous to tetramerous. In *Guiera* flowers are arranged in terminal capitula enclosed by four coloured bracts and resembling a flower bud. Individual flowers of *Guiera* lack bracts, which are present in other genera (Stace, 2007). Flowers have a well-developed saucer- to tube-shaped hypanthium with val-vate sepal lobes and stamens inserted on two different levels (obdiplostem-ony). Stace (2007) considered the hypanthium to be of two parts, a lower hypanthium (usually) fused to the ovary and an upper hypanthium bearing the perianth and stamens. This would make the ovary virtually superior. The ovary is mostly inferior to half-inferior in *Strephonema* (Stace, 2007) and is mostly topped by a broad nectary surrounding a single style. Petals are small and narrow, and often absent (e.g. *Terminalia*). One stamen whorl is occasion-ally staminodial or absent (antepetalous stamens in some *Combretum sp.*, ante-sepalous stamens in other *Combretum* and *Terminalia tetrandra*). Eichler (1878) mentions an increase of stamens by dédoublement in some *Combretum*. In *Lumnizera* the number of stamens is often lower than ten by partial reduction of antepetalous stamens (Fukuoka, Ito and Iwatsuki, 1986). Septa are lacking and five ovules are apically inserted in *Guiera*. The number of ovules varies from one to 20 (mostly two) but the number of carpels is difficult to determine. Eichler (1878) interpreted the flower of *Combretum* as five-carpellate because of ridges on the ovary; the five ovules of *Guiera* also point to five carpels, although the number is lower in flowers with fewer ovules. Tiagi (1969) found a correlation between the number of dorsal traces in the style and the number of ovules, and interpreted this as a reflection of the number of carpels (mostly two to three). However, Fukuoka, Ito and Iwatsuki (1986) showed that this correlation cannot be made in all cases.

10.2.2   *Remaining malvids: Malvales, Brassicales, Sapindales*

Malvales

The premolecular Malvales were expanded as to include four clades (e.g. Alverson *et al.*, 1998): core Malvales (Malvaceae), Thymelaeaceae, a bixalean clade and a dipterocarpalean clade, including Neuradaceae and Cistaceae.

Von Balthazar *et al.* (2006) interpreted Malvales as basically diplostemonous (present in Neuradaceae, Thymelaeaceae and Dipterocarpaceae), although several clades are characterized by a secondary stamen increase. However, indications for the existence of two stamen whorls were demonstrated for Cistaceae (Nandi, 1998) and Bixaceae (Ronse De Craene, 1989b).

Neuradaceae

Fig. 10.16A,B. *Neurada procumbens* L., based on Ronse De Craene and Smets (1996b)

✳ K5 C5 A5+5 -G- (10)

Inflorescences in *Neurada* are complex but built on a monochasial pattern (Ronse De Craene and Smets, 1996b). Flowers have a deep hypanthium with sepals, petals and stamens inserted on the inner slopes. The androecium is diplostemonous. The ovary is partly embedded in the receptacle and arises as ten free carpels. Two ovules are initiated in each carpel, but one aborts.

The Neuradaceae (with three genera) were included in Malvales on the basis of seed coat anatomy and chemistry and this was supported by molecular data (Bayer *et al.*, 1999), despite earlier associations with Rosaceae on the basis of striking convergences (e.g. Ronse De Craene and Smets, 1996b). The epicalyx of spinose appendages is not comparable to the bracts of Malvaceae but closely resembles the spines of *Agrimonia* (Rosaceae); it arises much later than the bracts (Ronse De Craene and Smets, 1996b). *Neurada* shares an increase of carpels with some Malvoideae.

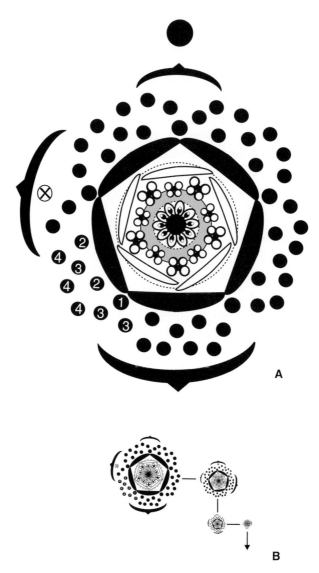

Fig. 10.16. *Neurada procumbens* (Neuradaceae): A. flower; B. inflorescence. Numbers refer to sequence of initiation of outer spines.

Cistaceae

Fig. 10.17A,B. *Cistus salviifolius* L.

✳ K5 C5 A5$^{\infty}$+5 G̲ (5)
General diagram: ✳ K3–5 C0/(3)5 A3-∞ G (2)3–5(6–12)

Fig. 10.17. *Cistus salviifolius* (Cistaceae): A. flower; B. partial inflorescence. Black dots, antepetalous stamens; white dots antesepalous stamens.

Flowers are solitary or grouped in unipartite cymes with flowers arising in a zigzag pattern. Flowers are small to large, polysymmetric and are subtended by a single bract. The calyx shows considerable variation between different species: it consists either of five equal sepals with a clear 2/5 arrangement, or the two outer sepals are significantly smaller and occasionally absent. Petals are imbricate-contorted, initially smaller and crumpled in bud. In species with only three sepals, two petals are opposite the sepals and the other three alternate with them (e.g. *C. ladanifer*). Eichler (1878) mentions a superposition of petals to sepals in some *Cistus* species. A similar arrangement is found in *C. creticus* and *C. salviifolius*, where the outer sepals are highly different from the inner and usually do not enclose the bud (pers. obs.). The position of petals is influenced by size differences of sepals and the more or less high degree of contortion in the flower.

The androecium is usually polyandrous with centrifugal stamen initiation. Five antesepalous stamens arise earlier and higher than the other stamens. They are followed by 5–13 antepetalous primordia and more primordia follow in a more or less regular arrangement on a ring primordium (Ronse De Craene and Smets, 1992a; Nandi, 1998). The androecium can be interpreted as complex haplostemonous or possibly diplostemonous with a secondary multi-plication of stamens from antesepalous primordia, as is common in Malvales. In large-flowered species, the number of primordia of the second whorl is usually much larger. The gynoecium consists of five (6–12) carpels in *Cistus*, but only of three carpels in other genera, with intruding parietal placentation. In *Cistus* the placentae are deeply intruding and appear axile (Fig. 10.17). Carpels alternate with petals and there is a single solid style. Ovules are inserted in pairs or in two rows on each placenta. No nectaries are found.

Some genera have a tendency for the flowers to become trimerous (*Lechea*) or petals are occasionally lost (*Crocanthemum*), and this was interpreted as a reduced state linked with cleistogamy (Nandi, 1998).

Malvaceae

General formula: $*$K5 C0/5 A5-∞ G(2–3)5-∞

Molecular data have shown that there is no support for the traditional delimitation of four families – Malvaceae, Sterculiaceae, Bombacaceae and Tiliaceae. Sterculiaceae and Tiliaceae are largely polyphyletic, while Malvaceae *sensu stricto* is the only natural clade (Alverson *et al.*, 1998, 1999; Bayer *et al.*, 1999). Nine well-supported evolutionary lineages have been identi-fied and these are best treated as subfamilies, as a subdivision into separate families is clearly unpractical (Bayer *et al.*, 1999; see comments on p. 60).

Core Malvaceae form a natural group that can be identified by a number of synapomorphies, including a typical dichasial inflorescence structure ('bicolour units' *sensu* Bayer, 1999), trichomatic floral nectaries on the perianth (Vogel, 2000), a valvate calyx and basically obdiplostemonous androecium (von Balthazar *et al.*, 2004, 2006).

The tricolour units consist of dichasial inflorescences subtended by three bracts. By reduction of lateral branches a single terminal flower remains, sur-rounded by three or more bracts, which form an epicalyx (Bayer, 1999).

Flowers of Malvaceae are often typically contorted; this contortion starts early in the development and leads to an oblique displacement of petal and stamen organs (van Heel, 1966). Sepals are not affected by contortion and are usually valvate and tubular with small lobes. Petals are often strongly contorted and basally fused to the androecium. In *Chiranthodendron* and *Fremontodendron* petals are apparently lost (von Balthazar *et al.*, 2006). The sterculioid clade has apetalous, unisexual flowers with ten to many stamens and an androgynophore (Venkata Rao, 1952; see below).

The androecium is extremely diverse in the family and is often difficult to interpret. Von Balthazar *et al.* (2004, 2006) and Venkata Rao (1952) interpreted an obdiplostemonous arrangement as ancestral for Malvaceae. The obdiplostemonous arrangement of stamens is linked with a sterilization of the antesepalous stamens, arising more towards the centre of the flower after the initiation of the antepetalous stamens. I interpreted this kind of obdiplostemony as different from other cases of obdiplostemony where the antepetalous stamens arise after the antesepalous stamens and show a tendency for reduction (see p. 9; Ronse De Craene and Smets, 1995b). In Grewioideae and Matisieae (van Heel, 1966; Brunken, pers. comm. 2008) both stamen whorls are developed and produce a few to many stamens on complex primordia.

The antesepalous stamens are seldom well developed (e.g. some Sterculioideae, Fig. 10.20A), more frequently sterile (Bombacoideae, Byttnerioideae, Fig. 10.19A, Matisieae) or almost undeveloped or absent (Malvoideae, Fig. 10.20B). This arrangement is reflected in the vasculature of the stamens throughout the family, with a single undivided bundle connecting the antesepalous unit and branching trunk bundles serving the other stamens (von Balthazar *et al.*, 2006; Janka *et al.*, 2008).

In one major lineage of Malvaceae, the Malvatheca clade (see Alverson *et al.*, 1999), more basal genera (e.g. *Fremontodendron*, *Ochroma*) develop triplets of one antesepalous and two antepetalous stamens; the antesepalous stamen is sterile and forms the tip of structures that contain one half of two separate antepetalous anthers, resembling a true stamen! The anthers become long twisted and fragmented at maturity. This arrangement is retained in other members of the clade with or without sterilization of antesepalous stamen primordia. *Ceiba pentandra* has a similar triplet arrangement (Janka *et al.*, 2008). Von Balthazar *et al.* (2006) derived the rows of half-anthers found in many Malvaceae from the subdivision of long anthers as in *Fremontodendron* and the subsequent development of stalks. Figure 10.18 shows the centrifugal development of stamens from two whorls of stamen initials.

There is a general tendency in Malvaceae for stamens to become monothecal. However, this has evolved along different, convergent routes. In clades outside the Malvatheca clade, stamens proliferate and monothecal anthers evolved by splitting of dithecal anthers; in the Malvatheca clade, monothecal anthers are the result of a compartmentalization of long anthers in rows (Alverson *et al.*, 1998; von Balthazar *et al.*, 2006; Janka *et al.*, 2008). The androecium of *Adansonia digitata* differs from other Malvaceae in the high number of small stamen primordia arising on a ring primordium (Janka *et al.*, 2008). This obviously represents a secondary proliferation, leading to a much higher stamen number.

In Malvoideae, there is a strong tendency for complete loss of the antesepalous sector, which is occasionally present as small teeth in some genera (e.g. *Pavonia*, *Hibiscus*) and could represent the connective of the lost unit or

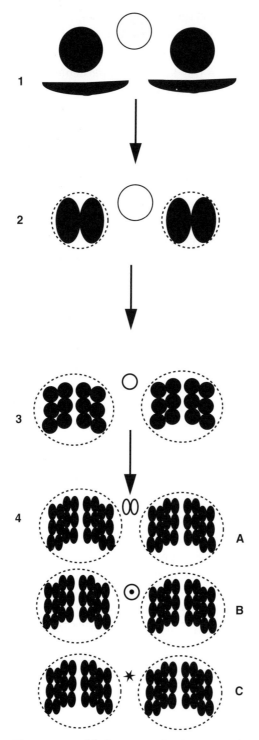

Fig. 10.18. Model of sequence of androecial development in the Malvatheca clade (Malvaceae), based on Janka *et al.* (2008), strongly modified. 1. Early stage with

sterile antesepalous stamens (van Heel, 1966; von Balthazar *et al.*, 2004). The antesepalous stamen is occasionally initiated. In *Althaea officinalis* (Ronse De Craene, unpubl. data), stamen groups appear in antesepalous position, while in species without the antesepalous stamens, groups are distinctly antepetalous with each half developing distinct rows of stamens.

All Malvoideae share a common early development of a primary ring primordium and the centrifugal initiation of two rows of secondary stamens opposite the petals (Fig. 10.18). The ring primordium is broken up in five or rarely ten sectors on which secondary stamen primordia arise in two alternating rows. In many taxa, secondary stamens tend to be split radially or longitudinally in tertiary half-stamens, which can be sessile or connected on a common stalk. The centrifugal proliferation of stamens is made possible by upward growth of the initial ring primordium developing into a tube around the ovary. This development is unique in angiosperms (Ronse De Craene, 1988) and leads to the formation of long staminal tubes from which the styles emerge centrally (e.g. *Hibiscus*).

The gynoecium is usually isomerous and develops five antepetalous carpels. There is the occasional proliferation of carpels in Malvoideae. In tribe Malopeae (*Kitaibelia*, *Malope*), numerous small carpels arise as five groups in a waveline and are squeezed in the confinement of the stamen tube (van Heel, 1995). Each carpel produces a single ovule and style. A second carpel whorl develops in Malvoideae–Ureneae but the inner whorl contains no ovules (e.g. *Pavonia*, *Urena*) (van Heel, 1978). Reductions to three (two) carpels are not infrequent in Malvaceae. Secondary apocarpy has evolved mainly in Sterculioideae and some other subfamilies (e.g. Endress, Jenny and Fallen, 1983; Jenny, 1988). Placentation is usually axile with ovules in two rows, often reduced to two lateral ascending ovules.

Nectaries of Malvaceae are present on either sepal or petal bases, and consist of secretory trichomes (Vogel, 2000).

Caption for Fig. 10.18. (cont.)

antesepalous stamen primordia (white dot) and antepetalous primordia (black dots);
2. lateral division of antepetalous primordia; 3. centrifugal differentation of
secondary primordia on the initial antepetalous primordia; 4. lateral division
of secondary primordia in tertiary primordia, developing into half anthers; A. the
antesepalous primordium produces two half anthers: e.g. *Pachira, Eriotheca, Adansonia*;
B. the antesepalous primordium is staminodial: e.g. *Bombax, Pseudobombax, Goethea*;
C. the antesepalous primordium is suppressed: most Malvoideae. Antesepalous
primordium drawn to scale relative to antepetalous primordium.

Byttnerioideae

Fig. 10.19A. *Theobroma cacao* L.

∗ K5 C5 A(5°+5²) G̲(5)

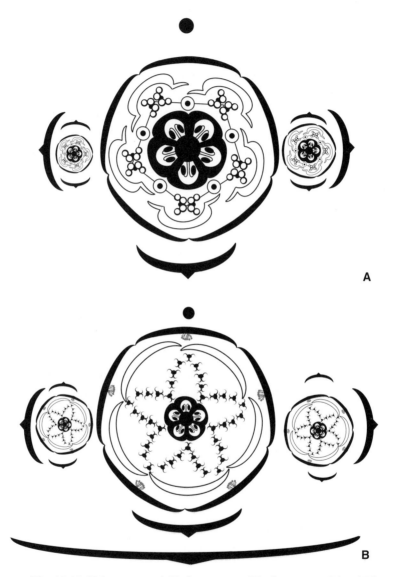

Fig. 10.19. Malvaceae: partial inflorescence of *Theobroma cacao* (A) and *Tilia × europaea* (B).

Byttnerioideae have retained the original obdiplostemonous arrangement. In *Theobroma*, antesepalous stamens develop into erect staminodes while antepetalous stamens divide in two lateral anthers facing away from each other. Stamens are rarely undivided, or more often in antepetalous triplets or with higher numbers (van Heel, 1966). Petals are usually cucullate and develop in a broad, cup-shaped basal part enclosing the anthers, while the petal tip is spathulate and recurved backwards (Fig. 10.19A; Bayer and Hoppe, 1990). Antepetalous stamens emerge before antesepalous staminodes and the carpels are arranged opposite the petals, increasing the obdiplostemonous appearance of the androecium.

Staminodes are occasionally lost and stamens can remain undivided (e.g. *Hermannia*). In *Theobroma* and allies, staminodes prevent selfing by mediating insect movement in the flower (Walker-Larsen and Harder, 2000). Flowers are rarely monosymmetric, with unequal petals and stamens and gynoecium grouped in an androgynophore (*Kleinhovia*: Bayer and Kubitzki, 2003).

Tilioideae

Fig. 10.19B. *Tilia* × *europaea* L.

✳ K5 C5 A0+5$^{\infty}$ $\underline{G}$(5)

Subfamily Tilioideae is restricted to two genera, much fewer than in premolecular treatments. The inflorescence peduncle is typically fused to the subtending bract in a conspicuous wing, favouring dispersal. Sepals are free and bear glandular trichomes at the base. Petals are narrow and slightly cucullate with weakly imbricate aestivation. The androecium arises centrifugally in antepetalous groups forming two rows connected with one upper stamen, which is occasionally staminodial (*Tilia tomentosa*). In other Malvaceae, staminodes are usually in antesepalous position. At the base neighbouring stamen fascicles are linked between the stamens in a waveline (van Heel 1966). The five carpels bear two ascending ovules per locule in *Tilia*.

Sterculioideae

Fig. 10.20A. *Sterculia coccinea* Jack

✳ K5 C0 [A(5+5) $\underline{G}$(3)]*
*shown as bisexual but with late abortion of one of the genders

Flowers are apetalous and unisexual, and lack an epicalyx or staminodes (Alverson *et al.*, 1999; Bayer *et al.*, 1999). Stamens and carpels are connected by an androgynophore. In staminate flowers, stamens are compressed in bud and overtop the pistillode, resembling *Nepenthes* buds (Ronse De Craene, unpubl. data). Venkata Rao (1952) described the androecium of *Sterculia foetida* as

Fig. 10.20. Malvaceae: A. partial inflorescence of *Sterculia coccinea*; *Abutilon megapotamicum*, flower (B) and partial inflorescence (C).

two-whorled with the alternisepalous stamens forming triplets. The androecium is variable in the subfamily, with various increases of the antepetalous stamens to 15 or more (e.g. *Brachychiton*, *Sterculia*) and loss of the antesepalous stamens (e.g. *Cola*, some *Sterculia*) (van Heel, 1966). In *Cola*, stamens are paired with bisporangiate anthers arranged in a radial line (van Heel, 1966; Bayer and Kubitzki, 2003). Carpels range from five to three in *Sterculia*. Secondary apocarpy evolved independently in the clade (Endress, Jenny and Fallen, 1983) and this is stressed at maturity by the development of individual carpophores. Trichomatic nectaries are variously spreading on the adaxial side of the sepal lobes and hypanthium, extending to the androgynophore.

Malvoideae

Fig. 10.20B,C. *Abutilon megapotamicum* (Spreng.) A. St. Hil. & Naudin

∗ K5 [C5 A(0+5$^\infty$)] G(5)

Malvoideae correspond to the classical delimitation of Malvaceae and were placed with part of Bombacaceae in a Malvatheca clade (Alverson *et al.*, 1999).

In *Abutilon* an epicalyx is absent, while it is well developed in several other Malvoideae and consists of two, three or more bracteoles (Pluys, 2002, unpublished thesis; Bayer and Kubitzki, 2003). In *Goethea* the flower is enclosed by four large petaloid epicalyx lobes surrounding the red flowers (pers. obs.). The calyx is valvate. The corolla is basally adnate to the stamen column. The long staminal tube enclosing styles and carpels is a typical feature of Malvaceae (Ronse De Craene, 1988; Ronse De Craene and Smets, 1992a). The number of stamens tends to be correlated with the extent of development of the stamen tube, ranging from few (e.g. *Wissadula*) to many (e.g. *Abutilon*, *Hibiscus*). Stamens are disposed in long rows in antepetalous position. An antesepalous staminode is occasionally developed (e.g. *Goethea*, *Hibiscus*), more often lacking (asterisks in Fig. 10.20B). Trichomatic nectaries are found in pockets on the calyx or opposite the petal lobes (e.g. *Abutilon*). Floral development of numerous species of Malvoideae was studied by several authors (e.g. Payer, 1857; van Heel, 1966, 1995; von Balthazar *et al.*, 2006).

Sapindales

The order forms a clade of nine families, forming three subclades: Sapindaceae, Anacardiaceae–Burseraceae and Rutaceae–Meliaceae–Simaroubaceae (Ronse De Craene and Haston, 2006). Sapindales share a syndrome of characters or interesting features that are not synapomorphic. Flowers are usually small and grouped in cymose inflorescences. They bear the typical characteristics of rosids: flowers are usually tetra- or pentamerous, polysymmetric and rarely monosymmetric (as a late event) with free, imbricate petals and a diplostemonous androecium. The gynoecium is superior, basically isomerous or reduced to two to three carpels with axile placentation. A prominent disc nectary is present and is either intrastaminal or extrastaminal. The superior ovary contains few ovules. Functionally unisexual flowers occur frequently and arise through late abortion during development (e.g. *Phellodendron*, Rutaceae: Zhou, Wang and Xiaobai, 2002; *Rhus*, Anacardiaceae: Gallant, Kemp and Lacroix, 1998).

Sapindaceae (incl. Aceraceae and Hippocastanaceae)

Fig. 10.21A,B. *Acer griseum* (Franch.) Pax

$\leftrightarrow$ K5–6 C5–6 A10–12 G̲(2)

Fig. 10.21C,D. *Serjania glabrata* HBK

↓ K5 C2:2 A5+3 G̲(3)*
*flowers unisexual with aborted stamens and ovary
General formula: $\leftrightarrow$ $*$↓ K(4)5 C0/(4)5 A4–10(-∞) G2–3

Sapindaceae are extended as to include Hippocastanaceae and Aceraceae (Judd, Sanders and Donoghue, 1994). A frequent combination of characters in Sapindaceae is: flowers functionally unisexual, oblique monosymmetry, petal appendages, eight stamens with papillose filaments, and trimerous–dimerous superior ovary surrounded by an extrastaminal disc. Eight stamens result from the loss of two antepetalous stamens (Fig. 10.21B; Ronse De Craene, Smets and Clinckemaillie, 2000). All ten stamens (original diplostemony) are rarely present in some *Acer* (Fig. 10.21A). There is a single ovule per carpel without funiculus but with a prominent obturator. Flowers of *Acer* appear secondarily disymmetric by the equal development of small petals and eight to ten stamens surrounding a bicarpellate ovary. Some Sapindaceae have a secondary stamen increase by lateral multiplication (e.g. *Deinbollia*: Ronse De Craene, unpubl. data).

Cardiospermum, *Aesculus* and *Serjania* (Fig. 10.21B) have elaborate monosymmetric flowers with a morphological distinction between adaxial and lower petals (see Endress and Matthews, 2006b). One lateral petal is lost, stressing the oblique monosymmetry. A similar orientation is found in *Moringa* or *Bretschneidera* (Brassicales), where cymose partial inflorescences facilitate lateral access to flowers (see p. 237). Petals have large ventral appendages. The extrastaminal nectary is strongly developed on the adaxial side of the flower, building a platform and pushing stamens and carpels to the abaxial side.

Acer griseum is unusual, compared with most *Acer*, which have only eight stamens as most Sapindaceae (Fig. 10.21A). In species of *Acer* with eight stamens, the stamens opposite the carpels are lost. In *A. griseum* there is a tendency for increase in merism by doubling of organs, as occurs in some other Sapindaceae.

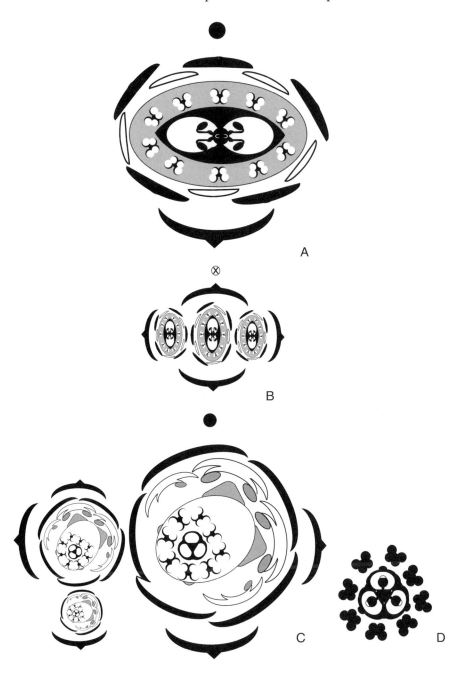

Fig. 10.21. Sapindaceae: *Acer griseum*, A. flower, B. inflorescence; *Serjania glabrata*, C. partial inflorescence with staminate flowers, D. reproductive organs of pistillate flower.

Flowers have small valvate sepals and petals that are morphologically similar. In some species, the corolla is absent. The number of carpels is always two in a transversal (or a more variable) position (Eichler, 1878). However, the genus is morphologically understudied. The nectary extends beyond the stamens, which appear to be immersed in it. The flattened flower appears disymmetric to polysymmetric. It is possible that *Acer* is derived from a predecessor with mono-symmetric flowers and three carpels, as they are nested in a clade with mono-symmetric flowers (Judd, Sanders and Donoghue, 2004). A floral diagram of *Koelreuteria paniculata* is given in Ronse De Craene, Smets and Clinckemaillie (2000).

Rutaceae

Fig. 10.22A. *Galipea riedeliana* Regel

↓ K(5) C5 A5 G̲(5)

Fig. 10.22B,C. *Boenninghausenia albiflora* (Hook.) Meissn.

✶ K(4) C4 [A4+4 G̲(4)]
General diagram: ✶/↓ K(3) 4–5 C(3) 4–5 A(4)5–10(-∞) G(1–2)4–5

Rutaceae is a large, variable family. Inflorescences are variable but partial inflorescences are cymose. Flowers are regular, weakly monosymmetric to strongly monosymmetric with two to three fertile stamens. Flowers are four- to five-merous, rarely trimerous (e.g. *Cneorum*), bisexual to rarely unisexual and dioecious. The calyx is often basally fused. The corolla is generally free or postgenitally adherent to stamens (*Galipea*), and imbricate. The androecium is mostly (ob)diplostemonous, occasionally with antepetalous staminodes or haplostemonous. An incomplete whorl of stamens is found in Galipeineae with two to three staminodes (e.g. *Erythrochiton*, *Galipea*) or *Ticorea* with five stami-nodes and two fertile stamens (three antepetalous stamens are missing: Baillon, 1871b). A prominent intrastaminal disc nectary is usually present. The ovary is isomerous and superior, mostly with a single stout style. The number of stamens and carpels is increased in some genera: stamens have undergone a lateral increase from antesepalous primordia (Payer, 1857; Moncur, 1988, accompanied with an increase of carpels in one whorl: *Citrus*, or both whorls: *Aegle*: Leins, 1967). There is occasionally a second carpel whorl ('navel oranges'). One to two ovules (rarely more) are arranged on axile placen-tae. Stamens and carpels are occasionally united on an androgynophore (Fig. 10.22B; *Cneorum*: Caris *et al.*, 2006b). In *Boenninghausenia* carpels appear secondarily apocarpous (pers. obs.), as in *Zieria* (Endress, Jenny and Fallen, 1983).

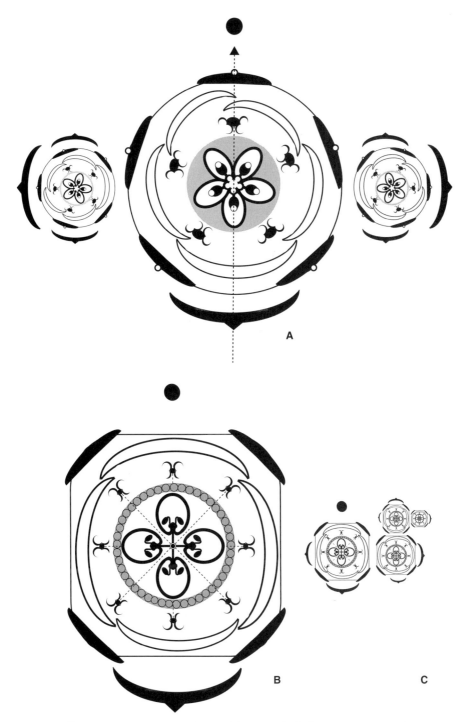

Fig. 10.22. Rutaceae: A. *Galipea riedeliana*, partial inflorescence; *Boenninghausenia albiflora*, flower (B) and partial inflorescence (C). Broken line in B shows attachment of stamens on androgynophore.

Anacardiaceae

Fig. 10.23. *Toxicodendron vernicifluum* (Stokes) F. A. Barkley

Fig. 10.23. *Toxicodendron vernicifluum* (Anacardiaceae). Particular flower with two staminodes (variable).

✳ K5 C5 A5+1–2° G̲(1:2°)

General formula: ✳/ Ẹ K5 C0/5 A5 or 5+5 or 1+9° G (1)3–13

Inflorescences are paniculate with all flowers maturing simultaneously. Bracts and bracteoles are present but early caducous. In several genera, flowers are functionally unisexual by the late abortion of carpels or stamens (e.g. Wannan and Quinn, 1991; Gallant, Kemp and Lacroix, 1998). The wind-pollinated *Amphipterygium* and *Pistacia* have reduced unisexual flowers, with a variable abortion in staminate and pistillate flowers and an unclear differentiation of bracts and sepals (Bachelier and Endress, 2007). Flowers are generally polysymmetric except for the gynoecium, which is pseudomonomerous. *Spondias* is isomerous in all whorls with carpels opposite the petals, or carpels are increased to ten in *S. pleiogyna* (Eichler, 1878); in *Pleiogynium* the carpel number ranges from 5 to 13 (Wannan and Quinn, 1991). The androecium is usually diplostemonous (not obdiplostemonous as mentioned by Eichler, 1878). Haplostemony is found in a few genera, such as *Pistacia*,

Amphipterygium, *Rhus* and *Toxidendron*; in the former stamens alternate with sepals (Bachelier and Endress, 2007); in *Toxidendron*, one to two antepetalous staminodes are occasionally present (Fig. 10.23). In *Anacardium* and *Mangifera*, the androecium is reduced to a single large fertile stamen (opposite sepal one) and other stamens are reduced in a descending order starting from the fertile stamen, resulting in a monosymmetric flower. The number of stamens is rarely secondarily increased (*Sorindeia*, *Poupartia*, *Gluta*: Eichler, 1878; Wannan and Quinn, 1991). A well-developed intrastaminal disc nectary is present in most species, although nectariferous trichomes on the corolla are reported for *Anacardium occidentale* (Bernardello, 2007). A distinctive character of the family is that two of the three carpels are sterile (except for *Spondias*), with variable degrees of reduction. The fertile carpel is situated in a latero-anterior position opposite sepal one, leading to oblique monosymmetry. The sterile carpels are often reduced to the styles (Fig. 10.23; Ronse De Craene and Smets, 1998b), or completely absent (*Anacardium*). A single basal-axile ovule develops in the fertile carpel. Styles are either free or fused and opposite the locule.

Brassicales

The order contains 15 families with a striking morphological divergence but a common chemical character (mustard oils, also present in Putranjivaceae of Malpighiales: Rodman *et al.*, 1998). No clear morphological synapomorphies exist for Brassicales. Several well-supported subclades can be recognized on a morphological and molecular basis (Tropaeolaceae-clade, Bataceae-clade, Moringaceae-clade and core Brassicales: Ronse De Craene and Haston, 2006).

Tropaeolaceae is associated with *Bretschneidera* and *Akania* (Akaniaceae) as a basal clade of Brassicales. Both families share oblique monosymmetry (at least in early stages of the development of *Tropaeolum*), an octomerous androecium, as well as a hypanthium (Ronse De Craene and Smets, 2001b; Ronse De Craene *et al.*, 2002). These characters are also found in Sapindaceae and it is not clear whether they represent a convergence (see Ronse De Craene and Haston, 2006). Moringaceae and Caricaceae are morphologically highly divergent, especially in their floral morphology, although they share a number of vegetative and floral characters (pentamerous flowers with a hypanthium, diplostemony, superior ovary with parietal placenta: Ronse De Craene and Smets, 1999a). Bataceae, Koeberliniaceae and Salvadoraceae represent a well-supported clade of small tetramerous, disymmetric flowers and a strong tendency for reduction (Ronse De Craene, 2005; Ronse De Craene and Wanntorp, 2009). Most core Brassicales (incl. the Bataceae-clade) share tetramerous flowers, a gynophore, curved embryos and campylotropous ovules, but there are several exceptions (Ronse De Craene and Haston, 2006).

The pentamery of Emblingiaceae and Pentadiplandraceae has been interpreted as a reversal from tetramery (Soltis *et al.*, 2005). A less parsimonious option is the repeated derivation of tetramery from pentamery in Setchellanthaceae, Bataceae-clade (*Koeberlinia* and *Salvadora* are occasionally pentamerous) and core Brassicales. This makes more sense on morphological grounds. *Pentadiplandra* functions as a morphological prototype for the deriviation of disymmetric Brassicaceae (see p. 242; Ronse De Craene, 2002). Brassicaceae *sensu lato* was recognized as a single family by APG II, although there is good morphological evidence to support a distinction between Capparaceae, Brassicaceae and Cleomaceae (Hall, Sytsma and Iltis, 2002; Ronse De Craene and Haston, 2006).

When unisexual, flowers of Brassicales often show a high dimorphism (e.g. Caricaceae, Bataceae, Gyrostemonaceae).

'Tropaeolaceae-clade'

Tropaeolaceae

Fig. 10.24. *Tropaeolum majus* L.

↓* K5 C3:2 A5+3 G̲(3)
*✓ in preanthetic flowers

Flowers of *Tropaeolum* arise singly in the axil of a bract. There is a conspicuous hypanthium lifting calyx and corolla. A large spur develops below the adaxial sepal within the hypanthial tissue, changing the symmetry of the flower from oblique to median (Ronse De Craene and Smets, 2001b). Sepals have a quincuncial arrangement or become valvate at maturity. The genus shows much variation in the development of the petals, which are strongly clawed; the two upper (adaxial) petals are often strongly divergent from the three lower, which are smaller (e.g. *T. peregrinum*) or occasionally absent (e.g. *T. pentaphyllum*: Eichler, 1878). In *T. umbellatum*, the posterior petals are reduced in size. The androecium consists of eight stamens with an unusual spiral initiation sequence. Ronse De Craene and Smets (2001b) discussed different interpretations for the origin of the octomerous androecium but concluded that two stamens opposite petals were lost, linked with spatial constraints imposed by the gynoecium (asterisks in Fig. 10.24). There is a striking analogy with the floral structure of *Koelreuteria* (Ronse De Craene, Smets and Clinckemaillie, 2000). The ovary is always trimerous, with a single large ovule per carpel. The spur can be variously developed, and is occasionally the most conspicuous part of the flower.

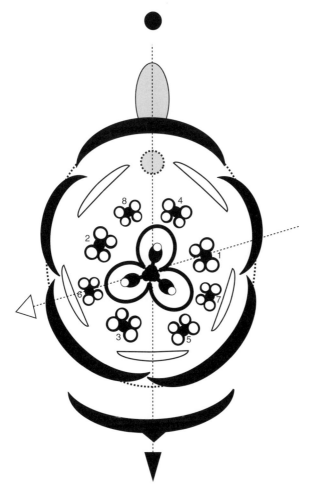

Fig. 10.24. *Tropaeolum majus* (Tropaeolaceae). Numbers give order of stamen initation. White arrow, symmetry in young bud; black arrow, symmetry at maturity; asterisks, lost stamens.

'Moringaceae-clade'

Caricaceae

Fig. 10.25A,B. *Carica papaya* L., based on Ronse De Craene and Smets (1999a)

Staminate: ✶ K5 [C(5) A5+5] G0
Pistillate: ✶K5 C(5) A0 G̲(5)

   Staminate and pistillate flowers are highly dimorphic and occur on the same or on different branches. Some cultivars are occasionally hermaphroditic. Staminate flowers are grouped in multiflowered dichasia with a strongly

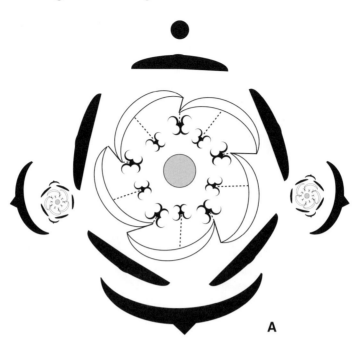

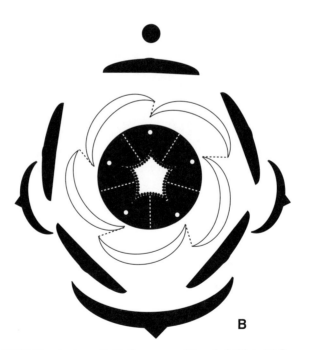

Fig. 10.25. Staminate partial inflorescence (A) and pistillate (B) flower of *Carica papaya* (Caricaceae). Note fusion of petals and stamen-petal tube.

contorted corolla fused with the two stamen whorls by the development of a stamen-petal tube. The gynoecium is sterile and is transformed in a nectariferous protuberance. Pistillate flowers are grouped in few-flowered dichasia. The flower development is similar to the staminate flowers up to the growth of the petals (Ronse De Craene and Smets, 1999a). No stamens are formed but the contorted petals are only loosely connected at the base. The massive gynoecium produces a high number of small ovules on five large parietal placentae. A short style supports a highly branched stigma. No nectar is produced and it is assumed that pistillate flowers trick insects by mimicry of the staminate flowers through their stigmas.

Moringaceae

Fig. 10.26. *Moringa oleifera* L., based on Ronse De Craene, De Laet and Smets (1998)

Fig. 10.26. *Moringa oleifolia* (Moringaceae): partial inflorescence. Note the orientation of half-anthers.

↘K5 C5 A5°+5$^{1/2}$ G̲(3)

Inflorescences are basically monotelic with a bract and two bracteoles (Olson, 2003). The monotypic genus is highly variable, with either strongly monosymmetric flowers with well-developed hypanthia to polysymmetric flowers with weak hypanthia (Olson, 2003). The androecium is diplostemonous with antesepalous staminodes resembling filaments; the antepetalous stamens bear only one theca and their orientation is specific (Fig. 10.26). In monosymmetric flowers, petals regulate the oblique monosymmetry by differences in size and curvature to one side (the petal between sepals three and five is larger) and anthers do this by a progressive maturation. This is occasionally enhanced by the suppression of the staminode opposite sepal four in some species (*M. concanensis*). Petals appear contorted in bud, running along the orientation of the anthers, although aestivation is quincuncial (Ronse De Craene, De Laet and Smets, 1998). While the largest petal occupies a posterior position, other petals are arranged pairwise as mirror images. As a result the flower resembles a pea flower (Ronse De Craene, De Laet and Smets, 1998; Olson, 2003). In polysymmetric species, the aestivation of petals and orientation of anthers is far less consistent. Illustrations of floral diagrams of different species were given by Olson (2003: 55). The lower slope of the hypanthium is covered with an inconspicuous nectary surrounded by filaments bearing trichomes. The ovary is superior, borne on a gynophore at the bottom of the hypanthium. Placentation is parietal with ovules in two rows, but appears to be opposite the carpel because of a groove running between the ovules. Olson (2003) demonstrated that monosymmetry is an early developmental process that is either accentuated later in development, or repressed.

'Core Brassicales'

Capparaceae

Fig. 10.27A. *Capparis cynophallophora* L.

✳ K4 C4 A8+4+4+8+4+4 G̲(2)

Fig. 10.27B. *Euadenia eminens* Hook. f.

↓ K4 C4 A5:6° G̲(2)
General formula: ↓/↔ K(2)4(6) C0/2(4)6 A(2)4-∞ G2(-12)

Capparaceae with 16 genera have a highly diverse floral morphology (see Endress, 1992; Ronse De Craene and Smets, 1997a). Cleomoideae were placed in Capparaceae, but they represent a morphological intermediate between Capparaceae and Brassicaceae and should be recognized as a separate family (Hall, Sytsma and Iltis, 2002; p. 61). Flowers are mostly tetramerous, (weakly) disymmetric to polysymmetric (*Capparis*), to strongly monosymmetric (e.g. *Euadenia*: Fig. 10.27B; Karrer, 1991). The calyx is well developed and often bears extrastaminal nectary scales at the base (developing on a weak hypanthium). Petals are generally clawed and caducous with

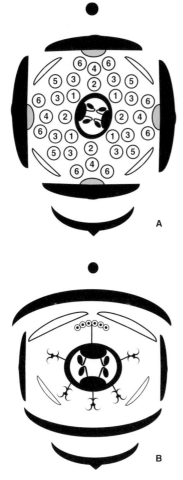

Fig. 10.27. Capparaceae: A. *Capparis cynophallophora*; B. *Euadenia eminens*. Numbers refer to order of stamen initiation.

the stamens. Petals are occasionally absent (e.g. *Boscia*, *Ritchiea*). The androecium is variable, although the number of stamens tends to be constant within a genus, such as *Capparis* (Ronse De Craene and Smets, 1997a). A multistaminate androecium arises centrifugally on a ring primordium (a torus) with a clearly whorled sequence, in cases where numbers are moderately high (Fig. 10.27A; Karrer, 1991; Ronse De Craene and Smets, 1997a). The gynoecium is mostly bicarpellate with parietal placentation and develops on a long gynophore. A style is absent or weakly developed. In some *Capparis* the number of carpels tend to be increased (e.g. *C. spinosa*: Ronse De Craene and Smets, 1997b).

Brassicaceae

Fig. 10.28. *Pachyphragma macrophyllum* Busch

Fig. 10.28. *Pachyphragma macrophyllum* (Brassicaceae). Note the absence of subtending bract.

↔ K4 C4 A2+4 G̲(2)

General formula: ↔ (↓) K4 C(0)4 A2(0)+(2)4(-24) G2

Inflorescences are generally racemose contrary to Capparaceae, which are more diverse. Bracts are generally missing. Brassicaceae (or Cruciferae) have one of the most readily recognizable floral diagrams shared by all members of the family (with very few exceptions enumerated by Endress, 1992). Flowers are tetramerous but appear to be dimerous and disymmetric. Because of a high similarity with the floral diagram of Papaveraceae, both families used to be associated into an order Rhoeadales until fairly recently, although the difference in chemical composition was a main reason to separate them.

Monosymmetry is rare and is caused either by petals (e.g. *Iberis*) or stamens. In *Iberis amara* the abaxial petals are much longer than the adaxial petals and this is regulated by late expression of *CYC* homologs (Busch and Zachgo, 2007). Monosymmetry corresponds with the arrangement of flowers in flattened racemes resembling umbels. Four sepals are arranged in a median and transversal position and are free (exceptionally fused). Four petals are clawed and arranged in diagonal position. They are rarely lacking (e.g. *Lepidium*). The androecium consists of six stamens, as in Cleomaceae, with two outer smaller stamens opposite the lateral sepals and four inner stamens not exactly opposite the petals, but shifted towards the median line. The genus *Lepidium* has the most variable stamen numbers: inner stamens are occasionally fused, or are missing, as are the outer stamens (Bowman and Smyth, 1998). As loss and fusion of stamens happen independently, the number of stamens fluctuates between two and six. Stamens are increased up to 24 in *Megacarpaea* but their position is currently unknown. The gynoecium is always bicarpellate with parietal placentation. A false septum often divides the ovary in two parts and contributes with the persistent placental ridge (replum) to fruit formation and seed dispersal. Nectaries develop at the base of the stamens, often in variable position, wherever there is sufficient space. Bernardello (2007) distinguished four types, which can be grouped into an annular nectary or a segmented nectary.

Flowers of Cleomaceae are either polysymmetric or show strong monosymmetry with unidirectional flower initiation (Karrer, 1991; Erbar and Leins, 1997). They share the same floral diagram as Brassicaceae, although stamen numbers can be laterally increased (e.g. *Polanisia*) or reduced (e.g. *Cleome*, *Dactylaena*). The adaxial side of the androecium is occasionally sterile, or stamens are missing. In *Dactylaena* only one abaxial stamen develops (Karrer, 1991). A nectary is usually developed on the adaxial side, often as a crest or scale on the androgynophore (e.g. *Cleome isomeris*: pers. obs.). Cleomaceae share the presence of a replum with Brassicaceae, but lack a false septum in fruit (Hall, Sytsma and Iltis, 2002).

Endress (1992) showed that the Brassicaceae are highly uniform, compared with the highly diverse Capparaceae. The floral Bauplan of Brassicaceae is considered to be ancestral for Brassicaceae and Capparaceae alike. At least some members of Cleomaceae, Capparaceae (e.g. *Steriphoma*) and Brassicaceae share the tetradidynamous androecium of two and four stamens and I believe this to be the plesiomorphic configuration. Hall, Sytsma and Iltis (2002) concluded on character reconstructions that an androecium of 7–15 stamens is plesiomorphic, but this is not helpful in understanding floral evolution, as they gave no indication of stamen position. There are two main interpretations for the origin of the peculiar flowers of Brassicaceae *sensu lato* (reviewed by Endress, 1992; Fig. 10.29). The first interprets the flower as basically dimerous with six whorls (K2+2 C2^2 A2+2^2 G2: Fig. 10.29A). This is based on the closer proximity of petals in a median plane and the occasional replacement of four inner stamens by two stamens (e.g. *Lepidium*, some *Cleome*). The double stamen and petal positions are interpreted as a result of dédoublement. The second hypothesis interprets the flower as basically tetramerous and five-whorled (K2+2 C4 A4+4 G2: Fig. 10.29B). The first hypothesis is mainly discredited by the absence of clear dédoublement and the phylogenetic position of core Brassicales (embedded in a

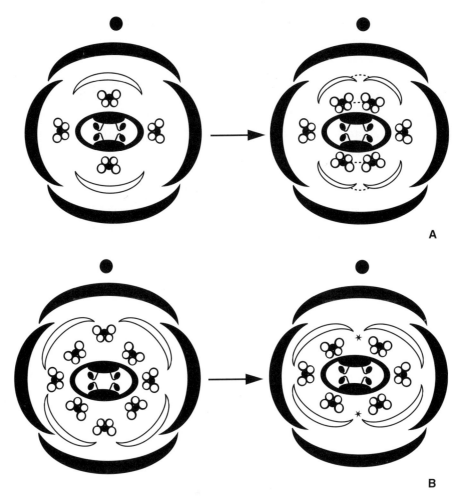

Fig. 10.29. Diagrammatic representation of the derivation of the Brassicales floral Bauplan. A. dimerous prototype; B. tetramerous prototype.

pentamerous, diplostemonous clade). The enigmatic genus *Pentadiplandra* can clarify the unusual floral diagram of Brassicaceae. Flowers of *Pentadiplandra* are pentamerous and diplostemonous. However, they are pressed between bracts and axis and appear disymmetric, with the initiation of lateral sepals preceding the median sepals. Brassicaceae may have arisen through a similar process leading to the loss of two median antesepalous stamens (Fig. 10.29B; Ronse De Craene, 2002). Androecial evolution of Capparaceae and Cleomaceae is more diverse, with a reduction of stamens up to one (e.g. *Dactylaena*) partial loss of stamens through monosymmetry (e.g. *Euadenia*, *Cleome*) and secondary increases of stamens (e.g. *Capparis*, *Polanisia*).

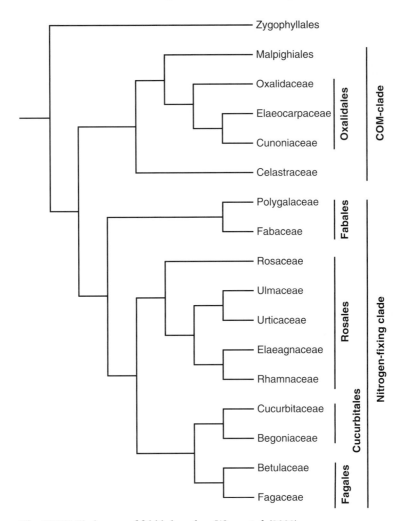

Fig. 10.30. Phylogeny of fabids based on Wang *et al.* (2009).

10.3   Fabids

Fabids consist of three subclades: a nitrogen-fixing clade of Rosales, Cucurbitales, Fagales and Fabales, a 'COM-clade' of Celastrales, Oxalidales and Malpighiales and a small clade of Zygophyllaceae and Krameriaceae not considered in this book (Fig. 10.30; Soltis *et al.*, 2005; Schönenberger and von Balthazar, 2006; Wang *et al.*, 2009). The common ancestor of the fabids had the ability to fix nitrogen. This was retained in at least ten families (four in Rosales) (Soltis *et al.*, 2005).

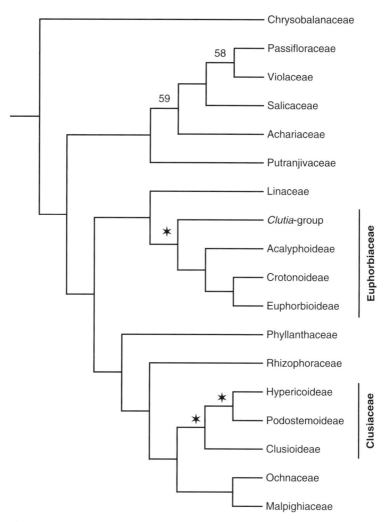

Fig. 10.31. Phylogeny of Malpighiales, based on Tokuoka and Tobe (2006). Asterisks denote branches with more than 90% support.

COM-clade: Celastrales–Oxalidales–Malpighiales

Malpighiales

The order is a huge, highly heterogeneous and mainly tropical assembly of 28 to ca. 37 families. Figure 10.31 is mainly based on Tokuoka and Tobe (2006). Affinities of several families are partly unresolved as support for branches is weak (e.g. Tokuoka and Tobe, 2006). Finding common morphological characters is challenging (Endress and Matthews, 2006a; Schönenberger and von Balthazar, 2006). Malpighiales are typical rosids with pentamerous flowers, free petals, diplostemony and a mostly superior three- to five-carpellate ovary with free styles. When present, petals are contorted or imbricate. A hypanthium is uncommon and nectaries are often distinct, rarely arranged as a disc. The ovary is either parietal or axile. Several Malpighiales are characterized by the presence of a corona, or various appendages on the petals. Oblique monosymmetry is found in Malpighiaceae and Chrysobalanaceae *sensu lato*.

There are some strongly supported associations: (1) Clusiaceae–Podostemonaceae–Hypericaceae–Bonnetiaceae, (2) Ochnaceae–Medusagynaceae–Quiinaceae and (3) Malpighiaceae–Elatinaceae. Other large families have been split up, such as Euphorbiaceae in five entities, and Flacourtiaceae in Salicaceae and Achariaceae. It will be important to better circumscribe subordinal groups in the future and attempts are currently being made to improve our knowledge of floral morphology in the order (e.g. Merino Sutter, Foster and Endress, 2006; Matthews and Endress, 2008).

Chrysobalanaceae

Fig. 10.32A,B. *Chrysobalanus icaco* L.

↘ [K5 C5 A5$^\infty$] G(1:1°)
General formula: ↘ K5 C(0)5 A3-∞ G1–3

Chrysobalanaceae *sensu lato* represents a grouping of four morphologically diverse families (Chrysobalanaceae, Euphroniaceae, Dichapetalaceae, Trigoniaceae). Most taxa have flowers with oblique monosymmetry and well-developed hypanthium. Monosymmetry is associated with the retardation/sterilization of the posterior part of the androecium and gynoecium (Fig. 10.32). In *Chrysobalanus* the gynoecium is pseudomonomerous with almost no trace of the second carpel at anthesis, although it is well visible in early stages, and the style is gynobasic through the extreme dorsal bulging of the fertile carpel. In other genera the gynoecium is tricarpellate and all carpels are fertile. Stigmatic lobes are carinal and two ovules are present in all taxa. A false septum develops occasionally between the two ovules (e.g. *Maranthes*).

Early stages of development of *Chrysobalanus* are regular (Matthews and Endress, 2008; pers. obs). A deep hypanthium places sepal lobes, petals and androecium as a rim around a cup-like depression. Sepals and petals have an imbricate arrangement and development. The numerous stamens in *Chrysobalanus* are arranged in a single ring, although they arise as five antesepalous groups (pers. obs.). Matthews and Endress (2008) interpreted the androecium as made up of two initial whorls (diplostemony with paired stamens arising opposite the petals after initiation of the antesepalous groups). I could see little indication for this as the primary primordia extend laterally in the petal zones. The stamens lying opposite the basal style are retarded and develop

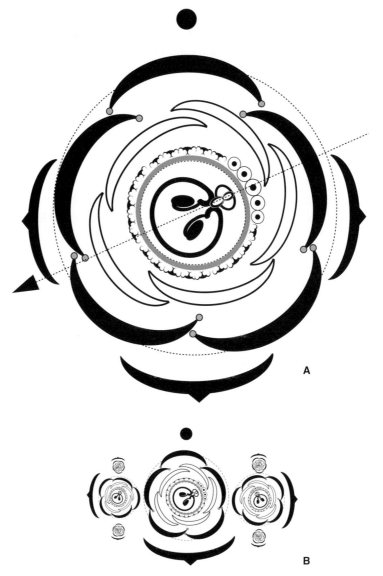

Fig. 10.32. *Chrysobalanus icaco* (Chrysobalanaceae): A. flower; B. partial inflorescence.

as staminodes. A similar development was described by Matthews and Endress (2008) for *Dactyladenia*. In *Hirtella* the number of stamens is reduced to three to seven, with a corresponding number of staminodes. A nectary develops on the inner slope of the hypanthium.

The four families share several morphological characters listed by Matthews and Endress (2008). The reader is referred to that paper for more information on the other families. The visibly most important synapomorphies are oblique monosymmetry (relatively rare in rosids) with a mixture of staminodes and stamens (fertile stamens being concentrated in the anterior part of the flower), strongly introrse anthers, a hypanthium, carpels well demarcated by a deep furrow and two parallel ovules on axile placentation.

Malpighiaceae

Fig. 10.33. *Galphimia glauca* Cav.

Fig. 10.33. *Galphimia glauca* (Malpighiaceae).

↘ (weakly monosymmetric) K5 C5 A(5+5) G̲(3)
General formula: ✱/↘ K5 C5 A(5)10(15) G2–3(4)

Inflorescences are indeterminate racemes with flowers subtended by a bract and bracteoles bearing two lateral glands (at least in *Galphimia*). Sepals often have large paired glands on their abaxial side (but not in *Galphimia*). Glands may occur on all sepals, or they may be missing on sepal three (*Stigmaphyllon*), or on all sepals except one (*Hiptage*: Eichler, 1878). The glands secrete oil or nectar. Petals are often fringed and strongly clawed, viz. paddle–shaped with a long narrow base. One of the latero-adaxial petals is more developed and becomes displaced in median position by a slight torsion of the pedicel. As a result the flower appears obliquely monosymmetric, with the symmetry line going through sepal three (Fig. 10.33; cf. Eichler, 1878). The androecium consists of two whorls of (basally fused) stamens with an obdiplostemonous arrangement and inserted on a torus around the tricarpellate ovary. Antepetalous stamens are slightly shorter than antesepalous stamens. Five antesepalous stamens are found in *Gaudichaudia* and *Aspicarpa*, two of which are fertile. Stamens may also be of unequal length (e.g. *Hiptage*, *Malpighia*) or the androecium may be partially sterile (*Stigmaphyllon*: antesepalous stamens, diagram in Niedenzu, 1897). Two carpels are fertile in *Acridocarpus* (Stevens, 2001 onwards). Each carpel with individual style contains one pendent ovule. There is no nectary.

Clusiaceae (incl. Hypericaceae, Podostemaceae)

Fig. 10.34. *Symphonia globulifera* L.f.

$\ast$ K5 C5 A(5^3) $\underline{G}$(5)

Fig. 10.34B–D. *Clusia sp.*

Staminate: $\ast$ K2+2 C2+2 A∞* G0
*arising as four weakly defined groups opposite the petals
Pistillate: $\ast$ K2+2 C2+2 A°8 $\underline{G}$(6)

Fig. 10.35A. *Garcinia spicata* (Wight and Arn.) Hook.

Staminate flower*: $\ast$K2+2 C4 A4^3 G0
*No material of pistillate flowers was available.

Fig. 10.35B,C. *Hypericum perforatum* L.

$\ast$ K5 C5 A3$^\infty$ $\underline{G}$(3)
General formula: $\ast$ K2(3)4–5(-20) C(0,3)4–5(-8) A4-∞ G1–5(-20)

The circumscription of the large and heterogenous family Clusiaceae *sensu lato* appears difficult on a morphological basis (Gustafsson, Bittrich and Stevens, 2002), as it includes Podostemaceae, a highly derived family of plants adapted to aquatic habitats. Four subfamilies (Hypericoideae, Podostemoideae, Kielmeyeroideae and Clusioideae) could be proposed.

Flowers are extremely variable, especially in Clusioideae, represented mainly by the heterogeneous *Clusia* and *Garcinia* (Figs. 10.34A–C, 10.35A; Gustafsson and Bittrich, 2002; Gustafsson, Bittrich and Stevens, 2002; Sweeney, 2008). A feature common to all Clusiaceae are the contorted or imbricate petals associated with antepetalous stamen fascicles (phalanges). When arising on a ring primordium, four groups can still be recognized in early stages (Fig. 10.34B, pers. obs.). Androecia are rarely obhaplostemonous (e.g. *Havetiopsis*, some *Clusia*: Gustafsson and Bittrich, 2002).

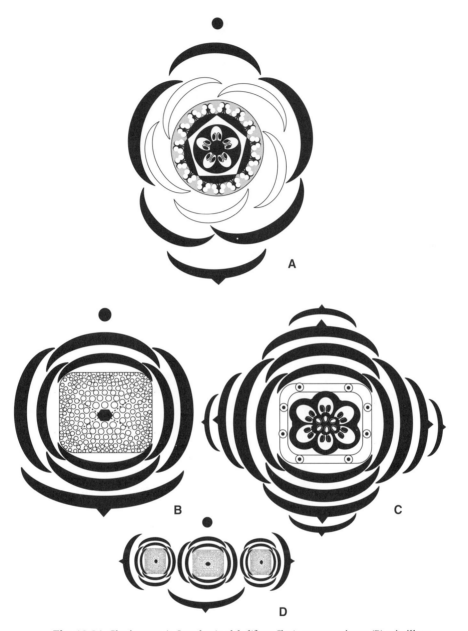

Fig. 10.34. Clusiaceae: A. *Symphonia globulifera; Clusia* sp.: staminate (B), pistillate (C) flowers and staminate inflorescence (D). Broken line in B shows delimitation of four stamen groups.

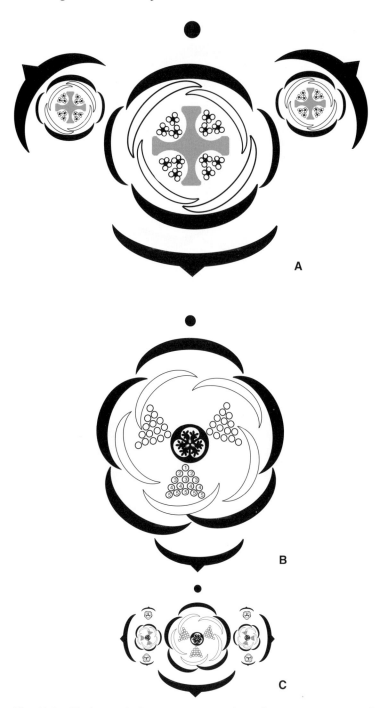

Fig. 10.35. Clusiaceae: A. *Garcinia spicata*, staminate flower; B. *Hypericum perforatum*. Numbers show initiation of stamens on common primordium; C. inflorescence.

In Clusioideae flowers are often unisexual (not in *Symphonia*), with a high divergence between staminate and pistillate morphs. While staminodes are usually present in pistillate flowers, pistillodes are either present or missing in staminate flowers (e.g. *Clusia*, *Garcinia*). The perianth and androecium are spiral throughout, often with decussate phyllotaxis and without clear boundaries between bracts, sepals and petals, an unusual condition in core eudicots (e.g. Gustafsson, 2000; *Clusia*, *Garcinia*). Bracteoles are usually present as a transversal pair not differing from the sepals. Petals are thick and imbricate (contorted in *Symphonia*), often four in number (or three to ten) and stamens arise centrifugally in antepetalous fascicles or on a ring primordium. Anthers may become highly modified in producing resin to attract small bees (Bittrich and Amaral, 1996). The petals may become fused into a tubular corona in *C. gundlachii* (Gustafsson, 2000). In *Garcinia* the androecium is extremely variable but develops mostly from massive fascicles (e.g. Baillon, 1871a; Leins and Erbar, 1991; Sweeney, 2008). The number of carpels is variable in *Clusia*, ranging from five to eight. The superior gynoecium is more stable than the androecium, with axile placentation and paired ovules. In staminate flowers of *Garcinia*, stamen fascicles alternate with broad disc-like nectaries (Fig. 10.35A), which correspond with antesepalous flap-like appendages alternating with staminodial phalanges in pistillate flowers (Sweeney, 2008). The nature of nectaries is unclear, as some authors considered them as staminodial, while Sweeney (2008) suggested that they are receptacular. In Hypericoideae stamen fascicles often alternate with nectaries that were interpreted as staminodial by Ronse De Craene and Smets (1991d: *Harungana*, *Vismia*) and as evidence for a diplostemonous ancestry. This may indicate that nectaries have evolved independently among different genera of Clusiaceae.

In *Hypericum* antepetalous stamen fascicles are distinct but they may coalesce into a ring (e.g. *H. androsaemum*). In some species the number of fascicles is reduced to three, alternating with three carpels, as the result of a lateral fusion of two fascicles in pairs (e.g. Fig. 10.35B; *H. olympicum*: Ronse De Craene, unpubl. data; *H. aegypticum*: Leins, 1964b; Prenner and Rudall, 2008). This development is similar to Cucurbitaceae.

Linaceae

Fig. 10.36. *Linum monogynum* G. Forst.

✳ K5 [C5 A(5+5°)] G̲(5)
General formula: ✳ K5 C5 A5 or 5+5 G2–5

In premolecular classifications (e.g. Cronquist, 1981) the family used to be broadly defined (including Humiriaceae, Ctenolophonaceae and Ixonanthaceae)

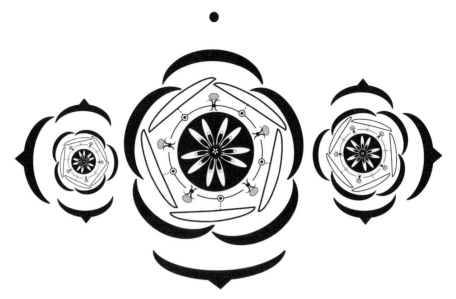

Fig. 10.36. *Linum monogynum* (Linaceae): partial inflorescence. Petal bases are adherent to the stamen tube.

and was related to Geraniaceae on the basis of superficial similarities. Indeed, the flowers show many similarities to the above-mentioned families. The current circumscription comprises Linoideae and Hugonioideae (Stevens, 2001 onwards), although affinities with other small families are unclear. The ovary fluctuates between three and five with six to ten pendent anatropous ovules, seemingly six- or ten-carpellate because locules are divided by false septa that do not reach the central axis in the upper part of the ovary (Fig. 10.36; Narayana and Rao, 1969. The five styles are twisted in *Linum monogynum*.

Flowers are diplostemonous, or haplostemonous with or without staminodes (e.g. *Linum lewisii*: Narayana and Rao, 1976). The report of obhaplostemony in *Anisadenia* (e.g. Cronquist, 1981) is erroneous. Stamens and staminodes are basally fused; staminodes are small and teeth-like in the species observed, but can be more developed. Staminodes are non-vascularized in *Anisadenia* and other Linaceae (Narayana and Rao, 1969, 1976). Stamens are of two lengths, the antepetalous being shorter (e.g. *Indorouchera*, *Rouchera*: Narayana and Rao, 1966, 1973, 1978).

In *Linum*, five inconspicuous nectaries are situated at the base of the staminal tube, abaxially of the fertile stamens. In *Anisadenia* only one gland is developed below the adaxial antesepalous stamen (Narayana and Rao, 1969), while only two glands are present in *Linum perenne* (Narayana and Rao, 1976). Other

Linaceae can have a prominent extrastaminal disc (e.g. *Philbornea*: Narayana and Rao, 1973). Petals are clawed and appear connected with the stamen tube by an erect basal portion situated against the tube. Sepals are leaf-like and equal in size.

The related Ixonanthaceae have highly similar ovaries with paired ovules but nectaries are intrastaminal (Narayana and Rao, 1966). Ctenolophonaceae share the extrastaminal disc with Linaceae.

Passifloraceae

Fig. 10.37A,B. *Passiflora vitifolia* Kunth

✳ K5 C5 [A5 G̲(3)]

General formula: ✳ K(3-)5(-8) C(3-)5(-8) or 0 A(4)5–8(-∞) G(2)3(-5)

The inflorescence of *Passiflora* produces a single flower at each node together with a leaf and a tendril. Each single flower of *Passiflora* is derived from a reduced cyme; the central axis develops into a tendril with a single lateral bracteole. The lateral bracteole encloses a flower together with two second-order bracteoles of the same size (cf. Eichler, 1878). *Passiflora* has a basically simple floral diagram. However, the flower is highly elaborate by the combined development of a hypanthium and a corona made up of several whorls of thread-like appendages. The corona is highly variable in development and shape. In *P. vitifolia* there are five series: three outer series of erect threads (radi and pali), a horizontally inserted rim (the operculum) consisting of fused threads and covering an inner chamber, and an inner rim with tiny hairs (limen) surrounding the base of an androgynophore. Nectar is secreted within the chamber. The outer series clearly mimic stamens in some species (e.g. *P. coriacea*). The five antesepalous stamens are initially introrse, but become versatile and pendent. The superior ovary has three massive styles alternating with parietal placentae. Sepals and petals are both petaloid; the sepals are recognizable by the development of a dorsal awn or crest. Petals are occasionally absent (e.g. *P. coriacea*: pers. obs.). Floral developmental evidence (Bernhard, 1999) has shown that the corona arises in later stages of development and has a receptacular origin, not staminodial as suggested by some authors. However, the limen is often regarded as being staminodial (e.g. De Wilde, 1974), although there is little evidence to support this. Other genera, such as *Adenia*, have a far less elaborate corona and flowers are occasionally unisexual. In some genera (*Barteria*, *Smeathmannia*) stamens proliferate laterally within a single whorl, increasing the stamen number to 25 (Bernhard, 1999). Some *Passiflora* have eight stamens through dédoublement of some stamens (Krosnick, Harris and Freudenstein, 2006).

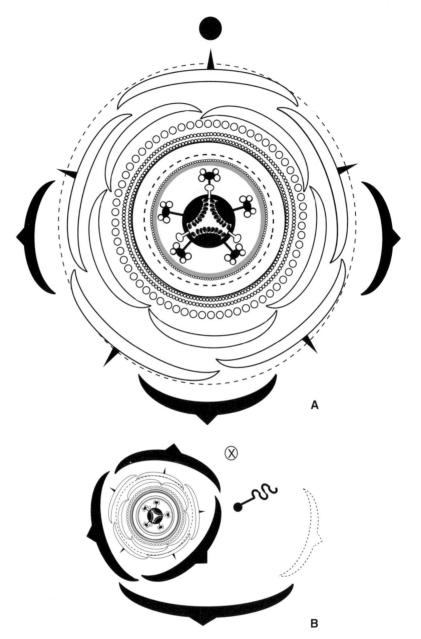

Fig. 10.37. *Passiflora vitifolia* (Passifloraceae): A. flower; B. position of flower on a node with accompanying tendril; bracteole on the right is lost.

De Wilde (1974) illustrated a series of floral diagrams of Passifloraceae. Passifloraceae, Malesherbiaceae and Turneraceae are closely related and optionally form a single family, Passifloraceae (APGII, 2003), sharing a hypanthium with corona, haplostemony and a trimerous superior ovary with parietal placentation.

Ochnaceae (incl. Quiinaceae and Medusagynaceae)

Fig. 10.38. *Ochna multiflora* DC.

Fig. 10.38. *Ochna multiflora* (Ochnaceae): partial inflorescence. Broken line delimits stamen groups.

✳ K5 C5 A5$^\infty$+0 G̲(5–6)
General formula: ✳ (↓) K5 C 5 A1-∞ G2–5(-7)

Flowers of Ochnaceae are variable, but constructed on a pentamerous Bauplan. Flowers are mostly regular and occasionally monosymmetric by the stamen orientation on the adaxial side of the flower. In *Testulea* the androecium is reduced to a single stamen. Patterns affecting the androecium are comparable to those in Dilleniaceae. Sepals and petals are both imbricate, or petals are contorted. The androecium is often polyandrous, more rarely diplostemonous (*Ouratea*) or with antepetalous staminodes (*Sauvagesia*). Staminodes can be conspicuous, sometimes fusing into a tube around stamens and ovary. A ring of outer structures is often described as staminodes in *Sauvagesia*, but is probably a receptacular emergence or corona (Eichler, 1878; Matthews and Endress, 2006b). Stamens have short filaments and poricidal anthers. Polyandry was described as centripetal by Pauzé and Sattler (1978), developing from five common primordia in *Ochna atropurpurea*, although I found the central stamen primordia to be initially longer in *O. multiflora*. Carpels are two to five (or more), with terminal style or seemingly apocarpous with a gynobasic style. Placentation is axile to parietal. Amaral (1991: 107) provided several floral diagrams without reference points.

The family was related with Quiinaceae and Medusagynaceae (e.g. Fay, Swensen and Chase, 1997), with whom they share contorted petals, polyandry, partial fusion of styles, a secondary increase of carpels and an absence of nectaries.

Salicaceae

Fig. 10.39. *Azara serrata* Ruíz & Pav. var. *serrata*

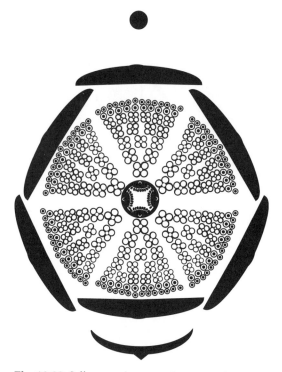

Fig. 10.39. Salicaceae: *Azara serrata* var. *serrata*.

∗ K4–6 C0 A4$^{\infty}$-6$^{\infty}$ G̲(4–6)
General formula: ∗ K0/(3)4–7 (8+) C0/3–8 A(2)4-∞ G2–9

Molecular evidence has demonstrated that the premolecular Flacourtiaceae are not a natural entity, that has to be split up in two clades belonging to two families with a previously much smaller circumscription, viz. Achariacae and Salicaceae (Chase *et al.*, 2002). Traditional Salicaceae (*Salix* and *Populus*) were shown to represent extremes in floral reduction and belong closely to members of former Flacourtiaceae. A morphological distinction between Achariaceae and Salicaceae was shown by Bernhard and Endress (1999), in that the former has centripetal polyandry while in the latter stamens arise centrifugally.

The family is highly variable and difficult to define, because major defining characteristics (superior ovary with parietal placentation) are shared by other families. However, a series of traits is common in the family. Flowers are generally small, often unisexual with variable merism (occasionally six- to eight-merous), with valvate calyx and rarely with corolla. When present, petals are small (e.g. *Homalium)* and they are frequently missing, as the attraction is taken over by stamens (e.g. *Azara*) or petaloid sepals (e.g. *Casearia*). An extreme reduction of the perianth occurs in *Salix* and *Populus*. The androecium is occasionally diplostemonous. When a single whorl, stamens are obhaplostemonous alternating with large antesepalous nectaries (e.g. *Homalium*). When multistaminate, stamens arise centrifugally on a ring primordium or on complex primordia (e.g. *Homalium abdessammadii* with obhaplostemonous triplets: Pauwels, 1993). The genus *Casearia* is highly variable in stamen number, ranging from five to eight to higher numbers in a single whorl. A hypanthium is often developed, placing the gynoecium in a half-inferior position (e.g. *Casearia*). The gynoecium always has parietal placentation. In cases where nectary tissue is present, it is variable in origin and position, ranging from large, receptacular outgrowths alternating with the stamens (e.g. *Casearia bracteifera, Azara microphylla*: pseudostaminodial structures: Ronse De Craene and Smets, 2001a) to an entire disc (e.g. *Dovyalis*: Ronse De Craene, unpubl.data), or glands replacing the perianth in *Salix*. When unisexual, flowers often have divergent morphs with absence of staminodes or carpellodes, as in *Dovyalis caffra* (Ronse De Craene, unpubl.data).

Euphorbiaceae

Fig. 10.40A–C. *Jatropha sp.* (Crotonoideae)

Staminate: *K5 C5 A5+3 G0
Pistillate: *K5 C5 A0 G̲(3)

Fig. 10.40D,E. *Clutia sp.* (Acalyphoideae)

Staminate: *K5 C5 A5 G0
Pistillate: *K5 C5 A0 G̲(3)
General formula: *K0/(3)5–6 C0/5–6 A1–5–15–∞ G(1)3(20)

The massive family Euphorbiaceae as previously circumscribed was divided in at least five families on the basis of molecular evidence (Tokuoka and Tobe, 2006: Pandanaceae, Euphorbiaceae *sensu stricto*, Phyllanthaceae, Picodendraceae and Putranjivaceae). Phyllanthaceae and Picodendraceae are sister families sharing biovulate carpels, while other families have a single ovule per carpel. However, all euphorbioid flowers share a combination of three characters not found elsewhere, which are epitropous, hanging ovules, a nucellar beak and an obturator (Sutter and Endress, 1995). Morphologically euphorbioid flowers

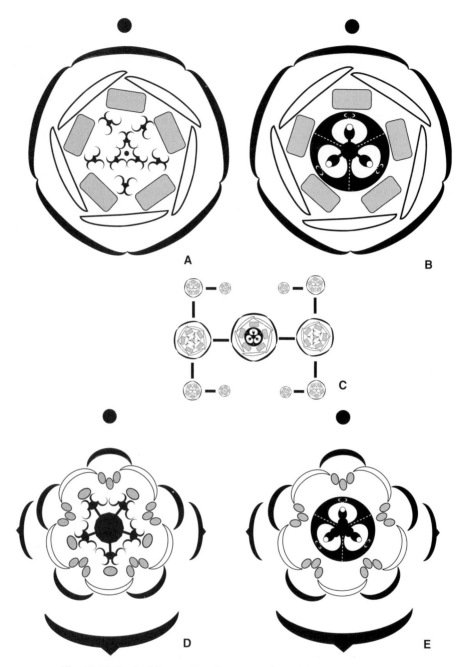

Fig. 10.40. Euphorbiaceae: *Jatropha* sp.: staminate (A) and pistillate (B) flower; C. partial inflorescence. *Clutia* sp.: staminate (D) and pistillate flower (E).

share many characters, including unisexual flowers with large separate nectar glands and a superior capsular ovary with bifid styles. However, this is not clearly reflected in the molecular phylogeny. Euphorbiaceae *sensu stricto* contain four clades that are not strongly supported.

Inflorescences are generally cymose or cymose-derived. However, floral diversity in the family is high, linked with unisexual flowers and a strong dimorphism between staminate and pistillate flowers. Merism can be highly variable, mostly tetra- or pentamerous or more variable (tri- to octomerous). When trimerous, flowers are simple and apetalous (e.g. *Mercurialis*). Floral nectaries are present, or in cases where flowers are reduced they are found on the inflorescence. Flowers generally have sepals with imbricate or valvate aestivation. Petals are present but rarely showy as in *Jatropha integerrima*, mostly equal to smaller than sepals, or frequently absent (e.g. *Manihot esculenta*). Aestivation of petals is contorted to valvate. The androecium of staminate flowers is highly variable, either (ob)haplostemonous, (ob)diplostemonous (e.g. *Jatropha aconitifolia*, *Manihot*) or in multiple whorls (e.g. *Croton*) (Baillon, 1874; Michaelis, 1924). When two-whorled, the antepetalous whorl is generally the outer and is shorter than the antesepalous whorl (e.g. *Jatropha*, *Cnidoscolus*). In some species the inner whorl consists of three stamens (e.g. Fig. 10.40A; Nair and Abraham, 1962) or a single stamen (e.g. *Croton sp.*: Gandhi and Dale Thomas, 1983). Stamens are generally monadelphous and fused into a column, and are often lifted on an androphore with clearly separate stamen whorls (e.g. *Caperonia*, *Ditaxis*: Michaelis, 1924). The gynoecium is generally strongly reduced, with small carpellodes (e.g. *Cnidoscolus*) or without evidence or carpels (e.g. *Clutia*). I suspect that stamens have replaced carpels in *Jatropha* or *Croton* through a process of homeosis, as one to three stamens occupy the expected position of the carpels (see also diagrams in Michaelis, 1924). In *Clutia* the five antepetalous stamens are fused into a column topped by a sterile hairy apex (Fig. 10.40D). When numerous, stamens develop in alternating whorls of five (e.g. *Codiaeum*, *Croton*: Baillon, 1874; Nair and Abraham, 1962; Gandhi and Dale Thomas, 1983). In some staminate flowers the stamen number can be very high and is linked with an increase in petal number (e.g. *Garcia*). *Ricinus* is exceptional by the development of highly branching fascicles (Prenner and Rudall, 2008). Nectar appendages are large and are situated in antesepalous or antepetalous position, occasionally in both (Michaelis, 1924). They are rounded or rectangular in shape, or resemble staminodial structures and have been interpreted as such (as in *Clutia*: Endress and Matthews, 2006b). Pistillate flowers usually resemble the staminate flowers up to the development of the perianth. In some taxa petals are present in staminate flowers but absent in pistillate flowers (e.g. *Croton sarcopetalus*: Freitas *et al.*, 2001); pistillate flowers have an extra whorl of nectaries in alternisepalous position with a vascular connection and may indeed be transformed petals. Staminodes may be variously present or are more generally absent. The gynoecium is generally trimerous, rarely dimerous and has a generalized morphology in the Euphorbiaceae. A single pendent, epitropous ovule is located on axile placentation in each locule. The globular ovary is topped by a short style with bifid stigmatic arms opposite the locules.

Euphorbioideae are characterized by pseudanthia with highly reduced staminate and pistillate flowers. Prenner and Rudall (2007) demonstrated the progressive reduction of the perianth in staminate and pistillate flowers, culminating in *Euphorbia*. Staminate flowers are reduced to a single stamen (often with a constriction on the pedicel), while pistillate flowers are reduced to three carpels. The inflorescence of *Euphorbia* is called a cyathium and is derived from a cymose inflorescence with terminal pistillate flower (cf. Fig. 10.40C). The cup-like inflorescence is formed by fusion of bracts and topped by large glands. Diagrams of inflorescences were shown in Stützel (2006). Similarly in *Dalechampsia* a cymose inflorescence gives rise to a complex pseudanthium enclosed by two petaloid bracts (e.g. Froebe and Magin, 1993).

Phyllanthaceae

Fig. 10.41A,B. *Leptopus chinensis* (Bunge) Pojark.

Staminate: $*$K5 C5 A5 $\underline{G}°(3)$
Pistillate: $*$ K5 C5 A0 $\underline{G}(3)$
General formula: $*$ K2–8 C0/(3)5(-9) A2–35 G(1)2–5(-15)

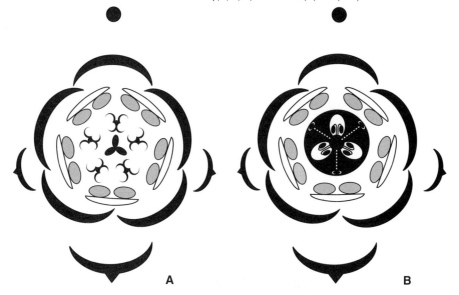

Fig. 10.41. *Leptopus chinensis* (Phyllanthaceae): staminate (A) and pistillate flower (B); note glandular appendages on the petals.

Although molecular evidence places Phyllanthaceae in a clade different from Euphorbiaceae flowers appear to be highly similar, except for ovule numbers, which are consistently two. Phyllanthaceae is sister to Picodendraceae and both families share a unique combination of characters with Euphorbiaceae (Merino Sutter, Foster and Endress, 2006) including unisexual, apetalous flowers (often trimerous with bifid stigmas), large extrastaminal nectary glands, epitropous ovules with large obturator, and explosive capsular fruits.

The perianth is mostly pentamerous in Phyllanthaceae (or three- to nine-merous; trimerous in Picodendraceae) with petals generally smaller than sepals or absent (Michaelis, 1924). The number of carpels is either three or two (always three in pentamerous flowers: Merino Sutter, Foster and Endress, 2006). The stigma is mostly bifid. Glandular tissue may be attached to the petals (Fig. 10.41), or in alternisepalous position. When a single whorl, the androecium is generally haplostemonous (more often obhaplostemonous in Euphorbiaceae: Michaelis, 1924). A pistillode may be variously developed.

Dichapetalaceae and Linaceae share a number of characters of ovules and placentation with Euphorbiaceae *sensu lato*. These include false septa in Linaceae and some species of Picodendraceae and Phyllanthaceae (Sutter and Endress, 1995; Merino Sutter, Foster and Endress, 2006).

Violaceae

Fig. 10.42. *Viola tricolor* L.

Fig. 10.42. *Viola tricolor* (Violaceae): partial inflorescence.

↓ K5 C5 A5 G̲(3)

General formula: ✱/↓ K5 C5 A(3)5 G(2)3(5)

Violaceae represents a heterogenous family with poly- and monosymmetric flowers. Common characters are haplostemony (free or basally fused stamens with broad anthers and protruding connective), and superior gynoecium with parietal placentation. *Leonia triandra* has three stamens (Cronquist, 1981). As demonstrated by Tokuoka (1998) actinomorphy is plesiomorphic in the family and symmetry has changed at least eight times.

Flowers of *Viola* are axillary, apparently single but with a small reduced inflorescence between flower and axis (pers. obs.). Sepals have a long abaxial appendage oriented downwards and resembling a second appressed sepal. The corolla is cochleate-descending and heteromorphic. Petals are clawed or sessile. In *Viola* the lateral and adaxial petals have a tuft of trichomes in the middle of the petal lobe; the base of the adaxial petal is prolonged as a spur in which two filiform nectaries are nested. Nectaries are connected to the two adaxial stamens between filament and anther. Only the tip of the appendage is nectariferous. Other Violaceae have their anthers connivent, often with a prolonged connective. A stamen tube is strongly developed in *Leonia*. Nectar scales develop on the back of all filaments, and are occasionally confluent in a tube (e.g. *Rinorea*, *Hymanthera*: Melchior, 1925).

Carpels are generally three in number with apical style; only *Leonia* has five carpels interpreted as a reversal (Tokuoka, 2008). The orientation of the ovary is inversed (cf. Eichler, 1878), with the odd carpel in an adaxial position.

Celastrales

The order contains the three families Lepidobotryaceae, Parnassiaceae and Celastraceae.

Celastrales share several synapomorphies or 'apomorphic tendencies' (summarized in Matthews and Endress, 2005). These include: a tendency to produce proterandrous or functionally unisexual flowers, large petals in bud with often quincuncial aestivation forming the protective organs, mostly haplostemonous flowers (diplostemonous only in *Lepidobotrys*), or with antepetalous staminodes, tricarpellate gynoecia with commissural stigma and solid styles. Nectaries are often well-developed discs extending around the base of the gynoecium and protruding between the stamens. A short hypanthium is occasionally present and this is linked with the semi-inferior position of the ovary.

Celastraceae (incl. Hippocrateaceae, Brexiaceae, Stackhousiaceae, Plagiopteraceae, Canotiaceae)

Fig. 10.43A. *Brexia madagascariensis* Thou. ex Ker Gawl

✱ K5 C5 A5+5° G̲(5)

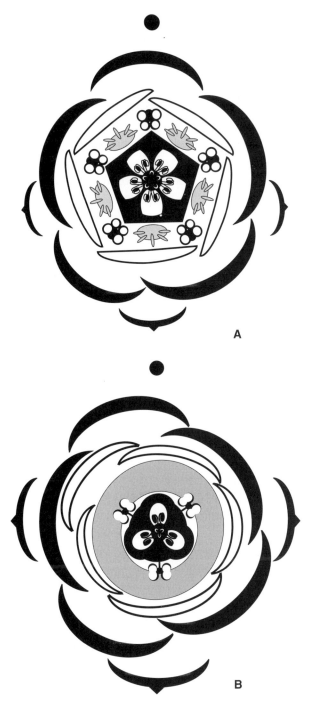

Fig. 10.43. Celastraceae: A. *Brexia madagascariensis*, B. *Tonteleia sp*. Note antepetalous staminodial (?) nectaries in A.

Fig. 10.43B. *Tonteleia sp.*

✳ K5 C5 [A3G(3)]

General formula: ✳ K(3)4–5(6) C(3)4–5(6) A(3)4–5(∞) G2–5(10)

The circumscription of the Celastraceae has been controversial for a long time. The family is currently broadly circumscribed (see e.g. Simmons, 2004) and contains Hippocrateaceae, along with some smaller monotypic families. Matthews and Endress (2005) described floral morphology and anatomy of several representative species.

A large receptacular disc nectary is present in several genera, and is sometimes extrastaminal (e.g. Hippocrateoideae). In Hippocrateoideae, stamen number is reduced to three, in alternation with three carpels (Fig. 10.43B). The number of stamens is strongly linked to the merism of the gynoecium, which is enhanced by the position of the stamens inside the disc nectary (Matthews and Endress, 2005). My observations on the position of the stamens correspond to the description by Payer (1857) for *Salacia viridiflora* but differ from the diagram presented by Eichler (1878: 367). Payer accepts congenital fusion for four of initially five stamens; this can explain the position of two of the stamens opposite petals (Fig. 10.43B). However, *Stackhousia* and *Tripterococcus* have three longer stamens and two shorter (the longer alternate with the three carpels). The condition in *Stackhousia* may be intermediate between the pentamerous and trimerous condition and trimery in the androecium could be the result of loss of stamens and a displacement of the remaining in alternation with the carpels (cf. Eichler, 1878). In other families (e.g. Cucurbitaceae, Clusiaceae) a trimerous androecium is obtained differently by fusion of four stamens or stamen groups in pairs (cf. p. 251, 269). The extrorse stamens of *Tonteleia* are fused with the gynoecium into an androgynophore. *Brexia* and *Siphonodon* are reported to have antepetalous staminodes, although these are mostly interpreted as a receptacular disc nectary in *Brexia* (e.g. Simmons, 2004). The nectaries of *Brexia* (Fig. 10.43A) resemble the staminodes of *Parnassia* and could be homologous with them, although they are initiated very late (Edgell, 2004). Both taxa share pointed extensions (more finger-like in *Parnassia* with glistening tips) and a thick nectariferous base, in *Brexia* extending around the base of the ovary. Likewise, staminodes of *Parnassia* might be receptacular emergences with nectariferous tissue on the ventral side and glistening pseudostaminodes on the appendages (Endress and Matthews, 2006b). However, presence of a vascular bundle at the base of the structures supports a staminodial nature. Bensel and Palser (1975a) interpreted the finger-like extensions of *Parnassia* and *Brexia* as individual staminodes and the whole structure as a fascicle and suggested that the androecium is derived from a fasciculate ancestor. Evidence for this are teratological cases with the replacement of the fingers by simple stamens (Wettstein, 1890 in

Matthews and Endress, 2005). However, this derivation is unlikely, as nearest clades are diplostemonous or haplostemonous. The only polyandrous member of Celastraceae is *Plagiopteron*, but the development of the androecium is unknown.

The ovary is often trimerous (pentamerous in *Brexia*) with axile placentation and two ovules per carpel, reduced to one in *Stackhousia*. The ovary is rarely pseudomonomerous (*Pleurostylia*).

Oxalidales

The Oxalidales consist of seven families. Flowers are generally tetra- or pentamerous in all whorls with the carpels in antepetalous position. Cephalotaceae is hexamerous throughout. Petals are either large in bud with imbricate (cochleate or contorted), rarely induplicate valvate aestivation (Cunoniaceae, Elaeocarpaceae), or are reduced or absent (e.g. Cunoniaceae). Two families, Cephalotaceae and Brunelliaceae, are apetalous. Petals are often three-lobed (Cunoniaceae, Elaeocarpaceae). Stamens are often basally fused, while petals tend to become postgenitally fused at the base (Oxalidaceae and Connaraceae). The androecium is basically obdiplostemonous (present in all families) and occasionally haplostemonous. A secondary stamen increase occurs in some Cunoniaceae and Elaeocarpaceae, with antepetalous stamens in pairs. In Elaeocarpaceae, groups of stamens are wrapped by petals that have three vascular bundles (Matthews and Endress, 2002). Filaments tend to be longer than anthers and erect in bud. Carpels are mostly superior and, when isomerous, are always alternating with the sepals, but they are reduced to two to three in Cunoniaceae and Elaeocarpaceae. Styles are always carinal and each carpel contains either two ovules (rarely a single ovule) or two rows of parallel or superposed ovules on axile placentation. A hypanthium, when present, is weakly developed. Nectaries, when present, develop as separate glands or as an intrastaminal disc.

Cunoniaceae (including Davidsoniaceae, Baueraceae, and Eucryphiaceae)

Fig. 10.44. *Caldcluvia paniculata* D.Don

✶ K4 C4 A4+4 G̲(2)
General formula: ✶ (3)4–5(-10) C0/(3)4–5(-10) A(3)4–5(-10)+ (3)4–5(-10)/∞
G2–3–5(-14)

Floral morphology of the family is diverse, with a mosaic of characters common to Oxalidales. Most Cunoniaceae have tetra- (penta-) merous flowers with a weak hypanthium and a valvate calyx. Petals are absent in more than half of the genera (e.g. *Davidsonia*, *Geissos*), while they are small in the other (Dickison, 1975;

Fig. 10.44. *Caldcluvia paniculata* (Cunoniaceae).

Moody and Hufford, 2000a; Endress and Matthews, 2006b). Most taxa are obdiplostemonous. *Eucryphia* is tetramerous throughout, with a secondary centrifugal stamen increase. Other Cunoniaceae occasionally have a lateral stamen increase from antesepalous primordia, while the antepetalous stamens are undivided (*Geissos*: Matthews and Endress, 2002).

Carpels are generally not isomerous, more frequently two (three) in median position, rarely more numerous in some *Eucryphia* (Dickison, 1978), variably in superior, half-inferior or inferior position. Fossil flowers of Cunoniaceae affinity have isomerous carpels in alternipetalous position (Schönenberger *et al.*, 2001).

Nectaries tend to be separate inter- or intrastaminal structures (Fig. 10.44; Dickison, 1975; Moody and Hufford, 2000a). Filaments are usually longer than petals at anthesis, resembling brush-flowers (e.g. *Cunonia capensis*).

Elaeocarpaceae (incl. Tremandraceae)

Fig. 10.45. *Crinodendron patagua* Molina

✳ K(3:2) C5 A5+5^2 G̲(3)
General formula: ✳ K4–5 C0/4–5 A4–300 G2–8(9)

Fig. 10.45. *Crinodendron patagua* (Elaeocarpaceae). Note extrastaminal nectary and position of anthers relative to petals; broken line refers to attachment of anthers to filament bases (white dots).

The family used to be placed in Malvales because of superficial similarities in flower structure (van Heel, 1966).

Sepals are valvate and basally united, breaking up in two unequal parts in *Crinodendron*. Petals are often folded and three-lobed and they enclose antepetalous stamens in bud in a comparable way to Rhizophoraceae (Fig. 10.45; Endress and Matthews, 2006b). In *Crinodendron*, petals have a ventral ridge at the base forming two tubular access channels to the nectary (Matthews and Endress, 2002). Petals are absent in *Sloanea*. The androecium tends to be obdiplostemonous (e.g. *Platytheca*), with antepetalous stamens in pairs (*Aristotelia*, *Crinodendron*: Fig. 10.45; Ronse De Craene and Smets, 1996a; Matthews and Endress, 2002). In *Tetratheca* the androecium consists of five pairs wrapped by the petals, but it is unclear whether the pairs are antesepalous or antepetalous stamens (Matthews and Endress, 2002). *Aristotelia fruticosa* has tetramerous haplostemonous flowers (van Heel, 1966). A multistaminate androecium with centrifugal development is found in *Sloanea* and *Vallea* arising from antesepalous primary primordia (van Heel, 1966; Matthews and Endress, 2002). A nectary is developed as a broad intrastaminal disc at the base of the ovary or surrounding the stamens, but is absent in *Sloanea* and Tremandraceae with specialized pollen flowers.

Oxalidaceae

Fig. 10.46. *Averrhoa carambola* L.

Fig. 10.46. *Averrhoa carambola* (Oxalidaceae): partial inflorescence. Note the antepetalous nectariferous staminodes.

✳ K5 C5 A(5+5°) G̲(5)

General formula: ✳ K5 C5 A5+5/5° G5

Aestivation of petals is imbricate ascending or descending, or contorted. The androecium is obdiplostemonous and antepetalous stamens are occasionally sterile (e.g. *Averrhoa*). Stamens are basally connate and nectaries occur as extra-staminal appendages of the antepetalous stamens or the staminodes are nectariferous at the base (pers. obs.).

Averrhoa has short- and long-styled flowers (heterostyly). The superior gynoecium has free styles and a fused ovary with two or a single row of ovules on axile placentation. In other genera two rows of ovules are formed, but in *Averrhoa* the constricted locules only allow fewer ovules to develop on either row (Matthews and Endress, 2002).

Floral developmental studies (e.g. Payer, 1857; Moncur, 1988) showed the petals to be retarded in their development, but they are large in bud (Matthews and Endress, 2002).

Oxalidaceae are closely related to Connaraceae, with whom they share several characteristics (see Matthews and Endress, 2002).

Flowers of Oxalidaceae superficially resemble Geraniaceae and Linaceae; they share pentamerous, pentacyclic flowers with isomerous carpels, contorted petals, fused stamens, obdiplostemony with occasional reduction of antepetalous stamens, and nectaries connected to the staminal tube. However, in Oxalidaceae the nectary is in antepetalous position, while in Geraniaceae it is connected with antesepalous stamens (cf. Rama Devi, 1991a), and there is a strict separation between style and ovary in Geraniaceae.

10.3.2   Remaining fabids: Cucurbitales, Fabales, Fagales, Rosales

Cucurbitales

The order contains seven families. Matthews and Endress (2004) provided several floral characters in support for the order, except for Anisophylleaceae, which has more morphological affinities with Cunoniaceae and might be misplaced (see Endress and Matthews, 2006a,b). While Corynocarpaceae and Coriariaceae appear to be more basal with more generalized rosid characters, several characters link Cucurbitaceae, Begoniaceae, Tetramelaceae and Datiscaceae. These include unisexual flowers, often with a strong dimorphism between staminate and pistillate flowers, an inferior ovary with intruding parietal placentation and a high number of ovules (see Judd *et al.*, 2002).

Cucurbitaceae

Fig. 10.47A,B. *Cucurbita palmata* S.Watson

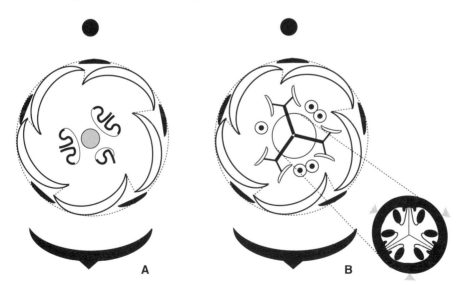

Fig. 10.47. *Cucurbita palmata* (Cucurbitaceae): staminate (A) and (B) pistillate flower.

Staminate: $\ast$ K(5) C(5) A1$^{1/2}$(2$^{1/2}$)(2$^{1/2}$)G0
Pistillate: $\ast$ K(5) C(5) A°5 Ğ(3)
General formula: $\ast$ K(3)5(6) C(3)5(6) A5 G2–3

Flowers are grouped in various inflorescences or are solitary. Flowers are unisexual with variable reductions of the other gender. In *Cucurbita palmata*, flowers are grouped in compound axillary racemes, consisting of a basal branch with a single pistillate flower only, and another branch with several staminate flowers. In *Cucurbita*, small staminodes are present in pistillate flowers, while no trace of the gynoecium is found in staminate flowers (Fig. 10.47A). Petals are free or fused, which is a rare phenomenon in rosids, although this is probably due to hypanthial growth as there is a connection with the sepals (at least in *C. palmata*). The androecium rarely consists of five free stamens opposite the sepals (haplostemony), which represents the basic condition in the family. Some genera have five monothecal stamens. Chakravarty (1958) demonstrated the existence of transitions between dithecal and monothecal stamens. In *Xerosicyos pubescens* four antesepalous stamens have either dithecal or monothecal anthers in variable proportions (Matthews and Endress, 2004). More often there are only three stamens; two are dithecal and inserted between two sepals, and the third is monothecal and opposite a sepal. Floral anatomical and developmental evidence (e.g. Payer, 1857; Chakravarty, 1958; Leins and Galle, 1971) has shown that the peculiar androecium is derived by the pairwise fusion of four monothecal stamens, while a single stamen remains free. Pollen sacs are characteristically twisted and convoluted and anthers appear asymmetrical. Stamens can be either free or variously fused at the base or in a synandrium (not in *C. palmata*). An extreme case is *Cyclanthera*, with the anthers forming a continuous horizontal dehiscence line, which is derived by fusion of three stamens (Chakravarty, 1958).

The ovary is inferior, (two-) three (–five)-carpellate and has three parietal placentae with two rows of ovules. Ovaries are unilocular or a septum runs to the centre of the ovary and the locules are filled with parenchymatous tissue. The style is often branched and twisted as in Begoniaceae.

Two types of nectaries are found in the family: mesenchymatous nectaries as an intrastaminal hypanthial disc, and trichomatic extrastaminal glands (Vogel, 1997); the latter were interpreted as substitutes for the ancestral disc nectaries. In *Cucurbita*, trichomes are present at the base of the corolla.

Begoniaceae

Fig. 10.48A,B. *Begonia albo-picta* Bull.

Staminate: ↓/↔* K2+2 A12 G0
* weakly monosymmetric to disymmetric
Pistillate: ✳ K5/2+3 A0 Ğ(3)
General formula: staminate ↓/↔ K2–4(5) C0(5) A4-∞ G0; pistillate
✳/↔K2–6 C0 G2–3(6)

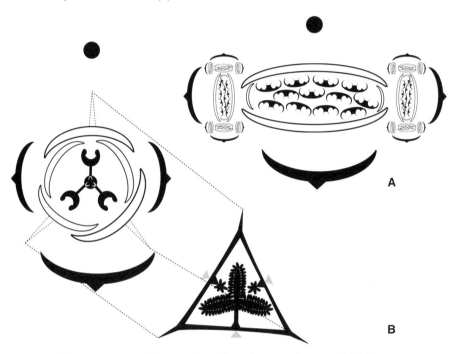

Fig. 10.48. *Begonia albo-picta* (Begoniaceae): A. staminate partial inflorescence;
B. pistillate flower; note the presence of abaxial crests on the inferior ovary and
apparently axile placentation by reduction of two placentae.

Flowers are arranged in cymes, with staminate and pistillate flowers on different inflorescences of the same plants (rarely dioecious), or are occasionally mixed. Bracteoles drop off early, leaving a scar. Merism is variable in the genus, even between staminate and pistillate flowers on the same inflorescence, ranging from dimerous to pentamerous or hexamerous. Petals are absent and the calyx is petaloid. Staminate flowers are mostly tetramerous (or apparently dimerous because of the difference between inner and outer calyx whorls (Ronse De Craene and Smets, 1990b). The calyx consists of an outer whorl of large petaloid parts, valvately arranged in bud and compressing the flower. The inner transversal sepals are much smaller and have a tendency to be lost (e.g. *B. heracleifolia*).

The calyx is rarely fused and tubular (*B. argenteomarginata*: pers. obs.). Stamens are numerous, or much lower in number (down to eight to four) and arranged on an inflated receptacle. Initiation is centripetal in regular to irregular whorls on an elliptical receptacle (Ronse De Craene and Smets, 1990b). The androecium is regular with extrorse to latrorse stamens, or monosymmetric will all stamens facing towards one of the median sepals. Staminate *Begonia* flowers offer pollen as reward and are without nectary, and stamens appear flattened with broad connective (Matthews and Endress, 2004). Pistillate flowers offer no reward and deceive visitors by mimicking stamens with their contorted stigmas. Only a few bird-pollinated begonias produce nectar in pistillate flowers (e.g. *B. ferruginea*: Vogel, 1998c). Merism in pistillate flowers is mostly five with a quincuncial arrangement, or hexamerous with an imbricate arrangement of an outer and inner whorl (e.g. *B. galinata*, *B. bogneri*). Arrangement of sepals in bud is similar to staminate flowers, with the outer sepals pressing the inner sepals in one plane. This arrangement is found in dimerous pistillate flowers (e.g. *B. incana*: Matthews and Endress, 2004). The ovary is inferior, mostly of three, rarely two carpels with parietal placentae. Ovaries have one to three wings as an extension of the locular space. The development of the wing is independent of the placental development. In *B. albo-picta* one of the placentae is more strongly developed than the two others with three extensions resembling an axile placentation (Fig. 10.48B). The placentation is frequently described as axile in the literature, which is erroneous. A high number of small ovules is produced. Styles are free opposite the carpels, and often have dichotomously branched stigmas that can be extensively twisted and contorted. No stamens are found in pistillate flowers.

The basal monotypic *Hillebrandia sandwicensis* differs from *Begonia* in having pentamerous staminate and pistillate flowers with reduced petals. The gynoecium is isomerous opposite the sepals (Matthews and Endress, 2004). Gauthier and Arros (1963) interpreted the petals as sterile stamens, because of their strong similarity.

Begoniaceae share the centripetal stamen development with Datiscaceae (Ronse De Craene and Smets, 1990b). Absence of a gynoecium in staminate flowers may be linked with the apparently unordered development, which is rare in rosids.

Fabales

The order contains four families of unequal size with unresolved affinities: Surianaceae, Quillajaceae, Polygalaceae and Leguminosae (Fabaceae). Surianaceae and Quillajaceae have retained the basic formula of rosids (K5C5A5+5G5), while zygomorphy has variously affected the other families. Floral diagrams for these two families were given in Bello, Hawkins and Rudall (2007).

Polygalaceae

Fig. 10.49. *Polygala* × *dalmaisiana* Dazzler

↓ K3+2 [C(5) A(4+4)] G̲(2)
General formula: K5 C3–5 A3–7(10) G2–8

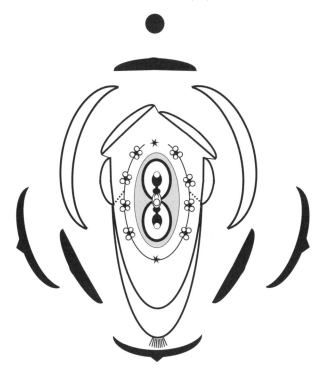

Fig. 10.49. *Polygala* × *dalmaisiana* (Polygalaceae). Broken line shows attachment of staminal ring to petals.

The Polygalaceae show a remarkable convergence with Leguminosae in developing monosymmetric pea-like flowers. However, the Bauplan of the flower differs in several characteristics. Inner lateral sepals are petaloid and contribute to floral display in a way comparable to the wing petals of Leguminosae. Petals are fused with the stamen tube in an elaborate compound system. The abaxial petal forms a protruding keel with a brush-like appendage, enclosing the stamens, while the four posterior petals are smaller, fused into lateral pairs and overlapping an entrance to the centre of the flower. Lateral petals are much smaller and are occasionally suppressed (Krüger and Robbertse, 1988; Prenner, 2004b). The stamens are fused except for the adaxial side, which is fringed with hairs; they are curved within the lip together with the style. The abaxial antesepalous stamen and the adaxial antepetalous stamen are lost (Fig. 10.49, asterisks). Anthers are poricidal and the stigma is asymmetric with only the adaxial lobe developed. There are two fused carpels on a gyno-phore and each bears an apical pendent ovule.

Other genera have more regular flowers. Prenner (2004b) mentioned devel-opmental similarities between Polygalaceae and Leguminosae, such as racemose inflorescences, pressure of bracteoles on floral development, suppression of organs and comparable acquisition of monosymmetry.

Levyns (1949) gave a floral diagram for *Muraltia* with seven stamens; one stamen is situated abaxially opposite the keel and three outer antesepalous stamens are missing, indicating a different evolutionary pathway from *Polygala*. Baillon (1860) mentioned a different position for stamens in *Muraltia*: one stamen opposite sepal two and two stamens opposite the latero-anterior petals are missing.

Leguminosae (Faboideae, Mimosoideae, Caesalpinioideae)

Most Leguminosae have 21 organs following the floral formula K5 C5 A5+5 G1. Main characters and differences between the three subfamilies are summarized in Tucker (2003a). All Leguminosae share the absence of fused petals (except Mimosoideae) and an androecium that is basically two-whorled (again with most variations in Mimosoideae). All Leguminosae are character-ized by the single superior carpel found in all taxa, except for *Archidendron* (Mimosoideae) with five or fewer carpels. Van Heel (1993) interpreted the pentamerous condition as plesiomorphic, but this is not supported by recent phylogenies where Mimosoideae represents a derived clade (Tucker, 2003a; Bello, Hawkins and Rudall, 2007). However, a pentamerous ovary is probably the plesiomorphic condition for all Fabales, as it is present in both Surianaceae and Quillajaceae. The androecium is basically two-whorled and diplostemon-ous. All families possess a broad intrastaminal nectary surrounding the ovary as a ring.

Leguminosae show a high level of variation in their flowers. Bracteoles can be part of the flower (e.g. *Isoberlinia*, *Afzelia*) or may be absent. Tucker (2001a, 2002b) described a progression in the reduction of petals through a stage of organ suppression, leading to their loss in some genera (see p. 48).

Molecular studies showed that Caesalpinioideae are paraphyletic, and one way to deal with this is to group Caesalpinioideae and Faboideae in one subfamily. Mimosoideae stands out as a separate group on the basis of several characters (see p. 275). Differences between Caesalpinioideae and Faboideae are minor and include partial fusion of petals and stamens in the latter. Faboideae and Caesalpinioideae differ from other pentamerous rosids in the position of the odd petal, which is adaxial and not abaxial. This may be linked with the adaxial-lateral position of the bracteoles altering the sequence of initiation of the petals (Endress, 1994). Petal aestivation is characteristically cochleate-ascending (Caesalpinioideae) or cochleate-descending (Faboideae).

Mimosoideae

Fig. 10.50. *Calliandra haematocephala* Hassk.

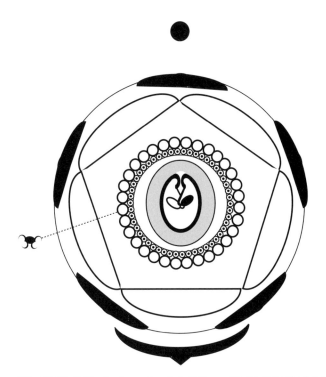

Fig. 10.50. *Calliandra haematocephala* (Mimosoideae). Note the inner ring of staminodes.

✳ K(5) C(5) A ∞ G1
General formula: ✳ K0–5 C0–5 A5-∞ G1(5)

Flowers are grouped in globular inflorescences and are subtended by a single bract. Flowers tend to arise synchronously (Tucker, 2003a). There is no pedicel. In *Calliandra* both calyx and corolla are fused for 2/3. The numerous stamens are showy with red filaments in at least two series curled up in bud. The inner whorl is staminodial (or inner trichomes?) curved over the disc-like nectary. Anthers are small and introrse. The single carpel is topped by a long style and dish-shaped stigma.

The subfamily differs from the other Leguminosae by a number of structural characters: absence of bracteoles, orientation of the flower, which has the odd petal in abaxial position as in most angiosperms, the radial symmetry of flowers (secondary?), which is usually associated with a secondary stamen increase and globular inflorescences, and the development of sepal and petal tubes. In Mimosoideae, there is much variation in the initiation sequence of sepals (simultaneous, unidirectional, helical), while petals arise generally simultaneously and have a valvate aestivation (Ramírez-Domenech and Tucker, 1990). Petals are generally much shorter and this is correlated with long showy filaments of the polyandrous flowers, analogous with some Myrtaceae (brush-flowers: Endress, 1994).

There is much variation in the number of stamens because of a secondary stamen increase. Stamen arrangement is generally diplostemonous or haplostemonous (e.g. *Mimosa*, *Neptunia*, *Calliandra*: Gemmeke, 1982; Prenner, 2004c). A secondary stamen increase is linked with alternisepalous primary stamen primordia (e.g. *Lysoloma*, *Accacia*) or the development of a ring primordium (e.g. *Albizia*, *Accacia*, *Calliandra sp.*) (Gemmeke, 1982; Derstine and Tucker, 1991). The stamens are increased in centripetal direction but also partly centrifugally (Derstine and Tucker, 1991).

Caesalpinioideae

Fig. 10.51A. *Afzelia quanzensis* Welw., based on Tucker (2002b)

↓ K5 C5 A5+2 G1

Fig. 10.51B. *Bauhinia divaricata* L.

↑ K(5) C2 A(1:4°+5°) G̲°1
General floral formula: ↓/ ✳ K0–5 C0–5 A1–10 G1

The subfamily shows the highest variation in floral diversity and this is linked with elaborate zygomorphy and open-access flowers with various adaptations for pollination. Flowers show some resemblance with Capparaceae in the development of spreading petals and long filaments.

Inflorescences are generally racemose. Bracteoles are usually well developed, occasionally enclosing the floral bud and replacing the reduced calyx (e.g. *Amherstia*: Tucker, 2000a,b, 2002b). Bracteoles are rarely absent (e.g. *Haematoxylum*, *Caesalpinia*). Flowers are generally bisexual, occasionally dioecious as in *Bauhinia malabarica* and *B. divaricata* (Fig. 10.51B, staminate flower; Tucker, 1988a). Flower orientation is with the odd sepal in abaxial position (except in *Ceratonia*: Tucker, 1992). Flowers are generally monosymmetric, except in some basal clades with regular flowers (e.g. *Ceratonia*). Asymmetric flowers occur in *Senna* mainly by curvature of the style and various modifications in size of floral organs (enantiostyly: Marazzi and Endress, 2008), or by reduction of organs (e.g. *Labichea*). Petals are often large and clawed but never

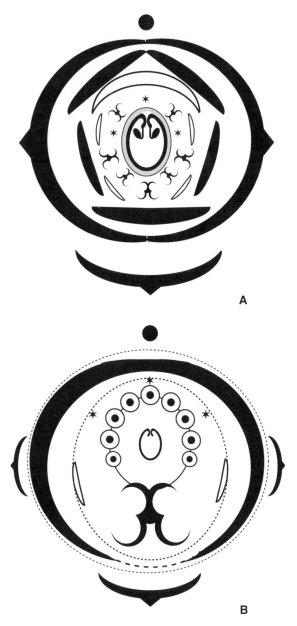

Fig. 10.51. Caesalpinioideae: A. *Afzelia quanzensis*; B. *Bauhinia divaricata*: staminate flower. The dehiscence line of the calyx tube is shown by a broken line. In A, asterisks refer to lost stamens; in B, asterisks refer to lost petals.

fused. The strongly monosymmetric flowers are differentiated early in floral development and influence the number of floral organs considerably. Initiation of organs is generally unidirectional. Some caesalpinioids have more or less irregular flowers linked with a reduction of petals (*Gleditsia*) or have evolved a papillionaceous shape (*Cercis, Haematoxylon*). In contrast to Faboideae the standard petal is the inner one, and several features of papillionoid flowers are absent. A hypanthium may be strongly developed, and various elaborations link androecium and petals in hiding access to the nectary (Endress, 1994).

In some taxa, adaxial sepals become fused during development (e.g. *Saraca, Schotia, Afzelia*: Tucker, 2000c, 2001b, 2002b; this may be a preliminary step for a transition to a tetramerous calyx. Prenner and Klitgaard (2008) considered that the tetramerous calyx of *Duparquetia* is formed by loss of one latero-adaxial sepal. In *Bauhinia divaricata* sepals are fused in a single calyptrate structure that dehisces abaxially under pressure of the single stamen (Fig. 10.51B). Sepals and petals can be frequently lost or suppressed. Petals are initiated but they are arrested in mid-development (e.g. *Crudia*: Tucker, 2001a); have completely vanished in *Ceratonia* (Tucker, 1992); all but one petal are initiated but suppressed (e.g. *Aphanocalyx, Monopetalanthus*: Tucker, 2000a); or all petals are completely lost (e.g. *Dialium*: Tucker, 1998). Petal rudiments can be variously present (e.g. *Bauhinia, Amherstia, Tamarindus*: Fig. 10.51B; Tucker, 2000b). In *Bauhinia divaricata* petal number ranges from two to three narrow appendages (Fig. 10.51B), although five petals are initiated (Tucker, 1988a). *Saraca* represents an unusual homeotic transformation of petals into stamens and this is accompanied by variable loss of antesepalous stamens (Tucker, 2000c).

The androecium ranges from one to ten stamens, with variable number of staminodes or missing stamens (e.g. *Afzelia, Amherstia, Bauhinia, Cassia, Duparquetia, Saraca, Senna*: Tucker, 1988a,b, 1996, 2000b,c, 2002b; Prenner and Klitgaard, 2008). Reduction of the androecium runs from the abaxial to the adaxial side of the flower and is often correlated with heteranthy, as in *Senna* with two pollinating stamens, four fodder stamens and three staminodes (Tucker, 2003a; Marazzi et al., 2006). *Bauhina divaricata* has a single functional stamen and nine staminodes (Tucker, 2003a). Reduction and loss of stamens affects the antepetalous stamens first (e.g. *Ceratonia, Tamarindus*: Tucker, 1992, 2000b). Stamens can be further reduced to five (3:2° in *Petalystylis*), four (in adaxial position, *Duparquetia*: Prenner and Klitgaard, 2008), two (*Dialium*: Tucker, 1998) or one (*Bauhinia divaricata*: Fig. 10.51B). In *Bauhinia galpinii* stamen reduction is correlated with lateral dédoublement of staminodes (Endress, 1994, 2008b).

The gynoecium is generally monocarpellate but can be displaced due to the monosymmetric development of flowers.

Faboideae (Papilionoideae)

Fig. 10.52A. *Swartzia aureosericea* R. S. Cowan, based on Tucker (2003b)

↓ K(5) C1 A∞ G1

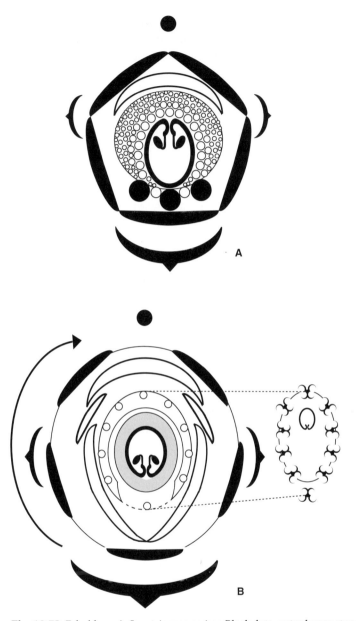

Fig. 10.52. Faboideae: A. *Swartzia aureosericea*. Black dots, outer larger stamens; white dots, inner smaller stamens. B. *Strongylodon macrobotrys*. The broken line represents the gap separating the odd adaxial stamen.

Fig. 10.52B. *Strongylodon macrobotrys* A.Gray

↓ K(5) C5 A1:(4+5) G1
General formula: ↓ K5 C5 A10-∞ G1

Inflorescences are mostly racemose (plesiomorphic condition), but cymose forms are occasionally found as a derived condition (Tucker, 1999). Flowers are nearly always subtended by a bract and two bracteoles.

Most Faboideae are median monosymmetric, with a surprisingly uniform ontogeny and rarely with loss or increase of floral organs (compared with Caesalpinioideae). The papilionate floral shape is generalized in the family and epitomizes a highly successful bee-pollination syndrome. The adaxial petal is mostly the largest (standard, flag or vexillum) and petals are arranged in cochleate descending aestivation. Anterior petals are usually connivent in the characteristic keel enclosing androecium and gynoecium. Main variation is found in fusion of members of the androecium, either all ten or as nine + one, where one adaxial stamen is separated from the tube by a slit allowing access to nectar at the inner side of the tube (e.g. Fig. 10.52, *Vicia*). All organs have a strong unidirectional development starting early in ontogeny, but there are several exceptions to this rule (e.g. Tucker, 1984), although the flag petal becomes the largest before anthesis. An obvious exception is *Cadia*, with radially symmetrical flowers. As several Faboideae have a regular symmetry at mid-development, Tucker (2002a) interpreted the condition in *Cadia* as a neoteny (a retention of the juvenile state), contrary to Citerne, Pennington and Cronk (2006) who saw this not as an evolutionary reversal, but as a homeotic transformation in which all petals have acquired the identity of the abaxial petal (see p. 50).

Flowers are mostly regular for perianth and androecium with two stamen whorls; reductions of organs are limited compared to Caesalpinioideae and are restricted to tribes Sophoreae and Swartzieae (Tucker, 1988b).

The genus *Swartzia* is basal in the order and resembles the Caesalpinioideae (Fig. 10.52A). Flowers are extremely heterogenous with one, two or no adaxial petals and a high number of stamens arising on a complete or partial ring meristem. In partial meristems a variable number of abaxial stamens arise before the ring meristem and are much larger (Tucker, 2003b). This phenomenon is best interpreted as a partial increase of the stamens, affecting the abaxial stamens to a variable extent. The androecium has thus partially complex polyandry linked with free adaxial stamens.

Fagales

The order consists of eight families of which Fagaceae, Betulaceae and Juglandaceae are the largest. Flowers are mostly highly reduced and well adapted to wind pollination. The character syndrome includes: unisexual flowers with separation of staminate and pistillate flowers on separate branches or on separate trees, reduction of the perianth and absence of petals, reduction of number of carpels and ovules (one to two per locule) and increase of size of styles. The order is characterized by intricate dispersal mechanisms involving fusions of bracts with flowers (e.g. Juglandaceae, Betulaceae) and associations of several bracts and flowers into cupules (e.g. Nothofagaceae, Fagaceae) (see e.g. Abbe, 1935, 1938, 1974; Oh and Manos 2008). Endress (1967, 1977) pointed out several similarities between Hamamelidaceae and Fagales, but these appear to be adaptive rather than a reflection of affinitiy.

Flowers of Fagales are well represented in the fossil record, with a mix of adaptations to insect and wind pollination (e.g. Friis, Pedersen and Schönenberger, 2006; Crepet, 2008). Fagales form the core of the Amentiferae, which were thought to be ancestral among flowering plants (e.g. Eichler, 1875). However, it appears that the reduced flowers are adaptive and derived.

Betulaceae

Fig. 10.53. *Carpinus betulus* L., based on Abbe (1935) and Endress (2008b)

Staminate: $\leftrightarrow$ K0 A4–6 G0
Pistillate: $\leftrightarrow$ K4–8 C0 $\underline{G}$(2)
General formula: $\leftrightarrow$/ $*$ K0/1–6 C0 A(1-)4(-6) G2(-3)

Pistillate flowers are arranged in pairs in reduced dichasia without central flower. A central flower is present only in *Betula*. Each flower is subtended by a bract and two bracteoles, which are postgenitally fused into a single unit. The calyx is reduced in size and has a variable number of parts; four larger sepals are found in median and transversal position, but with a variable number of smaller parts in between (Endress, 2008b). Sepal lobes are colleter-tipped and have strong bundles. Endress (2008b) argued that the variability of the sepals is linked with their reduction and change of function; the increase in parts may be the result of a subdivision of the four major sepals. The two carpels bear a single ovule each.

The staminate partial inflorescence is built on a dichasial system of three flowers (Abbe, 1935). There is much variation in the number of stamens of individual flowers and in the extent of development of the perianth. Staminate flowers lack a perianth in *Carpinus* as well as in *Corylus*, while two to four sepals are found in *Betula*, respectively *Alnus*. In *Carpinus*, the filament is branched and bears a pair of thecae topped with hairs.

Abbe (1935, 1938) interpreted the ancestral merism of Betulaceae as hexamerous. This relates merism of Betulaceae to the condition of Fagaceae, with a stronger floral reduction in the former.

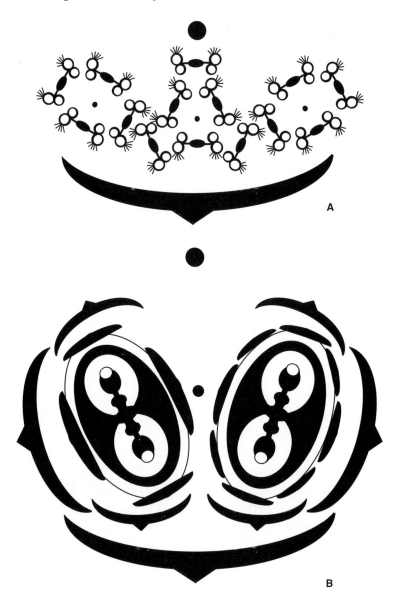

Fig. 10.53. *Carpinus betulus* (Betulaceae): staminate (A) and pistillate (B) partial inflorescence.

Fagaceae

Fig. 10.54A,B. *Castanea sativa* Mill.

Staminate: $*$ K3+3 A3^3+3 G(3°)
Pistillate: $*$ K3+3 A°3^3+3 Ğ(6)
General formula: $*$ K 6(-9) C0 A(3)6–12(-90) G2–6(-9)

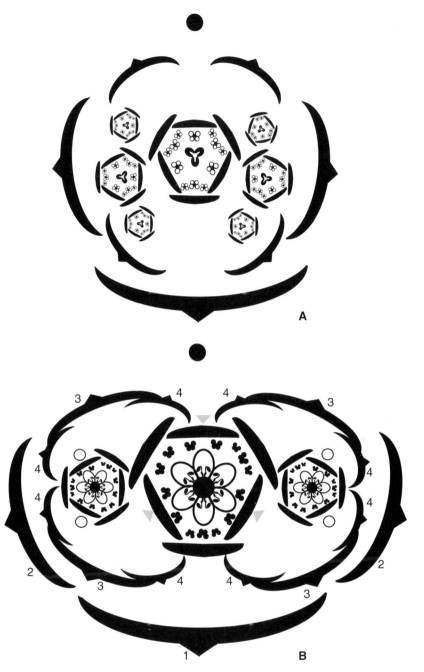

Fig. 10.54. *Castanea sativa* (Fagaceae): staminate (A) and pistillate (B) partial inflorescence. Numbers indicates order of initiation of bracts; white dots represent aborted flower buds.

Flowers in Fagaceae are often pentamerous to hexamerous, with an arrangement of parts in two whorls. There has been discussion whether flowers of Fagaceae are hexamerous or trimerous (see e.g. Endress, 1977; Okamoto, 1983), but it is reasonable to consider them as hexamerous because of the well-nested-position of the family within rosids (analogous to the condition in Polygonaceae). Staminate and pistillate flowers in *Castanea* are formed on separate catkin-like inflorescences or occasionally on the same with separate partial inflorescences. Staminate flowers are clustered in dichasia of up to seven to eight flowers, while pistillate inflorescences usually consist of three flowers surrounded by several bracts. Fey and Endress (1983) demonstrated that the four-valved cupules of *Castanea* are derived from complex highly contracted dichasia, consisting of suppressed lateral axes with their subtending bracts. Third-order bracts are fused with a pair of fourth-order bracts to form one of the valves (numbered in Fig. 10.54B). Staminate flowers in five genera of Fagaceae, including *Castanea*, usually have 12 stamens and the same number of staminodes in pistillate flowers (cf. Okamoto, 1983). The stamens opposite the outer sepals are arranged as triplets, although Baillon (1876a) argued that the position of the extra stamens is variable. Most other genera have six stamens only, opposite the sepals. In *Quercus* flowers are four- to five-merous with sequential initiation (Sattler, 1973).

There is high diversity in cupule morphology in Fagaceae, with considerable discussion on the origin and evolution of cupules as derived from complex dichasial systems (reviewed by Oh and Manos, 2008). There are two major types, the dichasium-cupule, as in *Castanea* or *Fagus*, enclosed by triangular valves, and the flower-cupule, as in *Quercus* or *Lithocarpus*, which is a single flower surrounded by a valveless cupule. The phylogeny demonstrates at least three independent derivations of valveless cupules from dichasial predecessors. In *Fagus*, the central flower of the dichasium is lost, leaving two lateral flowers enclosed by two bracts each, forming the valves (Fey and Endress, 1983). In *Quercus*, the dichasium is reduced to one terminal flower surrounded by several bracts of sterile flowers, forming scales on the acorn.

Nothofagaceae (southern beech) have often been included in Fagaceae. They occupy a basal position in Fagales.

Rosales

The order contains nine families and is strongly supported. The circumscription of Rosales has changed dramatically since Cronquist (1981). Soltis *et al.* (2005) enumerated a hypanthium and a reduction of endosperm as main synapomorphies. Stevens (2001 onwards) included the hypanthial nectary, petals that are clawed when present, valvate calyx and single apotropous ovule per carpel. The order is believed to be basically apetalous and wind-pollinated

(Ronse De Craene, 2003). This is reflected in families formerly placed in Urticales, which share a simple floral formula of K5 C0 A5 G(2). Within Urticales there is a progressive reduction of flowers linked with unisexuality, loss of a hypanthium, reduction of the perianth and pseudomonomery (e.g. *Urtica*, *Laportea*: Bechtel, 1921). Loss of a hypanthium occurred in the clade comprised of Cannabaceae, Moraceae and Urticaceae, while it is retained in Ulmaceae. Flowers are reduced to five or fewer stamens and bicarpellate unilocular ovaries with a single basal or apical ovule (Judd and Olmstead, 2004).

Rosaceae occupies a basal position in the order and this is linked with a unique kind of polyandry including petals of staminodial origin. I suspect that petals were secondarily derived from stamens on two occasions, once in Rosaceae and a second time in the Rhamnaceae and Dirachmaceae, corroborating earlier theories (e.g. Bennek, 1958). It would be interesting to study the genetic background for petal expression in Rosales.

Diplostemonous flowers are rare in the order (some Rosaceae, *Shepherdia* in Elaeagnaceae), although it is probably plesiomorphic. Obhaplostemonous flowers have mainly evolved in Rosaceae, Rhamnaceae and Elaeagnaceae, while 'Urticales' are haplostemonous (Ronse De Craene, 2003).

Rosaceae

Fig. 10.55A. *Potentilla fruticosa* L.

∗ K5 C5 A10+10+5+5 G∞

Fig. 10.55B. *Spiraea salicifolia* L.

∗ K5 C5 A10+5+5 G(5)

Fig. 10.55C. *Sanguisorba tenuifolia* Fisch. ex Link

∗ K4 C0 A4 G1
General formula: ∗ K(3-)5(-10) C0 or (3-)5(-10) A(1-)5–10-∞ G1–2–5-∞

Flowers are mostly arranged in cymose units or in racemose inflorescences. In racemes terminal flowers precede lower flowers (e.g. *Sanguisorba*, *Gillenia*, *Photinia*: pers. obs.; Evans and Dickinson, 2005).

Flowers are generally pentamerous, more rarely tetramerous (trimerous in *Cliffortia*: Eichler, 1878). Aestivation of sepals is generally valvate. An epicalyx is developed in several genera and alternates with sepal lobes. The epicalyx resembles the calyx closely (e.g. *Potentilla*: Pluys, 2002) or develops as spines with a centrifugal development (e.g. *Agrimonia*: Ronse De Craene and Smets, 1996b). The epicalyx has been interpreted as stipular in origin (e.g. Trimbacher, 1989), although homologies are not always clear (see p. 25).

Fig. 10.55. Rosaceae: A. *Potentilla fruticosa*: partial inflorescence; B. *Spiraea salicifolia*; C. *Sanguisorba tenuifolia*. Asterisks in A refer to empty stamen positions.

Rosaceae have either apetalous flowers or petaliferous flowers with a well-developed hypanthium (floral cup) reminiscent of Myrtaceae. Petals bear a short claw and have a variably imbricate aestivation. When petals are present, they are closely connected to the androecium and are interpreted as having a staminodial origin in the family and to represent the upper stamens of complex alternisepalous primordia (Ronse De Craene, 2003). In *Cecrocarpus* and *Neviusia* stamens occupy the position of petals in other genera (Kania, 1973; Lindenhofer and Weber, 2000). The androecium is extremely variable in the family by the presence of reductive patterns and secondary elaborations. The androecium of the Rosaceae appears to be unique in the angiosperms in the development of alternating decamerous or pentamerous whorls linked with the growth of the hypanthium (described as cyclic polyandry by Ronse De Craene and Smets, 1987). In most Rosaceae the androecium consists of several whorls (often 20–25 stamens) that develop thanks to the growth of the hypanthium. The upper (outer) stamen whorl is always decamerous and closely linked with the petal. Two more whorls of five stamens are generally formed in antesepalous and antepetalous position (Fig. 10.55B). However, the number of stamens formed can fluctuate strongly, even within a same species, and depends on reductions of the length of the hypanthium or an earlier onset of the development of the gynoecium (*Crataegus*: Evans and Dickinson, 1996). Kania (1973) interpreted the increase of stamens as secondary and derived from a diplostemonous ancestry. However, when the androecium is limited to ten stamens (e.g. *Crataegus*), these consist of the upper pair only and do not have a diplostemonous arrangement. Clearly diplostemonous androecia occur in *Agrimonia*, *Stephanandra* and *Plagiospermum*, although the number of stamens is variable and can be higher (Kania, 1973; Ronse De Craene, unpubl. data). *Filipendula* is also unique in the development of ten radial rows of stamens (Ronse De Craene, unpubl. data). In *Potentilla*, petals occupy the upper section of common primordia (ridges: Innes, Remphrey and Lenz, 1989), extending centripetally on the slopes of the hypanthium. The number of stamens can be increased dramatically, as in *Rubus*, through reduction of size of primordia and development of a ring primordium. Ronse De Craene and Smets (1992a) and Ronse De Craene (2003) argued that the cyclic development of stamen whorls is not different from a secondary multiplication of stamens on complex primordia. The common primordia are rarely well differentiated from the hypanthium (e.g. in *Potentilla*, *Crataegus*, *Fragaria*: Sattler, 1973; Innes, Remphrey and Lenz, 1989; Evans and Dickinson, 1996; Ronse De Craene, 2003). Fluctuation in stamen number and development of whorls is linked to the size of stamen primordia and the extent of growth of the hypanthium in a very similar pattern as found in Myrtaceae (Ronse De Craene and Smets, 1991a). An alternative interpretation was presented by Lindenhofer and Weber (1999a,b, 2000), that polyandry in Rosaceae is derived from a spiral androecium. They considered a spiraeoid pattern of 10+5+5 stamens as original and interpreted diplostemony as derived by reduction. They especially rejected the notion of dédoublement that was used to explain the paired position of

Fig. 10.57. *Ceanothus dentatus* (Rhamnaceae).

Ronse De Craene and Miller, 2004) and remain inconspicuous in the flower. In *Ceanothus* petals are hood-like. Some genera lack petals altogether and have a petaloid calyx and hypanthium (e.g. *Colletia*). The androecium is always obhaplostemonous. Most Rhamnaceae have little pigmentation and have greenish to yellowish flowers. The gynoecium contains two to three (rarely five) carpels with axile placentation. The ovary is superior or more frequently (half-) inferior with three carpels and a single, often deeply cleft, style. Each locule has one to two basal ovules per placenta. In *Ceanothus* one ovule per locule is developed (Fig. 10.57), while the number of fertile placentae can be variable in the family. For example, in *Colubrina* with a trimerous ovary, one septum bears two ovules, another one ovule and a third none (Medan and Hilger, 1992). The central part of the flower is mainly occupied by a large disc nectary, which is especially well developed in cases where the ovary is inferior.

Closest relatives of Rhamnaceae are Dirachmaceae and Barbeyaceae (Richardson *et al.*, 2000). Dirachmaceae shares several characters with Rhamnaceae, including obhaplostemony, although several unique features separate the family (Ronse De Craene and Miller, 2004).

Elaeagnaceae

Fig. 10.58. *Elaeagnus umbellata* Thunb.

✳ K4–5 C0 A4–5 G̲ 1

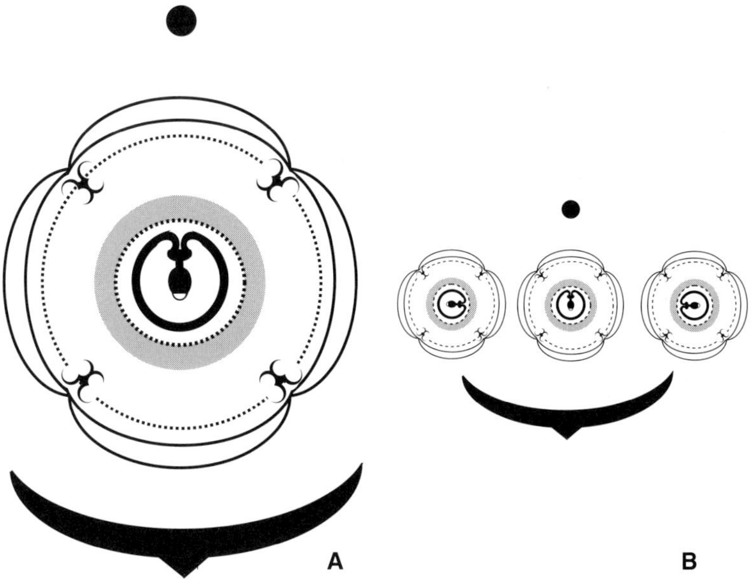

Fig. 10.58. *Elaeagnus umbellata* (Elaeagnaceae): A. flower; B. inflorescence.

The small family of three genera is characterized by the absence of a corolla, although this was questioned by Rao (1974) on anatomical grounds.

Flowers are tetra-, penta- or hexamerous and are grouped in clusters in the axil of leaves. Flowers are bisexual only in *Elaeagnus*. Sepal aestivation is always valvate. In *Hippophae*, lateral sepals are much smaller than median sepals and were described as bracteoles by Eichler (1878), who interpreted the flowers as dimerous. Stamens alternate with sepal lobes (*Elaeagnus*, *Hippophae*), or the androecium is diplostemonous (*Shepherdia*). The filament is very short. Stamens and sepals are connected by a long hypanthium that runs down below the ovary. In *Elaeagnus*, the ovary appears to be inferior but the ovary wall is not connected with the external wall. The appearance of an inferior ovary is stressed by the presence of a nectary on the hypanthial slope, surrounding the style. In *Shepherdia* well-developed nectary lobes alternate with the stamens or surround the carpel at the level of attachment of the sepal lobes (Baillon, 1870). The ovary is apparently monocarpellate with single, basal, anatropous ovule. Sepals, hypanthium and ovary are covered with peltate hairs.

Elaeagnaceae share many similarities with reduced flowers of Rosaceae, such as apetaly, valvate tetramerous calyx, alternate stamens and single carpel.

11

Asterids: tubes and pseudanthia

Asterids are subdivided into two main groupings: basal asterids (a grade) consisting of Cornales and Ericales, and euasterids (with Lamiids and Campanulids) (Figs. 11.1, 11.2; e.g. Judd and Olmstead, 2004). Spichiger *et al.* (2002) distinguished between archaic asterids, superior hypogynous asterids and superior epigynous asterids.

Characters common in a majority of taxa of asterids are sympetaly with adnate stamens (stamen-petal tube), unitegmic, tenuinucellate ovules, cellular endosperm formation, terminal endosperm haustoria, pollen that is released at the trinucleate stage, and the presence of iridoids (e.g. Judd and Olmstead, 2004; Soltis *et al.*, 2005). Apart from sympetaly, the other characters cannot be used in floral diagrams.

Most taxa share the development of a ring primordium in early stages of development (Erbar, 1991) indicating a generalized syndrome of petal development (see p. 34).

Core asterids (euasterids) share sympetaly, a bicarpellate gynoecium and haplostemony, while the basal orders Cornales and Ericales are much more variable with a basically diplostemonous androecium.

11.1   Basal asterids: Cornales, Ericales

Cornales

The basalmost order of asterids (Fig. 11.2) contains six to seven families, with three main families Hydrangeaceae, Cornaceae and Loasaceae. All share small sepals, an inferior ovary, epigynous disc nectary, and often drupaceous fruits (Soltis *et al.*, 2005). Soltis *et al.* (2005) argued that the half-inferior ovary of Hydrangeaceae evolved from epigyny common in the clade. Most Cornales have free petals, with

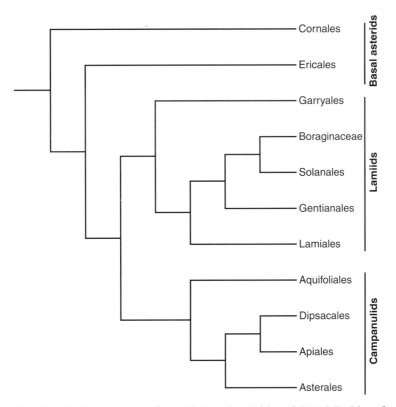

Fig. 11.1. Phylogenetic tree of Asterids, based on Soltis *et al.* (2005). Position of Boraginaceae is debatable.

the evolution of a corolla tube in a few Loasaceae. However, Hufford (1998) argued that Hydrangeaceae possess a 'corolline torus', which is comparable to early sympetaly described by Erbar (1991). Hydrangeaceae are strongly linked with Loasaceae in floral characters (see Roels, Ronse De Craene and Smets, 1997; Hufford, 1989a,b, 1992, 1998, 2001), with the same pattern of stamen increase, although the androecium is more complex in Loasaceae subfamily Loasoideae. Both families share highly polystemonous androecia, with a development that shifts between a centrifugal and centripetal direction (e.g. Hufford, 1990, 1998; Ge, Lu and Gong, 2007). Gynoecium and hypanthium development are highly similar with the development of a 'corolla ring primordium' (Erbar, 1991), also present in other Cornales. Development of a cup-like hypanthium in early stages of development in all Cornales links the clade with euasterids.

Hydrangeaceae

Fig. 11.3A,B. *Kirengeshoma palmata* Yatabe

$\ast$ K(5) C5 A5^3 -G- (3)

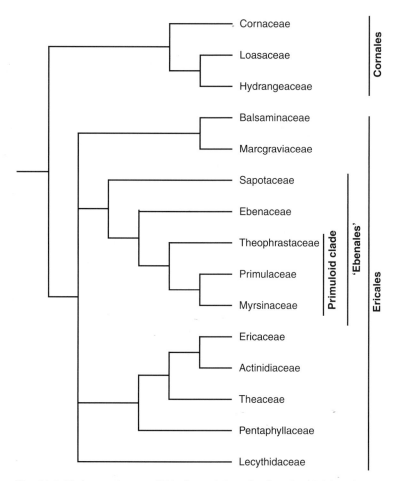

Fig. 11.2. Phylogenetic tree of Ericales and Cornales, based on Schönenberger *et al.* (2005).

Fig. 11.3C,D. *Dichroa febrifuga* Lour.

✳ K5 C5 A5+5 -G- (3)

General formula: ✳ K(4–5) C4–5 A10–∞ Ğ/-G- 2–5

Hydrangeaceae used to be associated with Saxifragaceae in subfamily Hydrangeoideae. Apart from molecular evidence, several morphological characters have indicated that both families are not related: e.g. rapid petal growth versus a retardation or loss of petals, diplostemony versus obdiplostemony (e.g. Klopfer 1973; Hufford, 1992a).

Inflorescences are usually cymose, often forming umbels with sterile peripheral flowers having expanded calyx lobes, or flowers are solitary. Floral merism ranges from four to five to six, rarely up to eight in *Deinanthe* and *Decumaria*,

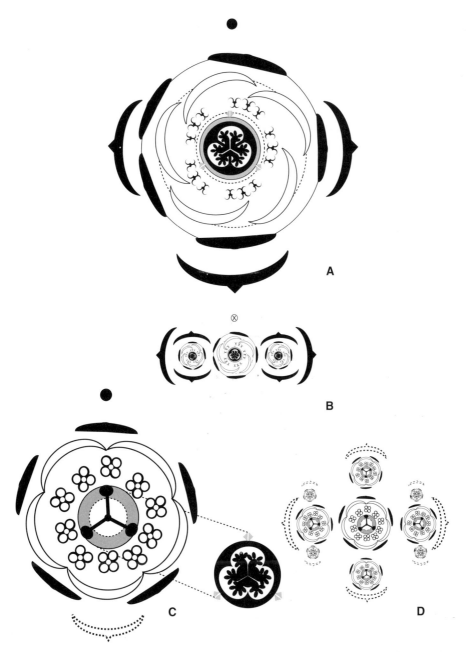

Fig. 11.3. Hydrangeaceae: *Kirengeshoma palmata*, A. flower, B. partial inflorescence; *Dichroa febrifuga*, C. flower, D. partial inflorescence.

leading to crowding of the organs in different whorls. Flowers are bisexual, rarely unisexual and dioecious in *Broussaissia*. The calyx is generally small and valvate (except for *Cardiandra* and *Deinanthe*, with imbricate sepals and spiral initiation of the perianth: Gelius, 1967; Hufford, 2001). Petals are generally valvate or contorted, rarely imbricate or calyptrate. The flower of *Kirengeshoma* is highly contorted, displacing the stamens into different rows (Roels, Ronse De Craene and Smets, 1997). The androecium is basically diplostemonous (e.g. *Deutzia*, *Dichroa*, *Hydrangea*), rarely haplostemonous (two species of *Dichroa*: Hufford, 2001), but is often complex polyandrous. The androecium of multistaminate Hydrangeaceae is basically haplostemonous, either by lateral division of ante-sepalous primordia in triplets and with more stamens extending in antepetalous position (e.g. *Kirengeshoma*, *Philadelphus*, *Decumaria*, *Cardiandra*: Roels, Ronse De Craene and Smets, 1997; Hufford, 1998), or by more complex increases involving a high number of stamens. This indicates that increase of antesepalous stamen primordia probably prevents the antepetalous stamens from developing. In *Carpenteria*, numerous stamens arise on complex U-shaped primordia extending in a waveline in antepetalous sectors of the flower (Hufford, 1998), resembling the development in Malvaceae. In *Deinanthe* and *Platycrater*, numerous stamens arise on a ring primordium, linked with identifiable primordia in antesepalous posi-tion, and covering the hypanthial slope (Hufford, 1998; Ge, Lu and Gong, 2007). Placentation is intruding parietal to axile, usually U-shaped with many ovules confined to the upper part of the ovary. A variable hypanthial growth is respon-sible for the position of the ovary, which is halfway emergent with usually free stylodes (Hufford, 2001). A disc nectary is intrastaminal to epigynous.

Loasaceae

Fig. 11.4. *Cajophora hibiscifolia* Urb. & Gilg, based on Leins and Winhard (1973) and Hufford (1990)

$$\ast \text{ K5 C5 A5}^{5°:\infty} \text{ Ǧ(3)}$$
General formula: $\ast$ K5 C5 A(3-)5-∞ G3–7

Inflorescences are generally cymose, rarely racemes (*Petalonyx*: Hufford, 1989a). Flowers are (tetra-) pentamerous, rarely six- to seven-merous (Urban, 1892). Sepals are valvate, rarely quincuncial (*Mentzelia*). Petals have a variable aestivation (valvate in most *Cajophora*), and can be relatively small in bud in Gronovioideae (Urban, 1892; Moody and Hufford, 2000b). There is a single style with apical stigmatic lobes, surrounded by a broad epigynous disc nectary. Placentation in Loasaceae is parietal and often five in number (with carpels alternisepalous and with a high number of ovules), although reduction to three or even a single subapical ovule occurs (e.g. Gronovioideae: Hufford,

Fig. 11.4. *Cajophora hibiscifolia* (Loasaceae). Antesepalous complex stamen primordia differentiate in staminodial scales and spread out opposite the petals.

1989a; Moody and Hufford, 2000b). Hufford (2001) argued that sympetaly has arisen at least two times in Loasaceae, either by congenital fusion (*Eucnide* section *Sympetaleia*) or by postgenital fusion (*Petalonyx*), although he also reported this for *Schismocarpus* (Hufford, 1989b).

Flowers of Loasaceae show the highest variability in their androecium, similar to Malvaceae. *Schismocarpus* is (ob)diplostemonous (Hufford, 1989b), while all members of subfamily Gronovioideae are haplostemonous, occasionally with some staminodial stamens (Urban, 1892; Moody and Hufford, 2000b), and most other taxa are polyandrous. The development of polyandrous flowers is highly complex, with continuity between antesepalous and antepetalous stamen groups, but the development can be linked to complex antesepalous (or more rarely antepetalous) groups of stamens (Hufford, 1990). A single stamen primordium followed by two lateral stamen primordia always precedes the initiation of more stamens. In *Mentzelia* and *Eucnide* development of the androecium runs centripetally from antesepalous forerunners, and stamens cover the slope of a deep hypanthium. In *Mentzelia* a variable number of first-formed stamens is staminodial, but this pattern is accentuated in other genera of subfamily Loasoideae (e.g. *Cajophora*, *Loasa*), with differentiation of antepetalous fertile stamens and development of the five initial stamen primordia into antesepalous staminodes (Fig. 11.4). Antepetalous stamens arise as an overflow of the antesepalous sectors and initiation of stamens is a mixture of centripetal and centrifugal development, comparable to some polyandrous Hydrangeaceae (e.g. *Platycrater*). Stamens in the antepetalous stamen fascicle consist of stamens of two adjacent mounds (e.g. Brown and Kaul, 1981; Hufford, 1990; contrary to the interpretation of Leins and Winhard, 1973), and this is comparable to the stamen development in Malvaceae. In *Blumenbachia* the pattern of initiation appears to be reversed, with antepetalous complex primordia arising before the antesepalous staminodes. One could possibly interpret this as an inversion of centres of initiation linked with the retardation of staminodial primordia.

The three outer staminodes (corresponding to the first formed stamens of the antesepalous common primordia) are connected into a single cucullate structure, while the two inner staminodes become elongated and cover the stigma (e.g. *Loasa*, *Cajophora*: Hufford, 1990). In other genera the two sets of staminodes are more variable in number (five to seven) and the shape of the staminodes can be variously elaborate (e.g. *Huidobria*, *Nasa*: Hufford, 2003). As the nectary is epigynous, staminodes may function as nectar containers (e.g. Brown and Kaul, 1981; Smets, 1988). Flowers of *Schismocarpus* are buzz-pollinated and lack nectaries (Hufford, 1989b). Distally forked filaments are found in some *Mentzelia* and are reminiscent of comparable structures in Hydrangeaceae (e.g. *Deutzia*: Hufford, 2003).

Ericales

Ericales contain about 23 families and were considerably extended compared with the premolecular classifications, incorporating several smaller orders (Theales, Ebenales, Sarraceniales, Primulales, Polemoniales, Ericales *sensu stricto*: Fig. 11.2; Anderberg, Rydin and Källersjö, 2002; Schönenberger *et al.*, 2005). Delimitation of Ericales, although well supported on a molecular basis, tends to be far less clear on a morphological basis, as some well-defined groups (such as the primuloid clade) are not well distinguished. Apart from three clades, the inter-familial relationships of Ericales are not well supported (Fig. 11.2; Schönenberger, Anderberg and Sytsma, 2005). Soltis *et al.* (2005) recognized two synapomorphies: gynoecia with protruding-diffuse placentae and theoid leaf teeth.

The order can be considered as a transitional group (together with Cornales) between asterids and rosids. Compared with euasterids, sympetaly is not universal, and the basic floral diagram is similar to rosids: K5 C5 A5+5 G5. Within Ericales this diagram tends to be retained in the Ericales *sensu stricto* (Ericaceae and satellite families). Some families (Polemoniaceae, Balsaminaceae) are haplostemonous. Sterile stamens occur in antesepalous position (Sapotaceae, primuloid clade), or more rarely opposite the petals (e.g. Diapensiaceae). A derivation of obhaplostemony (with or without sterile antesepalous stamens), while rare for the angiosperms, is common in families of 'Primulales', 'Theales', and 'Ebenales'. Except for Symplocaceae, increase of stamen number is usually from antepetalous stamens with stamens arranged as pairs (e.g. Fouquieriaceae, Ebenaceae, Styracaceae: Schönenberger and Grenhagen, 2005; Saunders, 1939; Dickison, 1993). Some families, such as Actinidiaceae, Lecythidaceae and Theaceae, have undergone a far-reaching specialization with secondary stamen multiplications on a ring primordium and a reversal to a spiral flower (van Heel, 1987; Tsou, 1998; Tsou and Mori, 2007). In many families of Ericales, sepals, petals and androecium arise in a spiral sequence leading to an imbricate arrangement of the petals (Ronse De Craene, 2008).

Schönenberger, Anderberg and Sytsma (2005) reconstructed morphological characters on their phylogeny and concluded that sympetaly has arisen on more than one occasion. They also assumed that diplostemony is derived and that haplostemony is plesiomorphic because it is present in basal groups. This assumption is not helpful and based on incomplete information. Morphologically there is no evidence to support the de novo development of a second whorl, even if the most parsimonious reconstructions indicate this to happen.

Monosymmetry is rare (some Ericaceae, Polemoniaceae, Balsaminaceae) and is expressed late. There is a strong tendency to develop sympetalous flowers, which are present in several clades (Ericaceae, Lecythidaceae, Ebenaceae, Sapotaceae, Primulaceae, Theaceae). This arises by zonal growth and includes petals with or without stamens. A corolla is never reduced or lost and this may be linked to the ubiquitous animal pollination in the clade. The gynoecium is generally syncarpous with five to three carpels and axile placentation. Comparison of gynoecium development in several Ericales shows an identical invagination of septa, which may or may not connect in the middle of the ovary (pers. obs.). In the primuloid clade, placentation is free central but this is probably derived by a reduction of the septa. Unitegmic ovules are frequent. A disc nectary is frequently present and closely linked with the base of the ovary (e.g. Ericaceae, Primulaceae) or nectaries are variously developed (e.g. Theophrastaceae) or absent (e.g. Actinidiaceae, Ebenaceae) (Bernardello, 2007).

Balsaminaceae

Fig. 11.5. *Impatiens platypetala* Lindsey

↑ K3C5A5G(5)

The family contains two genera, *Hydrocera* and *Impatiens*. Balsaminaceae forms a basal clade with three small families including Marcgraviaceae. There is little morphological evidence, except cryptic micromorphological and embryological characters, to understand the relationships of this clade (Kubitzki, 2004).

The floral diagram of *Impatiens* (Fig. 11.5) has a number of interesting features. Bracteoles are tiny and unequal (ignored by Eichler, 1878) and are absent in most

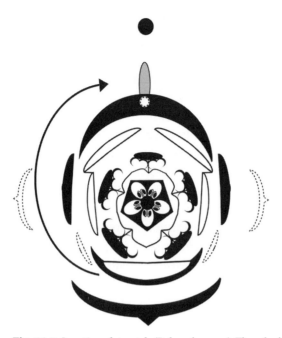

Fig. 11.5. *Impatiens platypetala* (Balsaminaceae). The adaxial petal is half sepaloid.

species. As a result, lateral sepals are the first to be initiated followed by the abaxial sepal (Caris *et al.*, 2006a). The flower is resupinate at maturity, adding confusion to the interpretation of the flower. At maturity, only three posterior sepals are present; two antero-lateral sepals are lost although they are occasionally initiated and are present in a few species (Caris *et al.* 2006a). The posterior sepal has a basal nectar spur arising late in development. Five petals are present; the laterals are basally fused into pairs, sometimes forming a single unit. The anterior petal is larger and appears to have a dual morphological nature of calyx and corolla. Rama Devi (1991b) thought the petal to have arisen by the congential fusion of the antero-lateral sepals and the anterior petal, while Grey-Wilson (1980) suggested that the antero-lateral sepals are reduced while their vasculature remains present at the base of the anterior petal. The flowers used for the present floral diagram had a dual nature in that the petal was half petaloid with a dorsal crest. It is possible that increased monosymmetry has broken down any clear delimitation of whorls, with gene activity of sepals invading the anterior petal. In some species, the anterior petal follows the spiral sequence of the sepals and behaves as a sepal (Caris *et al.*, 2006a). Interestingly, the petaloid sepals and all petals have a similar SEP3-like gene expression, which is normally only present in the three inner whorls of eudicot flowers (heterotopic petaloidy: Geuten *et al.*, 2006). A similar expression was also reported for *Marcgravia* (Marcgraviaceae), the putative sister group of Balsaminaceae.

The five antesepalous stamens are free but develop ventral appendages on their filaments that become connivent around the ovary. Monosymmetry is also expressed in the androecium, as the posterior stamen is smaller than the lateral stamens (possibly sterile?). Anthers open introrsely together with abundant secretion on the inside of the filaments, while pollen is loaded between the flaps and anthers (proterandry with secondary pollen presentation?). However, pollination mechanisms may be unknown and variable between different species. Anthers remain attached around the developing fruit while the filament breaks off. There are always five carpels with two rows of ovules, with a tendency for reduction to a single row (Caris *et al.*, 2006a). There is no distinct style but the upper part of the ovary is sterile.

The floral diagram of *Hydrocera* appears more regular and corresponds to the plesiomorphic condition of Balsaminaceae: five sepals are present, with the antero-lateral sepals of similar size as the laterals, and there is no fusion between the lateral petals (Eichler, 1878). The ancestral flower of Balsaminaceae was probably regular and haplostemonous.

Sapotaceae

Fig. 11.6. *Manilkara zapota* L., based on Moncur (1988) and Caris (1998)

Fig. 11.6. *Manilkara zapota* (Sapotaceae). Note trilobed petals.

$* K3+3 C(6) A6^o+6 \underline{G}(6–9)$
General formula: $* K5–12 C4–6(9) A4–6(12) G1–6(15–30)$

Flowers are generally bisexual or unisexual and arranged in fascicles. The calyx is mostly imbricate and in a single whorl but can be biseriate with two whorls of two, three, or four sepals, with the outer whorl valvate and the inner one imbricate in bud (Fig. 11.6: *Manilkara*). The corolla is always fused and the number of lobes can be either the same as sepals, or two to three times that of the sepals. Petal lobes are often divided in a median and two lateral segments (*Manilkara, Sideroxylon*). The lateral lobes spread horizontally while the median lobe remains erect and clasps the opposite extrorse anther (Pennington, 2004). Stamens are superposed to the corolla lobes (obhaplostemony) and are usually inserted on the hypanthial tube. In some tribes, the number of stamens is increased by forming antepetalous pairs or more stamens (e.g. *Bassia, Magodendron*: Saunders, 1939; Vink, 1995). Staminodes are occasionally formed opposite the sepals; they can take all sizes and are occasionally fertile in *Manilkara*. In *Mimusops commersonii* there are two outer series of 16 and eight staminodes, respectively (Endress and Matthews, 2006b). A nectary develops into a narrow disc around the ovary. The gynoecium is syncarpous with a single style, but the number of carpels is very variable. Each carpel contains one (rarely two) axile ovules with variable attachment (Ng, 1991). Pennington (2004) recognized five flower types in the family. The family has evolved different pollination syndromes related with an increase of the number of organs per whorl.

Sapotaceae were related with Lecythidaceae with weak support (Anderberg, Rydin and Källersjö, 2002), although Bremer *et al.* (2001) and Schönenberger, Anderberg and Sytsma (2005) linked the family with Primuloids. Based on floral morphology this is probably the best option.

Ebenaceae

Fig. 11.7. *Diospyros lotus* L.

$* K 4 [C(4) A(4+4)] \underline{G}(4)$
General formula: $* K3–5(-8) C3–5(-8) A(3-)12–20(-100) G2–8$

The family includes two genera, of which the large genus *Diospyros* is highly variable. Inflorescences are cymose and axillary. Flowers are mostly unisexual and heteromorphic, with a reduced pistillode in staminate flowers and a different merism with fewer staminodes in pistillate flowers. *Diospyros lotus* has bisexual flowers. The calyx is basally fused and accrescent; the corolla is sympetalous with reflexed lobes, and stamens are inserted in the lower part of the tube. Stamens are mostly diplostemonous, often with paired stamens in

Fig. 11.7. *Diospyros lotus* (Ebenaceae), inflorescence. Note the partially developed false septa.

antepetalous position, or stamens are in fascicles. The carpels have two ovules on apical placentae, generally with a false septum (Ng, 1991). There is a single short style or styles are free. A disc nectary is rarely developed (Wallnöfer, 2004).

'Primuloid clade'

The circumscription of the primuloid clade (previously Primulales) was revisited by Källersjö, Bergqvist and Anderberg (2000) and Anderberg, Rydin and Källersjö (2002) among others, who found that the traditional circumscription of Primulaceae was artificial, as Myrsinaceae appears to be nested within the family. They recognized Maesaceae, Theophrastaceae, Myrsinaceae, and a much smaller Primulaceae. These families represent a natural entity with strong morphological support and obvious synapomorphies. Main connecting characters are obhaplostemony (occasionally with a whorl of antesepalous staminodes, variously developed as petaloid appendages or as reduced stubs, sympetaly, and an ovary with free-central placentation (Anderberg and Ståhl, 1995). The development of a stamen-petal tube is highly comparable to other asterids. There is usually a lower tube common to stamens and petals and a fused corolla above the stamen insertion. The placentation is conspicuous as a broad central column with ovules in a single or several rows. The tip of the placenta is usually extended as a stalk penetrating in the common style. The stalk can be short and blunt (e.g. Myrsinaceae) or very long and filamentous (e.g. *Soldanella*: pers. obs.). The number of carpels making up the ovary is often difficult to verify, as no septa are present. However, the number of stylar lobes, fruit dehiscence patterns, ridges between the ovules, and vascular anatomy give indications of the number of carpels (Sattler, 1962; Ronse De Craene, Smets and Clinckemaillie, 1995). Apart from the primuloid clade, obhaplostemony is found in Sapotaceae as well. Filaments can be variously fused in a staminal tube or fused to the corolla, resulting from common zonal growth.

The division in families is reflected in the nectaries. Most Theophrastaceae have trichome nectaries (also found in *Aegiceras*, *Glaux* and *Lysimachia* of Myrsinaceae), which were interpreted as basal in primuloid taxa by Vogel (1997). Gynoecial nectaries are found in the basal Maesaceae (*Maesa*: Vogel, 1997), *Samolus* and Primulaceae (Vogel, 1998). Presence of stamen-petal primordia was interpreted as a derived condition by Sattler (1962) and Ronse De Craene, Clinckemaillie and Smets (1993). Floral evolution has proceeded independently in the different families, with several transitional genera (Källersjö, Bergqvist and Anderberg, 2000). The basalmost Maesaceae have bracteoles (absent in other primuloid families) but lack staminodes and have a semi-inferior ovary (Caris *et al.*, 2000).

Some Theophrastaceae (e.g. *Clavija*) and Myrsinaceae are dioecious, while heterostyly occurs mainly in Primulaceae.

Theophrastaceae (incl. Samolaceae)

Fig. 11.8. *Jacquinia macrocarpa* Cav.

∗ K5 [C(5)A5°]+(5) G̲(3)

Flowers are generally tetra- to pentamerous, arranged in cymose to racemose inflorescences. Sepals are lifted by a common hypanthium (Caris and Smets, 2004). All Theophrastaceae possess staminodes that arise with the petals on the margins of a ring primordium and become incorporated in the petal tube. Contrary to other primuloid families, stamens and petals do not arise from common primordia (except *Samolus*). Anthers are commonly extrorse (except *Samolus*: Källersjö, Bergqvist and Anderberg, 2000). Caris and Smets (2004) did not discuss carpel numbers as the gynoecium arises as a ring primordium. However, stylar lobes or fruit valves can be an indication of carpel number as two lobes were seen in *Deherainia*, three in *Jacquinia* and *Clavija*, and five valves in *Samolus*. A large number of ovules develop on the central placentation, except for *Clavija* with fewer ovules. Ovules are not embedded in placental tissue and the tip of the column is weakly developed.

Caris and Smets (2004) found the differences between *Samolus* and Theophrastaceae sufficiently important to separate *Samolus* as a separate family.

Fig. 11.8. *Jacquinia macrocarpa* (Theophrastaceae).

Primulaceae

Fig. 11.9A,B. *Soldanella villosa* Darracq

✳ K5 [C(5) A5°+5] G̲(5)

Flowers are grouped in cymose inflorescences, generally with a fused calyx and corolla. In *Soldanella*, sepal lobes are free and the corolla is divided in a variable but high number of segments. Petals are usually imbricate, although this is not visible in mature flowers of *Soldanella*. Obhaplostemony is the common androecial configuration, although staminodial antesepalous scales are present in some genera, including *Soldanella*. The scales stand opposite a separate vascular bundle in the corolla that was interpreted as remnant of the vascular supply of staminodes (e.g. Saunders, 1939). Scales on the corolla tube are present in *Androsace* and *Douglasia* but lack vascular tissue (Anderberg and Ståhl, 1995). The ovary has a basal nectary. The central placentation bears a high number of ovules that are not embedded in placental tissue.

Reflexed corolla lobes are present in *Dodecatheon*, resembling *Cyclamen* (Myrsinaceae) as an adaptation to a similar buzz-pollination mechanism (Källersjö, Bergqvist and Anderberg, 2000).

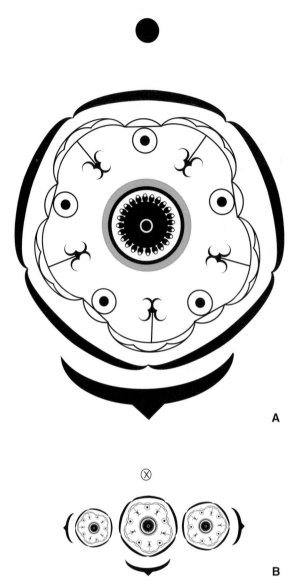

Fig. 11.9. *Soldanella villosa* (Primulaceae): A. flower; B. partial inflorescence. Note the formation of lobes on the fused corolla.

Clinckemaillie and Smets (1992) pointed to the high numbers of similarities between the floral diagrams of Plumbaginaceae and Primulaceae, which appears to be the result of homoplasy. The main difference is the basal placentation against the free-central placentation and location of nectary tissue, besides a different floral development.

Myrsinaceae (incl. Lysimachiaceae, Aegicerataceae, Coridaceae)

Fig. 11.10. *Anagalis arvensis* L.

Fig. 11.10. *Anagalis arvensis* (Myrsinaceae).

✳ K5 [C(5)A5] G̲(5)

General formula: ✳ (↓) K(3)4–5 C(3)4–5/0 A(3)4–5(+4°-5°) G5?

Flowers appear solitary in the axil of a single bract in *Anagalis*, although other Myrsinaceae have a racemose inflorescence. Flowers usually have a well-developed corolla (fused at the base), but it is absent in *Glaux*. Calyx and corolla are contorted or imbricate. In *Embelia* petals are exceptionally free (Anderberg and Ståhl, 1995). The filaments are hairy in *Anagalis* (trichome nectaries?). In *Lysimachia* glandular tissue consists of oil glands present on the corolla and stamens. Other members of Myrsinaceae have hanging flowers with poricidal anthers that are mainly buzz-pollinated. The ovules are generally embedded in the placental tissue without or with a short placental tip (e.g. *Lysimachia*, *Hymenandra*, *Myrsine*, *Embelia*: Caris, 1998; Ma and Saunders, 2003). Staminodes are generally absent (present in *Myrsine*). Stamens and petals arise on common primordia in all Myrsinaceae that were investigated (Ronse De Craene, Smets and Clinckemaillie, 1995; Caris, 1998; Ma and Saunders, 2003). *Coris* is unusual in Myrsinaceae in having an epicalyx and a monosymmetric corolla with imbricate aestivation (Ronse De Craene, Smets and Clinckemaillie, 1995).

Lecythidaceae

Fig. 11.11. *Napoleonaea vogelii* Hook. & Planch.

$\ast$K(5) [C(5)A80°+40°+5$^{2\circ}$+5^{2*}] $\breve{G}$(5)
*10 fertile half stamens in pairs and 10 staminodes in between
General formula: polysymmetric flowers: $\ast$ K4–5 C4–5(+4–5) A∞G(4–5); monosymmetric flowers ↑K6 C6 A∞ G(3)4–6

Fig. 11.11. *Napoleonaea vogelii* (Lecythidaceae). Curving of stamens over the nectary reflected by broken line connecting anthers to their insertion base. The pentagonal line delimits the stigma covering the anthers.

This large tropical family of mainly trees is highly diverse in floral structure. Flowers are usually large and merism tends to be variable, with four or six being the commonest number. Complexity is expressed mainly in the androecium, which is polyandric as a rule. Very often, the flower is strongly monosymmetric by the development of an abaxial staminodial hood covering the other stamens. The gynoecium is half-inferior to inferior as a rule, with axile placentation. Nectar is produced by an intrastaminal disc (e.g. *Napoleonaea*, *Barringtonia*) or occasionally from a modified part of the androecium, which acts as fodder or nectar for pollinators (several Lecythidoideae). Flowers of Lecythidaceae are adapted to a wide range of pollinators with specialized mechanisms mainly linked to the androecium (Tsou and Mori, 2007).

The androecium arises as a ring primordium with centrifugal initiation of a very high number of stamens. In polysymmetric genera such as *Gustavia* or *Barringtonia* (Endress, 1994; Tsou and Mori, 2007), stamens cover the ring primordium in an unordered mass. Monosymmetry is variously expressed in the family, including perianth, androecium and gynoecium to variable degrees. A flap of sterile tissue is detached from the anterior side of the ring primordium and grows over the fertile stamens in several genera. A diplostemonous androecium is found in *Cariniana micrantha*.

Napoleonaea, a West African genus at the base of the Lecythidaceae, is included as a representative of polysymmetric flowers in the family (Fig. 11.11). The androecium is highly complex with several series of thread-like or flap-like staminodes, surrounding ten fertile monothecal stamens inserted in five pairs and alternating with five pairs of staminodes. There are two outer whorls of staminodes, which progressively increase in number but decrease in size. The number of staminodes per whorl is fairly constant. Prance and Mori (2004) described the outer whorl as consisting of 60–70 strap-like appendages and the middle of 30–40 wider ones. They considered the 20 inner stamens to consist of five groups of four stamens, with the two outer fertile and the two inner sterile. My observations show that there are two whorls of paired stamens. The flower has often been interpreted as apetalous, and the petals as a corona or pseudocorolla, as no lobes are visible at maturity and the petals are little different from the staminodes in shape and colour (e.g. Endress, 1994). However, five petals are initiated and are fused into a crenellated rim, which is reflexed at anthesis (Ronse De Craene, unpubl. data). Petals closely resemble the staminodes in texture and colour. Therefore, I question the presence of a pseudocorolla in the closely related Scytopetalaceae (Appel, 1996). Floral developmental studies are necessary to elucidate this.

The stigma develops as a pentangular umbrella sheathing the anthers, which are bent in a hollow space between nectary and ovary. As the flower is hanging downwards, *Napoleonaea* probably shares the same pollination mechanism with Sarraceniaceae, another family in Ericales (Ronse De Craene, unpubl. data).

Theaceae

Fig. 11.12. *Stewartia pseudocamellia* Maxim., based on pers. obs., Erbar (1986) and Tsou (1998)

✳ K5 C5 A5$^\infty$ G(5)
General formula: ✳ K(4)5(–14) C(4)5–7 A5$^\infty$-∞ G(2)3–5(–10)

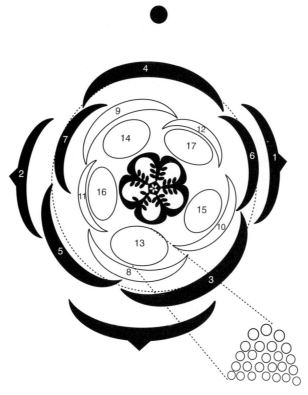

Fig. 11.12. *Stewartia pseudocamellia* (Theaceae). Numbers give order of initiation from bracteoles onwards; common stamen primordia divide centrifugally in many stamens (shown for stamen group labelled 13).

Theaceae was considered as a major family of Theales (Cronquist, 1981), an amalgamation of families now often referred to Malpighiales. Five subfamilies used to be recognized, but Theaceae is now restricted to three tribes, while Bonnetiaceae is placed in Malpighiales, and Tetrameristaceae and Ternstroemioideae are placed in other clades of Ericales (Stevens, 2001 onwards).

Flowers are solitary or grouped in axillary cymes. Tsou (1998) recognized two major groups on the basis of perianth phyllotaxis and differentiation. The first group, including *Camellia*, has a variable number of large sepals (10–14) and petals (five to seven) arising in a spiral sequence, with bracts, sepals and petals intergrading into each other (e.g. Sugiyama, 1991; Tsou, 1998). The multistaminate androecium arises centrifugally on a ring primordium and stamens are often basally connected into a tube, as are the petals (Payer, 1857; Tsou, 1998). The second group, where *Stewartia* belongs, is pentamerous with a better distinction between sepals and petals (with spiral initiation) and stamens are grouped in antepetalous fascicles (Fig. 11.12; Erbar, 1986; Tsou, 1998). The distinction between bracteoles, sepals and petals is unclear, with all organs arising in a spiral sequence. Stamens develop centrifugally on the five stamen fascicles. The gynoecium is tri- to five-carpellate with basal-axile placentation. The invaginating septa are weakly fused in the centre, at least for part of the genera. Styles are distinct and carinal. The development of the flower is accompanied by a central invagination leading to a half-inferior ovary. Based on the floral anatomy, Sugiyama (1991) suggested that the multistaminate androecium of *Camellia* is superimposed on an obdiplostemonous Bauplan, although this is not clear from the floral development. Complex obhaplostemonous flowers such as *Stewartia* could be derived by loss of stamens in antesepalous sectors.

Pentaphyllaceae (incl. Ternstroemiaceae)

Fig. 11.13. *Cleyera japonica* Thunb.

Fig. 11.13. *Cleyera japonica* (Pentaphyllaceae). Stamens develop laterally from common primordia (1); bracteoles with terminal gland.

✴ K5 C5 A5$^{5-7}$ $\underline{G}$(2–3)
General formula: K5 C5 A(5)10-∞ G(1)2–3

The family shares several morphological similarities with Theaceae, although it is placed in a different clade and is little known morphologically (Anderberg, Rydin and Källersjö, 2002). Flowers are single in the axils of leaf-like bracts. Sepals and petals arise in a spiral sequence, in continuation of two bracteoles, and are imbricate in bud. In some genera, petals and sepals are superposed, as petals continue the 2/5 sequence of the sepals (Ternstroemieae). The number of corolla lobes may vary with up to ten petals. In *Cleyera*, the first petal arises opposite sepal one, but becomes shifted in alternate position due to variable growth. Stamens are uniseriate in *Cleyera* but they multiply laterally from five antesepalous primordia (Ronse De Craene, unpubl. data). In some genera, flowers are unisexual with a non-functional or absent ovary in staminate flowers, and staminodes without anthers in pistillate flowers. An intrastaminal disc is present in *Cleyera*, but is absent in other genera such as *Ternstroemia*. *Pentaphylax* has only five stamens in antesepalous position. *Visnea* was reported as having obdiplostemonous flowers, although the floral development shows a succession of three pentamerous whorls with larger antesepalous stamens and pairs of antepetalous stamens (Payer, 1857). The ovary is superior and carpels initiate sequentially, often with one carpel conspicuously smaller. Placentation is axile and there is a single style, topped with carinal stigmatic lobes.

Ericaceae (incl. Epacridaceae, Pyrolaceae, Empetraceae, Monotropaceae, Vacciniaceae)

Fig. 11.14A. *Rhododendron tolmachevii* Harmaja

↓ K5 C(5)A5+5 G(5)

Fig. 11.14B. *Macleania stricta* A.C. Sm.

✳ K5 C(5) A5+5 Ğ(5)

General formula: ✳ (↓) K4–5 C(3)4–5 A5–10 G(3)4–5(10)

The circumscription of the family has considerably changed by the inclusion of several smaller families (e.g. Kron *et al.*, 2002). Flowers are subtended by a bract and two bracteoles, which may be reduced or absent. Flowers show a relative stability being pendulous and campanulate, with fused petals (free in *Pyrola*), two stamen whorls with (ob)diplostemonous arrangement, an intrastaminal disc nectary, and isomerous (five to four) carpels in a superior to inferior ovary. Ovaries contain a high number of small ovules on centrally protruding placentae. There is a single style with commissural stigmatic lobes. In *Gaylussacia* carpel number is increased to ten by the development of false septa (Eichler, 1875). Loss of antepetalous stamens leads to haplostemony in some *Rhododendron*. Anthers often bear appendages and are inversed with poricidal dehiscence. Flowers are occasionally slightly monosymmetric (e.g. *Rhododendron*), or strongly monosymmetric (e.g. *Elliotia*). The combination of a reduction of bracteoles and smaller sepals places the lateral petals outside the other petals in bud; flowers are

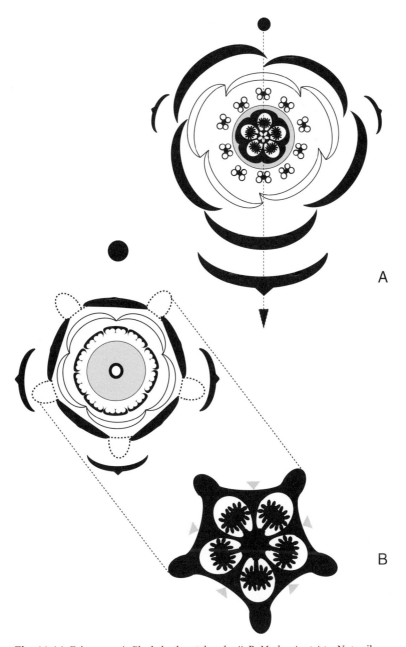

Fig. 11.14. Ericaceae: A. *Rhododendron tolmachevii*; B. *Macleania stricta*. Note ribs protruding from receptacle.

slightly to strongly monosymmetric with a difference in size between adaxial and abaxial petals and stamens, resembling Proteaceae (pers. obs.). In *Tripetaleia* the flower is partly trimerous by the progressive fusion of two of the petals in pairs and trimery of androecium and gynoecium (Nishino, 1988).

Anthers are often inverted with the morphological lower part on top. There are often appendages or awns, which are associated with a poricidal dehiscence common to the family. The obdiplostemonous arrangement of stamens is caused by a displacement during development (e.g. Leins, 1964a).

Rhododendron and *Enkianthus* have an unusual floral diagram in that the odd petal is not abaxial but adaxial (cf. Eichler, 1875).

11.2   Lamiids: Solanales, Gentianales, Lamiales

Figure 11.15 shows the phylogenetic tree of lamiids (euasterids I) based on Olmstead *et al.* (2001) and Judd *et al.* (2002). The clade is well supported and

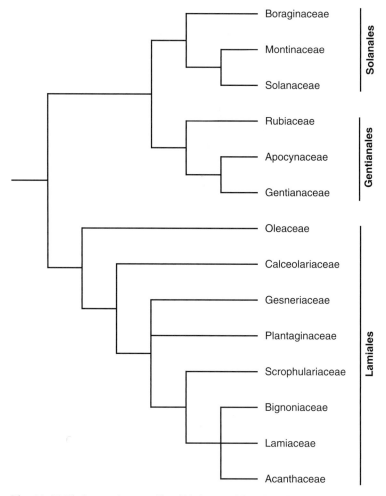

Fig. 11.15. Phylogenetic tree of lamiids (euasterids I), based on Olmstead *et al.* (2001) and Judd *et al.* (2002).

consists of three orders, Solanales, Gentianales and Lamiales. Apart from Lamiales, relationships of families are fairly well resolved (Stevens, 2001 onwards). Bremer *et al.* (2001) list the following synapomorphies for the lamiids: hypogynous flowers, late sympetaly, fusion of stamen filaments with the corolla tube, capsular fruits, opposite leaves and entire leaf margin. Lamiales have undergone an evolution to median monosymmetry, while the other orders are generally polysymmetric. Flowers are haplostemonous (exceptionally with higher numbers: Hoplestigmataceae, *Dialypetalanthus* and *Theligonum* in Rubiaceae) and an isomerous or bicarpellate gynoecium. A nectary is generally developed as a disc around or on top of the ovary.

Solanales

The order consists of five families (Solanaceae, Convolvulaceae, Hydroleaceae, Montiniaceae, Sphenocleaceae). The position of Boraginaceae is uncertain (maybe in Lamiales) with which it occasionally shares nuts with false septa. However, Lamiales are mostly monosymmetric (or polysymmetry is derived).

Synapomorphies are generally poor (e.g. radial symmetry, a plicate corolla tube, stamen number equaling the petals, persistent calyx: Judd and Olmstead, 2004). All Solanales have late sympetaly, which is correlated with the occurrence of superior ovaries. Although there is a tendency in some families for a secondary loss of petal tubes (e.g. Montiniaceae: Ronse De Craene, Linder and Smets, 2000), petal tubes are often elongated with short petal lobes or are trumpet-shaped. Ovaries are generally superior and consist of two carpels (occasionally more through secondary increase: *Lennoa*).

A possible synapomorphy for Boraginaceae and Solanaceae is oblique monosymmetry, which is occasionally expressed in both families (see diagrams in Stützel, 2006).

Boraginaceae (incl. Hydrophyllaceae, Lennoaceae)

Fig. 11.16A,B. *Echium hierrense* Webb ex Bolle

↖K(5) [C(5) A5] G(2)
General formula: ✳ (↖) K(4)5(-8) C(4)5(-8) A(4)5(-8) G2

Boraginaceae is a large family, consisting of four subfamilies that are sometimes recognized as families on their own.

The flowers are characteristically arranged in scorpioid cymes with a displaced bract relative to the flower. Flowers are mostly regular and pentamerous (up to eight-merous in some *Cordia*), occasionally monosymmetric with four staminodes (*Caccinia*: Eichler, 1875). Sepals are imbricate or valvate. Aestivation of the corolla is imbricate (in 2/5 arrangement: e.g. *Bourreria*), contorted (e.g. *Cordia*) or cochleate

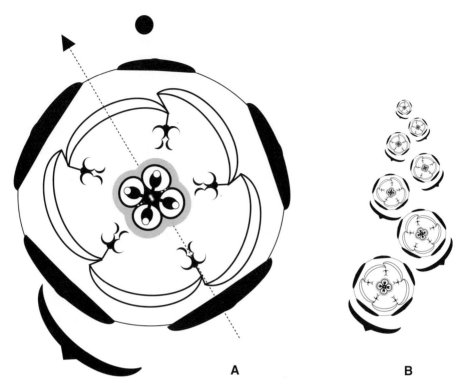

Fig. 11.16. *Echium hierrense* (Boraginaceae): A. flower; B. partial inflorescence. Broken line in the ovary shows the false septum.

ascending (e.g. *Echium*). The corolla tube sometimes produces invaginations that have occasionally been interpreted as staminodes. The gynoecium is superior and surrounded by a disc nectary. The carpels range from one (*Rochelia*, pseudomonomerous: Hilger, 1984) to two, or occasionally many more. A false septum develops in the same way as in Lamiaceae by invagination of a partition leading to four locules (e.g. Baillon, 1862; Gottschling, 2004; p. 331). The ovules are two per carpel with basal placentation.

Echium is slightly monosymmetric through the size difference of sepals and stamens and oblique orientation of the style. Eichler (1875) and Leredde (1955) reported an unequal length of the stamens in the genus with a displacement to the upper part of the flower.

Solanaceae

Fig. 11.17. *Cestrum parqui* L'Hérit.

✳ K(5)* [C(5) A5] G(2)
*Presence of long abaxial slit
General formula: ✳ (↓) K5 C5 A5 G2(5)

Inflorescences are cymose and flowers generally have bracts and bracteoles (tendency for reduction in *Cestrum*). Flowers of Solanaceae are generally

Fig. 11.17. *Cestrum parqui* (Solanaceae). Note the valvate petals with involute margins.

polysymmetric except for the oblique gynoecium, more rarely obliquely mono-symmetric with the symmetry line running through sepal one (e.g. *Schizanthus*, *Salpiglossis*: Eichler, 1875; Knapp, 2002). Knapp (2002) argued that monosymmetry has arisen several times in the family, with great variety of monosymmetric forms affecting several or a single organ. In *Salpiglossis* and related genera both corolla and androecium are monosymmetric with the abaxial stamen sterile in *Browallia*. Petal tubes are often very long with shorter lobes, or short with reflexed lobes (*Solanum*, linked with buzz pollination). Aestivation is variable, ranging from quincuncial to contorted or valvate. Stamen number is generally five (haplostemony). Stamens can be unequal in length with one shorter stamen (e.g. *Nicotiana*), or stamens of variable length (the one opposite sepal one shortest, the two neighbouring longest, and the two posterior intermediate in size: e.g. *Petunia*, *Physalis*); in *Schizanthus*, only the stamens alternating with the carpels are fertile (Eichler, 1875; Knapp, 2002). The gynoecium is characteristically inserted obliquely, mostly of two carpels, rarely five (*Nicandra*). Placentation is axile with many ovules. The single style has an undifferentiated stigma, or if differentiated, stigmatic lobes are carinal. A nectary is generally developed at the base of the ovary, except in the buzz-pollinated species. Several floral diagrams are shown in Knapp (2002).

Gentianales

The order Gentianales is a well-supported clade, which consists of five families, Rubiaceae, Gentianaceae, Loganicaceae, Gelsemiaceae and Apocynaceae (Backlund, Oxelman and Bremer, 2000).

Characters in common include interpetiolar stipules (with colleters), sympetaly with usually contorted corollas (rarely imbricate or valvate), a scarcity of (slight) zygomorphy and unisexual flowers, near absence of wind pollination, and plasticity in the gynoecium (Nicholas and Baijnath, 1994; Backlund, Oxelman and Bremer, 2000).

Rubiaceae (incl. Dialypetalanthaceae, Theligonaceae)

Fig. 11.18A,B. *Rondeletia laniflora* Benth.

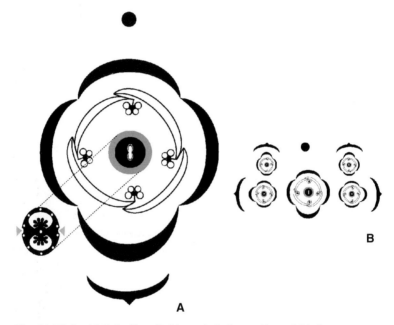

Fig. 11.18. *Rondeletia laniflora* (Rubiaceae): A. flower; B. partial inflorescence.

✳* K(4) [C(4) A4] Ǧ(2)
*weakly monosymmetric
General formula: ✳ K0/(3)4–5(-15) C0/(3)4–5(-15) A(3)4–5(-∞) G2–5(-9)

Rubiaceae differ from other Gentianales in early sympetaly linked to an almost ubiquitous inferior ovary. A secondarily superior ovary has arisen occasionally and in different ways, in *Mitrasacmopsis* by development of a prominent beak (Groeninckx *et al.*, 2007), or in *Gaertnera* by differential zonal growth within the ovary tissue (Igersheim *et al.*, 1994).

Inflorescences are generally cymose and flowers are subtended by a bract and two bracteoles. Inflorescences occasionally develop as pseudanthia with petaloid bracts (e.g. *Geophila*: Robbrecht, 1988). Flowers may occasionally fuse by their ovaries and calyces (e.g. *Mitchella* with paired flowers). Flowers are generally polysymmetric, rarely monosymmetric by unequal development of the calyx or corolla. In *Mussaenda*, one calyx lobe is extensively developed. Merism is mostly four to five, although it ranges from three to 15. Flowers are bisexual, except for several genera, which have evolved wind pollination: flowers are reduced and strongly heteromorphic (e.g. *Coprosma*, *Theligonum*, *Galopina*: Robbrecht, 1988; Rutishauser *et al.*, 1998; Ronse De Craene and Smets, 2000). The calyx is either well developed or reduced to a weak rim (e.g. *Galium*, *Galopina*). The corolla is generally long and tubular with short lobes and variable aestivation, rarely distinct (e.g. *Dialypetalanthus*, *Mastixiodendron*). Some Australian *Spermacoce* have elaborate petal appendages (Vaes *et al.*, 2006). These appear to have evolved in relation to specific pollination mechanisms. Petal tubes or lobes are often hairy on the inside (e.g. *Paederia*: Puff and Igersheim, 1991). Stamens are equal in number and opposite sepals; they are rarely numerous (*Dialypetalanthus*, *Theligonum*). The insertion of anthers in the tube is variable, either at the corolla mouth, or at its base, more rarely at different levels of the tube (e.g. *Paederia*). The inferior ovary is topped by an epigynous disc nectary surrounding a single style. Two carpels occur in the majority of taxa, although five carpels are found in several genera (e.g. *Hamelia*; rarely up to 16 in *Praravinia*: Robbrecht, 1988). Placentation is generally axile to basal. The gynoecium develops in two parts, a basal part producing the placenta and an apical part developing as a septum (e.g. Svoma, 1991; Ronse De Craene and Smets, 2000). Ovules range from several on protruding placentae to a single basal ovule. The ovary of *Theligonum* is pseudomonomerous with a single basal ovule curving over a low septum and filling the empty locule (Rutishauser *et al.*, 1998).

Gentianaceae

Fig. 11.19. *Exacum affine* Balf. f. ex Regel

✳ K5 [C(5) A5] G̲(2)

Clearly defined synapomorphies do not exist for Gentianaceae (Albert and Struwe, 2002), although diagnostic characters fit well with the tribes.

Loss of bracteoles leads to displacement of the two outer sepals in lateral position (Eichler, 1875). In *Erythraea*, the flower is inversed with sepal two abaxially. Flowers are generally penta- (tetra-)merous, rarely six- to 12-merous (*Blackstonia*). Petals as well as sepals are usually strongly contorted. There is much variation in the extent of development of a petal tube and a basal corona

Fig. 11.19. *Exacum affine* (Gentianaceae): partial inflorescence. Note dorsal appendages on sepal lobes, giving a contorted appearance to the calyx. Oblique orientation is caused by curvature of style and stamens.

is occasionally present (e.g. *Symbolanthus*: Struwe *et al.*, 2002). In *Exacum* (and most Exaceae), the calyx lobes have a prominent keel. The androecium is haplostemon-ous with equally long stamens, or these are rarely unequal (e.g. *Lisianthus*). Anthers of *Exacum* are poricidal, suggesting buzz pollination, but pollination syndromes are manifold in the family with an absence of wind pollination. Flowers of *Exacum* and *Orphium* are weakly monosymmetric by orientation of style and anthers. This is caused by the inversed position of flowers on the inflorescence linked with curving of the pedicel. Flowers of some genera such as *Canscora* have median monosymmetry with an unequal insertion of heteromorphic stamens (Struwe *et al.*, 2002). Nectaries either develop as a disc at the base of the ovary or as separate glands on the inner surface of the corolla tube (Lindsey, 1940; Struwe *et al.*, 2002). The petal tube of *Halenia* bears spurs with nectaries. Placentation is either axile or parietal. The ovary is occasionally semi-unilocular because the placenta is not developed distally (Struwe *et al.*, 2002).

Apocynaceae (incl. Asclepiadaceae)

Fig. 11.20A. *Nerium oleander* L.
Fig. 11.20B. *Periploca graeca* L.

$\ast$ K(5) [C(5) A5] $\underline{G}$(2)

Although Asclepiadaceae were considered a separate group, recent research has shown that they represent a level of complexity derived within the Apocynaceae

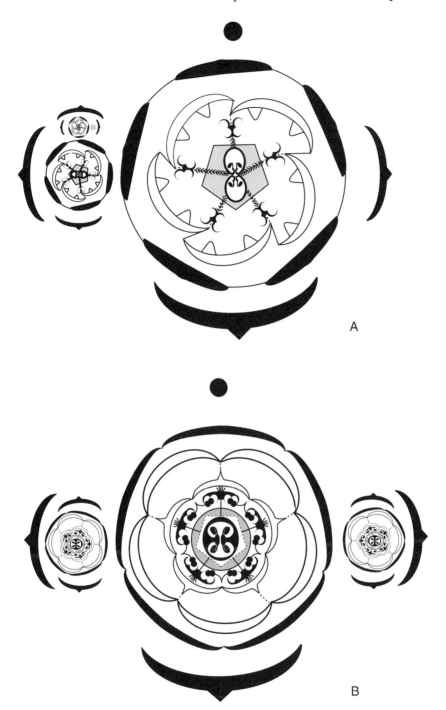

Fig. 11.20. Apocynaceae: A. *Nerium oleander*: partial inflorescence; note the connectives twining around the style; B. *Periploca graeca*: inflorescence.

(Endress *et al.*, 1996). A synapomorphy is the distinct carpels connected by their styles and an expanded apical portion of the style (Judd and Olmstead, 2004).

The floral Bauplan is generally well conserved within the family, contrary to other Gentianales. Floral adaptations to diverse pollinators are highly diverse and are superimposed on the highly synorganized flowers (Endress, 1994). While the flower is pentamerous, the ovary is bicarpellate and superior (very rarely up to eight). The corolla has variable aestivation, but is often contorted as are Gentianaceae, hence the old name 'Contortae' used for Gentianales. The corolla may bear a prominent corona. In *Asclepias* or *Periploca* this becomes a prominent petaloid organ that functions as a nectar recipient (Fig. 11.20B; Endress, 1994). Stamens alternate with petals and have a short filament, which can be distinct or fused into a tube around the ovary. In Asclepioideae the style develops as a pentagonal shield, which is connected postgenitally with the anthers into a gynostegium. In *Periploca* and other Asclepioideae, each theca (with pollen sacs fused into a pollinium) is connected with that of a neighbouring anther through a clasp-like organ (translator). A visiting insect will pull out the pollinia. In *Nerium* the connectives are extended in long plumose appendages curling around the style (Fig. 11.20A).

The ovary is surrounded by a disc nectary or two nectaries alternating with the carpels (Bernardello, 2007). Placentation is intruding parietal or axile, mostly with numerous ovules. In some genera the ovary becomes secondarily apocarpous (Endress, Jenny and Fallen, 1983), as is generally the case for the fruits. A floral diagram of *Asclepias* was presented in Endress (1994).

Lamiales

The order consists of 21 families and almost 18 000 species. The internal relationships of Lamiales are not well resolved, but families such as Oleaceae, Calceolariaceae and Gesneriaceae occupy a more basal position in a grade. The differentiation of families is very difficult to make on a morphological basis, as major distinctive characters sweep through the order and the relationships of families remain largely unresolved (Stevens, 2001 onwards). Reeves and Olmstead (1998) and Olmstead *et al.* (2001) recognized that the large family Scrophulariaceae is polyphyletic and proposed its subdivision between at least eight lineages in a strongly altered Lamiales. However, the lack of any clear morphological synapomorphies to differentiate large clades, such as Plantaginaceae, from Scrophulariaceae remains an issue. The same problem that existed for Lamiaceae and Verbenaceae was partly solved by merging most genera of Verbenaceae in Lamiaceae (e.g. Cantino, 1992). A more global family approach, with well-defined subfamilies, such as for Rubiaceae or Malvaceae, appears to be more appropriate.

Figure 11.21 shows the major steps in floral evolution of Lamiales. The calyx is generally well developed and often basally fused. A characteristic feature of

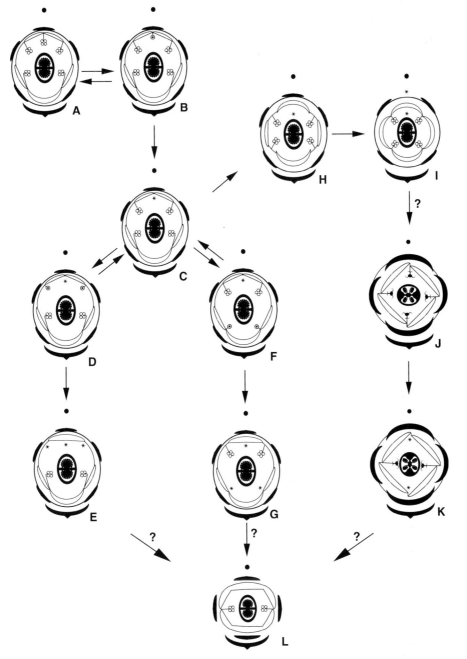

Fig. 11.21. Floral evolution in Lamiales. A. weakly monosymmetric flower with five stamens; B. monosymmetric flower with adaxial staminode; C. adaxial staminode is lost; D. sterilization of lateral adaxial stamens; E. loss of adaxial stamens and fusion of posterior petals; F. sterilization of lateral abaxial stamens; G. loss of abaxial stamens and fusion of posterior petals; H. fusion of posterior petals in a single structure; I. tetramerous flower with loss of adaxial sepal; J. tetramerous flower with four stamens; K. Oleaceae with loss of median stamens and reorientation of the ovary; L. flower of Calceolariaceae with two stamens.

the order is the presence of a monosymmetric bilabiate corolla mostly constructed on a 2/3 pattern, and a superior bicarpellate ovary. The androecium varies from one to five stamens, with various reductions and losses occurring in a median plane of the flower. The basic floral formula for the order is K(5) [C(5) A1–5] G̲(2).

Although median monosymmetry is the most common feature, there are several reversals to polysymmetry (see p. 328; Fig. 11.21H–K; e.g. Endress, 1998; Soltis *et al.*, 2005: 235). These flowers are tetramerous and have probably been derived by the fusion of the two posterior petals after the loss of the posterior stamen (e.g. Ronse De Craene and Smets, 1994; Donoghue, Ree and Baum, 1998; Endress, 1999; Bello *et al.*, 2004), although more studies should indicate whether this is the only process. Calceolariaceae and Oleaceae are more basal to other Lamiales, although their floral structure is clearly derived from more elaborate precursors with tetramerous flowers (Fig. 11.21K,L; Mayr and Weber, 2006).

The corolla is initiated at the same time or after the stamens (late sympetaly, cf Erbar, 1991); this development tends to be common for all Lamiales and is linked with a superior ovary. The reduction of the androecium generally starts with the adaxial stamen, which is staminodial or disappears (Fig. 11.21B,C). The four fertile stamens are arranged at two levels and are often connected by their anthers in pairs (didynamy). Further reductions lead to two stamens, either in latero-adaxial or in latero-abaxial position (Fig. 11.21D–G). Walker-Larsen and Harder (2000) argued that the tendency for partial loss of stamens in monosymmetric Lamiales is a reversible process. They concluded that some Lamiales have become secondarily polysymmetric and that this is linked with the restoration of the lost stamen (e.g. *Verbena*, *Verbascum*, *Oroxylum*, as these taxa occur in derived clades). Endress (1998) gave an overview of taxa where polysymmetric flowers occur. He argued that the limited extent of reduction of the staminode makes a reversal possible, as in some basal clades of Lamiales which have retained five stamens, or where the odd staminode is well developed. Gesneriaceae have the largest proportion of polysymmetric flowers with five stamens (Endress, 1998). It is the only family of Lamiales where the odd stamen is not missing from a particular genus (except for an odd species). In Bignoniaceae, the odd staminode is mostly present and often larger than the stamens (only absent in *Tourettia*: Endress, 1992, 1998). The possibility of a reversed process of staminode fertility should be critically investigated in the order.

Flower development has been investigated in several representative species, indicating an early onset of monosymmetry (e.g. Endress, 1999). Depending on the direction of development, aestivation of the corolla is highly variable within a family (e.g. Plantaginaceae, Acanthaceae: Armstrong and Douglas, 1989; Scotland, Endress and Lawrence, 1994).

Oleaceae

Fig. 11.22. *Osmanthus delavayi* Franch.

Fig. 11.22. *Osmanthus delavayi* (Oleaceae).

✱ K(4) [C(4) A2] G̲(2)

Inflorescences are cymose. In contrast to most other Lamiales, flowers are polysymmetric and built on a tetramerous merism, although the origin of tetramery is unclear. Merism is rarely higher in some *Jasminum*. Most taxa have two stamens in transversal position. It is implied that these were derived from a tetramerous condition as is found in some *Chionanthus* and a few other Oleaceae. Sepals are usually four, although occasionally five (as on Fig. 11.22, with two smaller lateral sepals). The superior ovary has a single style with carinal stigmatic lobes. Placentation is axile with two ovules per carpel. A nectary if present develops at the base of the ovary (gynoecial nectary). In the wind-pollinated species of the genus *Fraxinus*, the perianth is reduced or absent and flowers are unisexual (Eichler, 1875).

Calceolariaceae

Fig. 11.23. *Jovellana violacea* (Cav.) G. Don

↓ K(4)[C(2)A(2)] G̲(2)

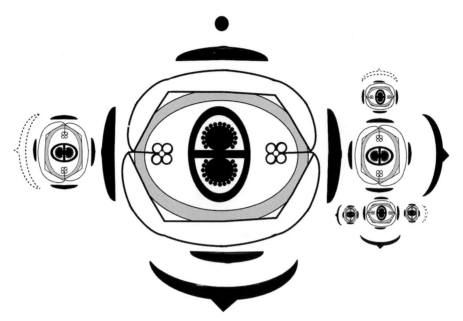

Fig. 11.23. *Jovellana violacea* (Calceolariaceae): partial inflorescence.

Flowers are arranged in compound dichasia. There are four sepals in median and transversal position and only two broad median petals alternating with two stamens. The ovary is bicarpellate with numerous ovules on axile placentation. Mayr and Weber (2006) have clearly demonstrated that the dimerous perianth is derived from a tetramerous condition, as in Oleaceae, and not from pentamery. In Veroniceae a similar reduction to tetramery is linked with loss of the median adaxial sepal and fusion of two petals in a single structure (see p. 328). There is no evidence for this in Calceolariaceae. An odd third stamen develops in an adaxial position in *Stemotria triandra* or as a teratology in *Calceolaria* and is linked with a bilobed lower lip. In *Jovellana*, the lower petal lip arises as a bilobed structure. Together with the presence of two lateral vascular bundles in the lower petal, this is evidence that the lower petal represents a composite structure linking Calceolariaceae to Oleaceae with four lateral petals (Mayr and Weber, 2006).

Plantaginaceae (incl. Callitrichaceae, Globulariaceae *p.p.*, Scrophulariaceae *p.p.*, Hippuridaceae)

Fig. 11.24A,B. *Penstemon fruticosus* (Pursch.) Greene var. *scouleri* (Lindl.) Cronq.

↓ K(5)[C(5)A4:1°] G̲(2)

Fig. 11.24C,D. *Hebe cheesemanii* (Buchanan) Cockayne & Allan

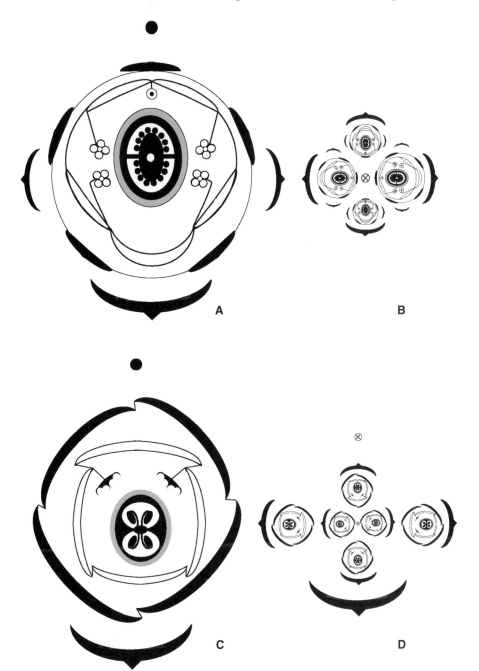

Fig. 11.24. Plantaginaceae: *Penstemon fruticosus* var. *scouleri*, A. flower, B. partial inflorescence; *Hebe cheesemanii*, C. flower; D. partial inflorescence.

↓ K(4)[C(4)A2] G̲(2)
General formula: ↓(✳) K(0)4–5 C(0)4–5A(1)2–4 (:1°) G(1)2

As mentioned on p. 322, it remains problematic to distinguish different families within the Scrophulariaceae–Plantaginaceae complex. Orobanchaceae appears more easily distinguished by the combination of a parasitic habit and constant absence of an odd staminode (Endress, 1998).

Secondarily polysymmetric flowers are clearly derived from monosymmetric flowers in *Aragoa* and some other Plantaginaceae (e.g. Endress, 1999; Bello *et al.* 2004). The calyx and corolla arise unidirectionally from the abaxial to the adaxial side (Bello *et al.*, 2004). In *Aragoa* and some *Veronica*, a small fifth adaxial sepal is present, while it is lost in other species of *Veronica, Hebe* and *Plantago*. Sepals and petals are completely lost in *Callitriche* and much reduced in *Hippuris* (Leins and Erbar, 1988).

The androecium fluctuates between five (always with an abaxial staminode: *Antirrhinum, Digitalis*: Walker-Larsen and Harder, 2000), four and two stamens; when two in number stamens are in a latero-abaxial position (e.g. *Veronica, Hebe*). In *Gratiola*, fertile stamens are latero-adaxial with two latero-abaxial staminodes (Bello *et al.*, 2004).

The bicarpellate ovary is relatively uniform as for most Lamiales. In a few genera (e.g. *Gratiola, Hippuris*) it is unilocular or tetralocular by the development of false septa (*Callitriche*).

Flowers of some Plantaginaceae tend to combine nectar secretion (as a disc surrounding the ovary) with oil secretion on the corolla (e.g. *Monttea*: Sérsic and Cocucci, 1999). This is comparable to *Diascia* in Scrophulariaceae (see Fig. 11.25 below).

Studies of Hufford (1992b, 1995) on *Besseya* and *Synthyris* in tribe Veroniceae illustrated the divergence of evolutionary patterns in Plantaginaceae. The number of calyx lobes fluctuates between two and five through connation and reduction of lobes. There is a tendency for the abaxial sepals to become reduced or vanish. A similar pattern affects the corolla where a stamen-corolla tube tends to become shorter or fails to develop at all in some species of *Besseya*. The adaxial petal lobes become fused into a single adaxial lip separate from a trilobed abaxial lip.

Scrophulariaceae (incl. Buddlejaceae, Myoporaceae, Selaginaceae, Oftiaceae)

Fig. 11.25. *Diascia vigilis* Hilliard & B. L. Burtt

↓ K5[C(5)A4] G̲(2)
General formula: ↓(✳) K(2)4–5(-8) C(0)4–5(-8) A2–4–5(-8) G2

Fig. 11.25. *Diascia vigilis* (Scrophulariaceae). Elaiophores shown as openings in the stamen-petal tube, with underlying spores.

Scrophulariaceae were considerably reduced compared with earlier circumscriptions of the family. Weakly monosymmetric flowers are found in *Verbascum*, with the occasional presence of an adaxial fertile stamen. Most other members of the family are two-lipped with four or less fertile stamens and 2:3 is the most common petal arrangement (exc. *Mutisia*: 4:1).

The abaxial staminode is elaborate in *Scrophularia* (reflected in *Penstemon* of Plantaginaceae: Endress, 1998). Flowers of *Diascia* have two petal spurs containing oils and no nectaries (Fig. 11.25). The bicarpellate ovary resembles Plantaginaceae to a great extent. The tetramerous and polysymmetric flowers of *Buddleja* are probably derived in a similar way as *Plantago* (see Bello *et al.*, 2004).

Armstrong (1985) used floral anatomical characters to distinguish Scrophulariaceae from Bignoniaceae. Scrophulariaceae have a simple axile placentation and only two large dorsal bundles in the ovary, while Bignoniaceae consistently have two to four distinct placental ridges and four large bundles, two opposite the septum and two dorsals. This character should be studied further in the context of renewed phylogenetic relationships.

Acanthaceae

Fig. 11.26A–C. *Odontonema strictum* Kuntze

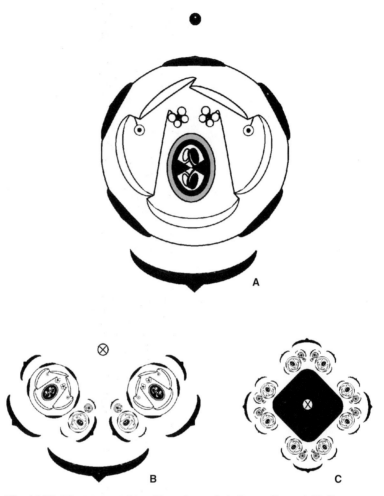

Fig. 11.26. *Odontonema strictum* (Acanthaceae): A. flower; B. partial inflorescence; C. view of partial inflorescences at level of node.

↓ K(5) [C(5) A2:2°] G̲(2)
General formula: ↓(✳) K(0)4–5 C4–5 A2–5 G2

The family is characterized by the presence of showy bracts and flowers arranged in decussate terminal inflorescences.

The calyx is generally well developed and basally fused. It is strongly reduced or absent in *Thunbergia*, where attraction and protection is transferred to bracts (Rao, 1953; Schönenberger and Endress, 1998; Endress, 2008b). The calyx of *Thunbergia* can develop as a lobed rim resembling the calyx lobes in some

Caprifoliaceae. The adaxial sepal is occasionally reduced or two sepal lobes are laterally fused. Flowers are rarely regular, more often monosymmetric. The monosymmetric flowers have a long petal tube with elaborate lower lip. Aestivation of the corolla tends to be highly variable (Scotland, Endress and Lawrence, 1994). Petals are generally arranged as 2:3, rarely 0:5 (e.g. *Sclerochiton*). In a few genera belonging to Strobilanthinae, flowers are resupinate through twisting of the bud and petals are arranged as 3:2 (Moylan, Rudall and Scotland (2004). The posterior staminode is often absent, although it can be initiated and pervade in the vasculature (Rao, 1953), and the androecium has four fertile didynamous stamens (e.g. *Acanthus*, *Thunbergia*), two stamens with two staminodes (Fig. 11.26, *Sanchezia*), or two latero-adaxial stamens (e.g. *Nelsonia*). An odd staminode is occasionally present and very rarely five stamens (Endress, 1998). The ovary is superior, surrounded by a disc nectary, with relatively few ovules compared with Plantaginaceae. *Mendoncia* has a pseudomonomerous ovary (Schönenberger and Endress, 1998). There is a single style with funnelform stigma or with two stigmatic lobes (occasionally reduced to one).

In *Hemigraphis* and other Strobilantheae, an internal partition formed by an extension of the stamen-petal tube ('filament curtain') divides the internal flower in two compartments and leads pollinators to an abaxial chamber containing nectar (Moylan, Rudall and Scotland, 2004).

Lamiaceae

Fig. 11.27A,B. *Westringia fruticosa* (Willd.) Druce
Fig. 11.27C. *Clerodendrum petasites* (Lour.) A. Meeuse

↓ K(5) [C(5) A4] G̲(2)

Lamiaceae (or Labiatae) used to be easily recognized by a combination of square stems with opposite leaves, bilabiate flowers with four stamens, a bicarpellate gynoecium with gynobasic style, and a fruit with four nutlets. The circumscription of the family has become much extended following molecular systematics, as several genera of Verbenaceae were displaced to Lamiaceae (Cantino, 1992). This makes an identification of Verbenaceae flowers easier, but adds complexity to the circumscription of Lamiaceae. A calyx is well developed and variously fused. The corolla is monosymmetric and often bilabiate (Fig. 11.27A), or less clearly so (*Clerodendrum*) with cochleate descending aestivation. The orientation of the lobes is mostly 2:3 (see also Donoghue, Ree and Baum, 1998), more rarely 1:4 (e.g. *Perovskia*), or 0:5 (e.g. *Teucrium*). The corolla tube is very long in *Clerodendrum* with imbricate lobes; the adaxial corolla lobe appears to be the upper one because of hanging flowers. Stamens are mostly four in number and didynamous, and there is never a trace of an abaxial staminode. In *Salvia*, there is further reduction

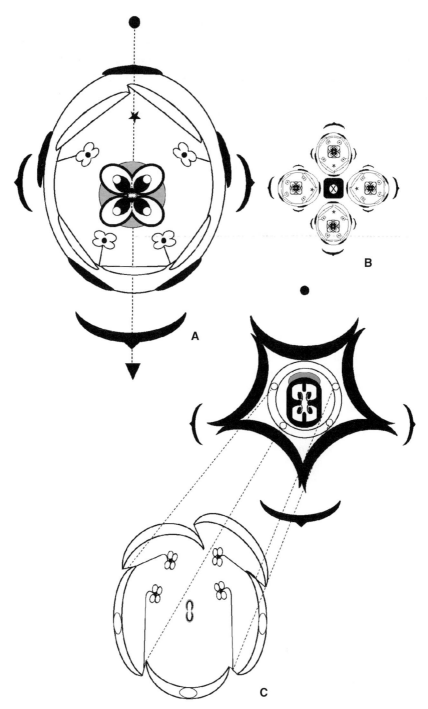

Fig. 11.27. Lamiaceae: *Westringia fruticosa*, A. flower, B. partial inflorescence; *Clerodendron petasites*, C. flower shown at lower and upper level. In A asterisk represents lost stamen position.

to two latero-adaxial stamens with various elaborations, including a lever mechanism and the development of only one fertile theca (see Stützel, 2006, and review by Classen-Bockhoff, Wester and Tweraser, 2003). The gynoecium is mostly bicarpellate (rarely five) and superior, with either a terminal style or a gynobasic style. Each locule is usually subdivided by a false septum separating the two erect ovules. The false partition is either incomplete (e.g. *Clerodendrum*), or complete, and the latter is correlated with a gynobasic style. Cantino (1992) argued that the gynobasic style must have evolved more than once in the family. There is an annular disc nectary present below the gynoecium, which is occasionally displaced in unilateral position.

11.3 Campanulids: Dipsacales, Apiales, Asterales

Figure 11.28 shows the phylogenetic tree of the campanulids (euasterids II), based on Soltis *et al.* (2005). Besides the basal order Aquifoliales the clade contains

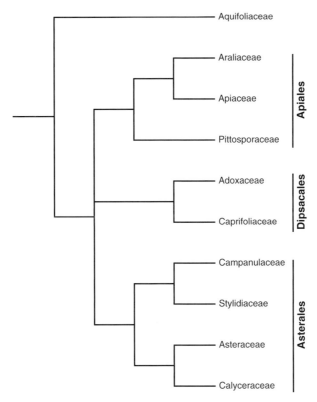

Fig. 11.28. Phylogenetic tree of campanulids (euasterids II), based on Soltis *et al.* (2005).

three well-defined orders. Bremer *et al.* (2001) listed epigynous flowers, early sympetaly with distinct petal primordia, and alternate, dentate leaves as distinctive
characters of campanulids. Several characters are recurrent, such as inferior ovaries
with loss of ovules and carpels, tendency for reduction of the sepals, reduction of
the stamen-petal tube, and arrangement of flowers in pseudanthia.

Dipsacales

Two families make up the Dipsacales. The circumscription of
Caprifoliaceae and Adoxaceae is broad, containing several smaller families
from previous classifications (Judd, Sanders and Donoghue, 1994). Dipsacales
share several characters with other asterids, such as a stamen-corolla tube,
monosymmetry with loss of adaxial stamens, and inferior ovary.

Common characteristics of Caprifoliaceae are the particular trichome nectaries, which are absent from Adoxaceae (Wagenitz and Laing, 1984). Roels and
Smets (1994, 1996) implied that there is no early sympetaly *sensu* Erbar (1991) in
Dipsacales. Floral organs arise on the margin of a depression on a ring-like zone
and a petal tube is formed by upward growth of the marginal area. The septate
ovary develops an apical placentation with pendent ovules curving into five to
three locules (Roels and Smets, 1994; 1996; Erbar, 1994). The carpel wall curves
as a roof above the placentation without connecting to it.

Caprifoliaceae (incl. Dipsacaceae, Morinaceae, Valerianaceae, Triplostegiaceae)

Fig. 11.29A,B. *Lonicera giraldii* Rehder

↓ K5 [C(5) A5] Ğ(3)

Fig. 11.29C,D. *Valeriana repens* Host

↙ K∞ [C(5) A3] Ğ(1:2°)
General formula: ↓ /↙/✱ K(0)5–∞ C5 A1–5 Ğ2–3(5–8)

Molecular phylogenies have shown that the traditional circumscription of
Caprifoliaceae is paraphyletic. Important synapomorphies of Caprifoliaceae
(against Adoxaceae) are: monosymmetry (occasionally asymmetry), a long style,
capitate stigmas alternating with locules, presence of specific unicellular trichome
nectaries, inferior ovary with a tendency for pseudomonomery, and tubular corolla
(see Wagenitz and Laing, 1984; Judd, Sanders and Donoghue, 1994; Howarth and
Donoghue, 2005), although the latter two characters are also present in *Viburnum*.

Different approaches to the intrafamilial classification have been presented,
which are not satisfactory (overview in Donoghue, Bell and Winkworth, 2003).
Based on more recent phylogenies, it is possible to recognize four clades or

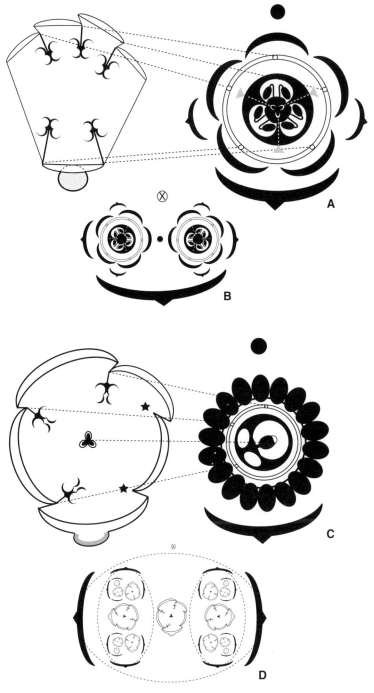

Fig. 11.29. Caprifoliaceae: *Lonicera giraldii*, A. flower, B. partial inflorescence; *Valeriana repens*, C. flower, D; partial inflorescence.

subfamilies: Diervilleae, Caprifolieae, Linnaeaeae and Valerianeae. Inflorescences are built on a monotelic pattern and consist of racemes with cymose lateral axes (thyrses). Partial inflorescences are dichasial with three flowers (e.g. *Leycesteria*), or are two-flowered in *Lonicera* and *Linnaea* by abortion of the central flower; in *Symphoricarpos* only the central flower is formed. Other genera have a more complex pattern with a higher number of small flowers (e.g. *Patrinia*, *Valeriana*: Fig. 11.29D), or forming capitula-like inflorescences in Dipsaceae.

An epicalyx is frequently present (in Dipsaceae, Morineae and *Triplostegia*) and is six- to four-lobed, consisting of two to three pairs of alternating bracts (*Hochblätter*: Hofmann and Göttmann, 1990). Roels and Smets (1996) interpreted the epicalyx as derived from lower bracts of a thyrsoid inflorescence as is found in *Centranthus* or *Lonicera*. Reduction of all flowers of a lateral branch, except for the top flower, leads to typical Dipsaceae flowers with an epicalyx. Each flower on a capitulum would be interpreted as a remnant of a complex thyrsoid inflorescence (Roels and Smets, 1996). However, assuming that the ancestral condition consisted of three pairs (as in *Linnaea*), it is not possible to know whether there are two separate origins of an epicalyx or a single origin followed by loss, although the second option is unlikely (Donoghue, Bell and Winkworth, 2003).

The calyx is rarely well developed; it is mostly cupulate or as a small rim, knobs (e.g. *Valeriana*) or bristles, or is absent (in analogy with the Asteraceae). Reduction of the calyx is accompanied by the development of the epicalyx, which takes over the protective and dispersal functions in combination with the calyx, or at the expense of calyx and ovary wall (see Donoghue, Bell and Winkworth, 2003).

Monosymmetry is median (e.g. *Lonicera*: Fig. 11.29A) to transversal (e.g. *Valeriana*: Fig. 11.29C) or the flower is asymmetric by reduction of stamens and carpels (*Centranthus*). Although symmetry patterns tend to be similar to those of Lamiales (Donoghue, Ree and Baum, 1998), changes in stamen number become decoupled from petal evolution (Donoghue, Bell and Winkworth, 2003). Stamen number is rarely five (e.g. *Patrinia pentandra*), mostly reduced to four (e.g. *Dipsacus*, *Patrinia*: loss of the adaxial stamen), three (e.g. *Valeriana*), two (e.g. *Fedia*) or one (e.g. *Centranthus*). In *Morina* the posterior pair is fertile, while the anterior is staminodial. Reduction of stamens tends to follow the phylogeny with clades characterized by a given stamen number (Donoghue, Bell and Winkworth, 2003).

Petal arrangement is usually as 2:3 with three anterior petals, rarely 1:4 (e.g. *Lonicera*, *Triostemum*). Most Dipsaceae are secondarily polysymmetric with four petals (by fusion of two posterior petals: e.g. *Dipsacus*, *Knautia*). In *Scabiosa*, five corolla lobes and four stamens are formed (Eichler, 1875; Roels and Smets, 1996). Howarth and Donoghue (2005) mentioned this for *Symphoricarpos* but tetramery was not detected in that genus, except for the ovary.

Trichome nectaries are inserted on the corolla tube, often concentrated in one area and occasionally in a spur (e.g. *Lonicera*: Fig. 11.29A, *Centranthus*).

The gynoecium is always inferior arising in a depression. *Leycesteria* has five carpels in antepetalous position with ovules in two rows in each locule (Eichler, 1875). In *Symphoricarpos* the ovary is four-carpellate but the median carpels are positioned lower than the transveral carpels and bear single ovules instead of pairs in the latter (Eichler, 1875; Roels and Smets, 1996). The ovary is mostly tricarpellate and inferior with three apical ovules but two ovules are frequently sterile or absent (pseudomonomery). Eichler (1875) presented the orientation of the gynoecium of *Valeriana* with the odd carpel in adaxial position; I found the opposite pattern (Fig. 11.29C). There is a complete transitional series between well-developed carpels and sterile carpels that are barely visible (Hofmann and Göttmann, 1990). In Dipsaceae it is not clear how many carpels make up the ovary, as the position of the single ovule is variable (see Hofmann and Göttmann, 1990; Roels and Smets, 1996). The single style has three stigmas in a commissural position.

Several species of Caprifoliaceae were studied ontogenetically by Roels and Smets (1996) and Hofmann and Göttmann (1990).

Adoxaceae (incl. Viburnaceae)

Fig. 11.30A,B. *Viburnum grandiflorum* Wall. ex DC.

✳ K(5) [C(5) A5] Ǧ(1:2°)

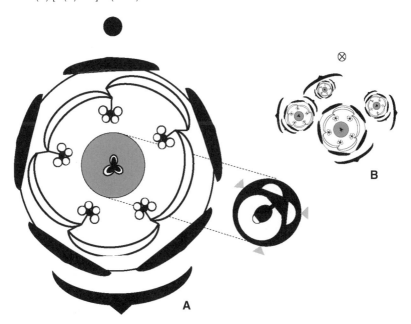

Fig. 11.30. *Viburnum grandiflorum* (Adoxaceae): A. flower; B. partial inflorescence.

The family is restricted to *Adoxa, Viburnum* and *Sambucus*, which have the following characters in common: polysymmetry, a short style with lobed stigma (lobes carinal), a rotate corolla, three to five carpels and drupaceous fruits (e.g. Judd, Sanders and Donoghue, 1994; Howarth and Donoghue, 2005). Flowers contain five stamens in a haplostemonous arrangement. Sepals are short, valvate and fused at the base. The arrangement of lateral flowers of *Adoxa* is inversed with the odd petal adaxially (Erbar, 1994), and this was also reported for *Sambucus ebulus* by Eichler (1875). The corolla is imbricate. *Sambucus* and *Adoxa* have extrorse anthers and a short stamen-petal tube. The stamens of *Adoxa* arise as common primordia but are split to the base forming pairs of half-stamens. In the sister genera *Sinadoxa* and *Tetradoxa*, the filament is divided to the base, respectively to the middle (Donoghue, Bell and Winkworth, 2003). In *Viburnum*, stamens are occasionally of different length (e.g. *V. grandiflorum*). The ovary is inferior with apical insertion of ovules. In *Viburnum*, the ovary is pseudomonomerous by reduction of two locules. Vestigial arechesporial tissue is present at the base of the style of all Adoxaceae. This was interpreted as the remnant of a second ovule present in ancestral groups (Erbar, 1994).

In *Adoxa*, glandular multicellular trichomes are formed in groups at the base of petals. In *Viburnum*, a disc-like nectary sits on top of the ovary (Erbar, 1994). *Sambucus* lacks nectaries and this is interpreted as a loss, but Vogel (1998b) mentioned the presence of glands associated with inflorescences of *S. javanica*.

Some species of *Viburnum* (e.g. *V. opulus*) have a tendency to develop complex inflorescences with marginal sterile flowers that have increased calyces, analogous to Hydrangeaceae.

A floral diagram of *Adoxa moschatellina* was provided by Erbar (1994).

Apiales

Apiales contain eight to ten families (Chandler and Plunkett, 2004) of which the largest are Apiaceae and Araliaceae. Araliaceae and Apiaceae are difficult to keep separated and some authors implied that Apiaceae evolved from within a paraphyletic Araliaceae (e.g. Judd, Sanders and Donoghue, 1994). Hydrocotyloideae are included in Araliaceae and Pittosporaceae is considered to be the closest sister group, although the internal relationships within the large Apiaceae and Araliaceae remain unresolved.

Apiales share a combination of characters. The complex inflorescences are probably derived from cymes. There is a tendency for sepals to become reduced or absent. Apiales usually do not have a stamen-petal tube although they develop early sympetaly (Erbar, 1991). It is assumed that a stamen-petal tube failed to develop and that apopetaly is derived. The androecium is mainly haplostemonous, except in Araliaceae with a higher variation in numbers.

Diplostemony may be a plesiomorphic condition in the family, linking Apiales with basal clades of the asterids. Plunkett, Soltis and Soltis (1996) implied that bicarpellate gynoecia are plesiomorphic in the Apiales, and that five-carpellate and pluricarpellate gynoecia are derived. Ovaries are generally inferior, except for Pittosporaceae. The nectary of Apiales is not a disc (as erroneously mentioned in the literature), but a gynoecial nectary (Erbar and Leins, 1988a). Main differences between Araliaceae and Apiaceae are in fruit structure and habit.

Pittosporaceae

Fig. 11.31. *Pittosporum tobira* (Thunb.) Aiton

Fig. 11.31. *Pittosporum tobira* (Pittosporaceae): partial inflorescence.

✻ K5 C5 A5 G̲ (2-)3(-5)

Flowers are grouped in compact panicles, with two bracteoles that may enclose additional flowers. Sepals and petals are erect and free, but petals do not spread by the tight constriction of sepals. Placentation is broadly parietal with few ovules inserted in two irregular rows. A gynoecial nectary develops at the base of dorsal carpel flanks and is visible as an area not covered by trichomes. Erbar and Leins (1995a) reported characters linking Pittosporaceae with Apiales, including early sympetaly and the gynoecial nectary in *Pittosporum tobira*. Plunkett (2001) mentioned the basal fusion of petals in some Pittosporaceae and Araliaceae – it is to be questioned whether this fusion is postgenital by compression between the erect sepals. He also implied that the superior ovary of Pittosporaceae is secondary, although this is not supported by floral ontogeny (Erbar and Leins, 1995a).

Araliaceae

Fig. 11.32. *Scheffleria* aff. *elliptica* (Blume) Harms

Fig. 11.32. *Scheffleria* aff. *elliptica* (Araliaceae).

✳ K5 C5 A5 -G(5)-
General formula: ✳ K0–5 C5–10(-12) A5–10(-120) G(1)2–5–10(-200)

Araliaceae flowers tend to be easily recognizable. Inflorescences are capitula or umbels with small flowers. Erbar and Leins (1988a) demonstrated the tendency for bracts to be lost in *Aralia elata* and *Hedera helix*. The position of outer sepals depends on presence/absence of the two bracteoles in *Hedera*, while it is more variable in *Aralia*. The calyx is small or develops as a rim. Sepals are absent in *Meryta* and *Hydrocotyle* where they are not initiated (Erbar and Leins, 1985). Petals are free with a generally valvate (e.g. *Hedera*), occasionally imbricate (e.g. *Panax*) aestivation. Petals are usually five in number, rarely ten or more arising by lateral division (Philipson, 1970). Contrary to other haplostemonous Apiales, the androecium can be highly variable. Secondary polyandry is present in some genera (*Plerandra*, *Tetraplesandra*: Philipson, 1970) with lateral stamen increase. In *Tetraplesandra*, the androecium varies from haplostemony, diplostemony, to eight times the petal number (Costello and Motley, 2004). Stamens are inflexed in bud, as in Apiaceae.

Merism is highly variable in Araliaceae. There is frequently a lateral increase of carpels, ranging from six to up to 200 in *Tupidanthus* (Sokoloff *et al.*, 2007). The multiplication of carpels is often linked with a lateral increase of stamens and petals, while sepals remain pentamerous.

The ovary is usually inferior, rarely superior. While most *Tetraplesandra* species have inferior ovaries, Costello and Motley (2004) reported an upward expansion of the ovary resulting in a secondary superior position in *T. gymnocarpa*. When isomerous, carpels are antepetalous.

In *Seemannaralia*, the bicarpellate ovary becomes unilocular by formation of an incomplete septum (Burtt and Dickison, 1975). Some taxa are pseudomonomerous (e.g. *Diplopanax* and *Eremopanax*: Philipson, 1970). Only one apical hanging ovule develops in each locule. The other ovule on the placenta develops in a modified tapering organ, which is also found in Apiaceae (Philipson, 1970; Erbar and Leins, 1988a). It was considered as vestigial, although Philipson (1970) believed it has another function, maybe acting as an obturator. As in Apiaceae, a stylopodium is generally formed on top of the ovary bearing a nectary.

Apiaceae

Fig. 11.33. *Aegopodium podagraria* L.

$*$ K5 C5 A5 $\breve{G}(2)$
General formula: $*/\downarrow$ K0–5 C5 A5 G2

The general floral structure is very similar to some Araliaceae. Inflorescences are umbels or capitula, occasionally with showy basal bracts, forming pseudoflowers (e.g. *Astrantia*). Flowers are either polysymmetric or bilaterally symmetric by unequal development of petal lobes, especially in the peripheral flowers of umbels. The calyx is usually reduced to teeth or is absent at maturity, Although the calyx is initiated in most cases (Leins and Erbar, 1985), it has a strong tendency for reduction with a retarded initiation (e.g. *Anthriscus*: Sattler, 1973). Petals are valvate and free, usually clawed, divided and with inflexed tip (enclosing the anthers in bud); there is usually only a single vein per petal. The five stamens are haplostemonous and inflexed in bud. The ovary of two carpels bears a stylopodium with nectary that is confluent with each style. Each locule has one apically hanging ovule, while the other is reduced.

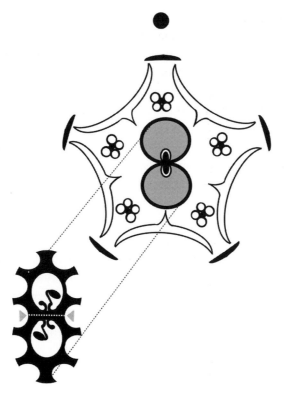

Fig. 11.33. *Aegopodium podagraria* (Apiaceae).

Asterales

The order contains 11 families but internal relationships are not well resolved (Stevens, 2001 onwards).

All studied members of the order share early sympetaly *sensu* Erbar (1991). Flower structure and development have been extensively studied in the order by Erbar (1992, 1993), Leins and Erbar (1987, 1989, 2000) and Erbar and Leins (1988b, 1989, 1995b).

Secondary pollen presentation is generalized and is linked with proterandry of flowers. Pollen is deposited on the outside of the style by growth of the style through a ring formed by the dehisced anthers. Erbar and Leins (1995b) and Leins and Erbar (2007) discussed the different secondary pollen presentation mechanisms in the order.

Flowers are mostly pentamerous and either polysymmetric or monosymmetric. In cases where the corolla is monosymmetric, it is usually deeply split on one side with the stamens exserted towards the slit (e.g. Goodeniaceae, Stylidiaceae, Lobelioideae of Campanulaceae). A similar construction is rarely found in some Lamiales, such as *Sclerochiton* (Acanthaceae). Stamens are five in number and haplostemonous, rarely fewer (two in Stylidiaceae). The ovary contains two (Asteraceae, Lobelioideae of Campanulaceae, Stylidiaceae), or three to five carpels (Campanuloideae of Campanulaceae, Calyceraceae) with axile, basal or apical placentation. With fewer fertile carpels there are clear indications of pseudomonomery in Calyceraceae, Brunoniaceae and Stylidiaceae (Erbar and Leins, 1988b; Erbar, 1992, 1993). An epigynous disc nectary is common. A single style is formed, occasionally with stigmatic lobes reflecting carpel number (Asteraceae). Corolla aestivation is generally valvate (excluding Stylidiaceae–Donatiaceae). In Menyanthaceae and Goodeniaceae, petals are induplicatively valvate. The petal apex is reduced and has two lobes (Endress and Matthews, 2006b). This is reflected in the petal vasculature of Asteraceae, where the median vein is also reduced. Gustafsson (1995) gave an overview of petal venation in Asterales and related families; he found that Asteraceae, and the closely related families Calyceraceae, Menyanthaceae and Goodeniaceae share an unusual petal vasculature in which prominent marginal veins meet in the apex of the petals. Stamens are strongly associated with the style, often forming a tube of connivent anthers surrounding the protruding style.

Campanulaceae (incl. Lobeliaceae)

Fig. 11.34. *Lobelia tupa* L.

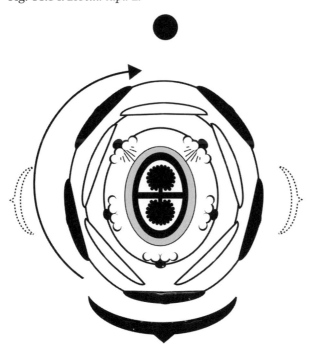

Fig. 11.34. *Lobelia tupa* (Campanulaceae). Arrow points to resupination of 180°.

↓ K(5) C(5) A(5) Ğ(2)

General formula: ∗/↓ K5 C5 A5 G2–5

Lobelioideae differ from Campanuloideae in the bicarpellate gynoecium versus three to five carpels, and the monosymmetric flowers with unusual orientation.

The odd petal of *Lobelia* is placed in adaxial position (cf. Eichler (1875). However, the flower is resupinated at maturity, restoring the condition found in most other core eudicots. The corolla is split on the side where the fused stamens protrude. Anthers are broad and closely packed. The two adaxial anthers bear long hairs missing in the other anthers (cf. Payer, 1857). A broad septum divides the ovary in two locules and bears a placenta in its upper part covered with a high number of ovules. A nectary covers the ovary roof (Erbar and Leins, 1989).

In *Campanula* and related genera the base of the filament is strongly inflated and covers the top of the ovary as a nectar chamber (Erbar and Leins, 1989). Access to the nectary is through spaces between the stamen bases. When isomerous, the ovary is either antesepalous or antepetalous according to Eichler (1875). In *Clermontia*, the outer (sepal) whorl is identical to the corolla through a process of homeosis (Albert, Gustafsson and Di Laurenzio, 1998).

Nemacladus (Nemacloideae) appears intermediate between Campanuloideae and Lobelioideae in having fused filaments attached to the corolla tube. Two pseudonectaries are found in adaxial position at the base of the filament tube (Stevens, 2001 onwards).

Stylidiaceae (incl. Donatiaceae)

Fig. 11.35A,B. *Stylidium graminifolium* Sw.

↙ K(5) C(4:1) [A2 Ğ(2)]

Flowers of *Stylidium* are arranged in short racemes. Stamens and style are connected in a single tubular structure (called gynostemium or filament-style tube), with two extrorse anthers inserted laterally of the stigmatic opening. The stamens are those situated opposite sepals four and five (cf. Eichler, 1875; Baillon, 1876b). In *Donatia fascicularis*, a third stamen may occur (Carolin, 1960). At maturity, flowers of *Stylidium* are obliquely monosymmetric with the combined style and anthers pointing to the left of the flower. In bud, monosymmetry of the flower is median, with a shorter abaxial petal (Fig. 11.35A). Early developmental diagrams for two species of *Stylidium* were shown by Erbar (1992). At the level of detachment of the petal lobes two erect appendages (auricles) are found. During development, the common stamen-style curves forward and fits

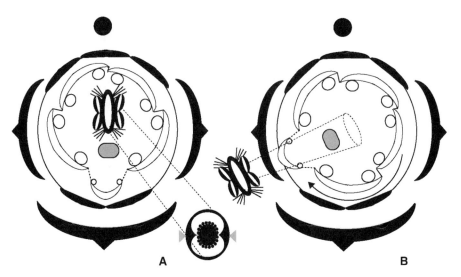

Fig. 11.35. Stylidiaceae: *Stylidium graminifolium*, A. flower in bud; B. flower at anthesis. Note corona of globular protuberances opposite petals. Arrow shows resupination of 45°.

between the two auricles of the shorter petal. This corresponds with a torsion of the pedicel of 45° (not 90° as suggested by Erbar, 1992) to the left (Fig. 11.35B). As the small petal diverges much lower on the tube, a short slit is present. The ovary has septa only in the basal part of the ovary, connecting a globular central placentation. One of the carpels is sterile in *S. adnatum*, while the posterior locule can be shorter in *S. graminifolium* (Erbar, 1992). In *Donatia* septa are almost completely formed (Carolin, 1960). Two nectaries are reported to be in the median plane of the flower with the adaxial larger (Erbar, 1992). In my material only the adaxial nectary was well developed. Below the anther–stigma complex long hairs are present, functioning in the reception and distribution of pollen. A complete picture of the flower structure and pollination mechanism of *Stylidium adnatum* and *S. graminifolium* was given by Erbar (1992).

Except for Erbar, other authors have not mentioned oblique monosymmetry in the family. Baillon (1876b), Sattler (1973) and Erbar (1992) studied the floral development of *Stylidium*. The anthers differentiate earlier than the corolla and push the style rapidly upwards. In *Donatia* the anthers are not connected to the ovary. The origin of the stamen-style tube appears difficult to resolve because no clear carpel primordia are visible in *Stylidium* (only a narrow slit), although clear primordia are visible *in Levenhookia*. Erbar (1992) interpreted the tube as receptacular, lifting up carpels and stamens. A similar filament tube surrounding the style was seen in Lobelioideae and Calyceraceae (Erbar, 1993), suggesting a fusion of style and filaments that would preferably be described as a filament-style tube.

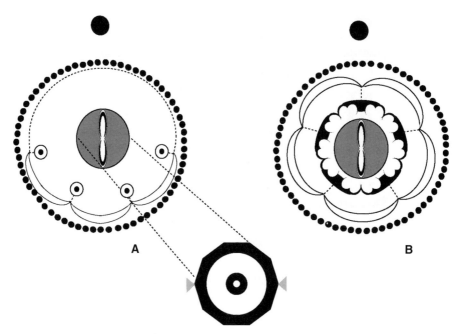

Fig. 11.36. Asteraceae: *Ligularia dentata*, A. ligulate pistillate flower; B. tubular bisexual flower. The ovary is the same for A and B.

Asteraceae

Fig. 11.36. *Ligularia dentata* (A. Gray) Hara

Fig. 11.36A. Ligulate (pistillate) flower:

↓ K∞ [C(3) A4°-5°] Ğ (2)

Fig. 11.36B. Tubular (bisexual) flower:

✱ K∞ [C(5) A5] Ğ (2)
General formula: ↓/ ✱ K0-∞ C3–5 A0–5 Ğ (2)

The largest family of core eudicots (ca. 20 000 species) has one of the most readily recognizable inflorescences. Flowers are grouped in heads or capitula, which are condensed racemose inflorescences enclosed by involucral bracts (phyllaries). The entire capitulum functions as a pseudoflower, which may consist of one (e.g. *Ligularia*) or several capitula (e.g. *Echinops, Achillea*: Harris, 1995). Flowers (florets) are generally small, ranging from a few per capitulum to over 1000. Subtending bracts (palea or 'chaff') can be present or are absent. When present, they arise with the flower on a common primordium, which may be due to condensation of the inflorescence (Harris, 1995). Flowers of Asteraceae are much homogenous throughout the family. Flowers are mostly bisexual – peripheral flowers may be pistillate (Fig. 11.36A) or sterile. Unisexual capitula are rare (e.g. *Xanthium*). A distinction can often be made between peripheral ray florets (ligulate flowers) with monosymmetric petals and central polysymmetric tubular disc florets. Some capitula only have tubular florets (e.g. *Echinops*), or only ligulate florets (e.g. *Hieracium*). Sepals are generally small and reduced and replaced by scales, teeth or hairs (pappus). The ovary is bicarpellate and inferior, with a single basal ovule (situated on the adaxial carpel. Therefore, the ovary is basically pseudomonomerous. Petals are fused with valvate aestivation and are connected with a whorl of stamens. Anthers are connivent around the style and this arrangement is linked to the secondary pollination mechanism. The stigma spreads open as two lobes above the stamen ring and is covered abaxially with pollen. Ligulate flowers which function as petals in the pseudanthium can have three fully developed and two reduced lobes (2:3), three lobes only (Asteroideae, Fig. 11.36A), a 0:5 (Cichorioideae), or 1:4 arrangement (e.g. *Barnadesia*). Staminodes are present in *Ligularia* (Fig. 11.36A), but they are rarely noticeable in mature flowers of Asteraceae (Harris, 1995).

Flowers develop acropetally along a helix. The ray florets lag behind in development compared with disc florets. Flower parts arise on a concave apex (early sympetaly: Leins and Erbar, 1987, 2000; Erbar, 1991; Harris, 1995). The corolla ring primordium is the first organ to arise and the calyx emerges from five primordia alternating with the corolla, randomly or on a circular rim. In ray florets, adaxial lobes are initiated but are generally suppressed (Harris, 1995).

There are strong similarities between Calyceraceae on the one hand, and Goodeniaceae on the other, with Asteraceae (Erbar and Leins, 1988b; Erbar, 1993; Leins and Erbar, 1989). Together with Menyanthaceae, the four families form a well-supported clade with clear morphological synapomorphies.

PART III   CONCLUSIONS

12

Distinctive systematic characters and cryptic apomorphies

Floral diagrams can describe flowers with a high degree of detail. Although it is difficult to represent specific three-dimensional depth, the amount of information provided is considerable and may include developmental and anatomical evidence (see p. 37).

Floral diagrams support recognition of plant species, as major groups can be identified by their floral diagram, but are also a reflection of the evolution of flowers in angiosperms. Major changes in the floral Bauplan can be stressed by representing floral diagrams in the context of the phylogenetic tree of angiosperms. Floral diagrams make a comparison between divergent characters possible by stressing the positional relationships of floral structures.

Floral diagrams are not rigidly fixed in time, but are an expression of the developmental plasticity in flowers. Important morphological changes are often the result of subtle shifts in the primordial body during development, and this is also reflected at the genetic level. One aspect of morphological observations is that changes are often gradual, without clear-cut boundaries in characters between different clades. Several plant groups are characterized by 'tendential features' or 'apomorphic tendencies' (see Endress and Matthews, 2006a), namely characters that may not be generalized in a clade, but occur on a much more frequent basis than in any related clades. I call these characters 'cryptic apomorphies'. These characters can either be synapomorphic (present in a clade but not in all members), non-synapomorphic (a predisposition to evolve a character in a clade) or represent several independently derived features (autapomorphies) in a clade (Endress and Matthews, 2006a). Suites of morphological characters can be used to identify clades, even if they are not present in all taxa. A precondition for certain characters can be present in ancestors but is not expressed in all descendants. Characters also appear to be correlated because one character change can have an effect on another character. Syndromes of correlated characters can be detected through floral diagrams. These syndromes can be linked with pollination mechanisms. An example is the syndrome of synandria with extrorse anthers in fully unisexual flowers (e.g. Triuridaceae, Nepenthaceae, Myristicaceae, Menispermaceae). Syndromes are intimately linked with the Bauplan of the flower. For example there is a correlation between early sympetaly and inferior ovaries and between late sympetaly and superior ovaries which reflects coherence between several interdependent characters linked through floral development.

However, there are also characters that appear to be unrelated and these have more systematic value (Endress and Matthews, 2006a,b). The gain or loss of structures may characterize subclades within major groups, even if these changes are not synapomorphic. A good example is the unexpressed predisposition for halving anthers that is characteristic for Malvaceae. Monothecal anthers have arisen convergently in *Durio* and Malvoideae (Bayer *et al.*, 1999); it is not a synapomorphy for the clade and has arisen several times independently. Another cryptic apomorphy is the tendency for loss or reduction of sepals or the transformation of a calyx in bristles or hairs in several campanulids, correlated with elaboration of compact, small-flowered inflorescences (umbels or capitula). Endress and Mathews (2006b) emphasized that sharing of characters in major clades could represent homoplasious tendencies based on shared genetic preconditions. Such preconditions could be considered as characters on their own right. Rasmussen, Kramer and Zimmer (2009) demonstrated for the Ranunculales that a genetic programme for petal identity is present in most families, even if petals are not developed or are highly divergent.

Character mapping and its interpretation on phylogenetic trees remains controversial, especially in basal angiosperms, as we do not know how extensive morphological variability was at the lower nodes. Extant families represent tips of millions of years of evolution and extinction that can only be partially resolved by the fossil record. A good example is Proteales, grouping three families with extremely different morphologies. Because of the principle of parsimony and because of lack of intermediate taxa, character evolution is often referred as 'reversal', in cases where more basal characteristics are found at the end of nodes. This can be reasonable in a cladistic context, but does not always help in understanding evolution of morphological characters.

13

Floral diagrams and major angiosperm groups

This book is not exhaustive, as the total number of families has not been covered, although it is comprehensive in reflecting floral diversity of the majority of angiosperms. Major groups of angiosperms have been covered to reflect important floral evolutionary patterns. Sections below summarize most important floral attributes of major clades of angiosperms.

13.1 Basal angiosperms, monocots and early diverging eudicots

Floral evolution of basal angiosperms has been studied in a phylogenetic context by several authors (e.g. Doyle and Endress, 2000; Ronse De Craene, Soltis and Soltis, 2003; Zanis *et al.*, 2003; Endress and Doyle, 2009). Mapping of floral characters on phylogenetic trees implies that cyclic flowers have evolved rapidly from a helical flower with relatively few stamens and carpels as in *Amborella* and the ANA-grade (Austrobaileyales, Nymphaeales, Amborellales), with frequent reversals to a spiral phyllotaxis in different orders. The most recent addition in Nymphaeales, Hydatellaceae, raises questions about the flowers of the earliest angiosperms (see Rudall *et al.*, 2007; Endress and Doyle, 2009).

Flowers of basal angiosperms have a low synorganization with high plasticity in organ number and position (Endress, 1990, 2001, 2008c). The transition between vegetative shoot and flower is often gradual with a progressive change from leaves to bracts and tepals within the inflorescence or at the base of the flower (Endress, 2003b; Remizowa and Sokoloff, 2003; Buzgo *et al.*, 2007). Several taxa, also in early diverging eudicots and basal monocots, have an unclear distinction between bracts and perianth. Taxa with spiral flowers look remarkably similar in the gradual transition of bracts to floral organs and transitions between different organ categories (e.g. Austrobaileyaceae, Eupomatiaceae, Calycanthaceae). Transitional organs include staminodes between fertile stamens and carpels, or between tepals and stamens, and have intermediate characteristics, explaining the difficulty in differentiating bracts from tepals, or tepals from staminodes (e.g. Endress, 1980c, 1990; Staedler and Endress, 2009). This is in contrast with core eudicots, where an abrupt transition from the vegetative shoot to the flower is commonplace.

Changes in phyllotaxis appear to be complex in basal angiosperms and early diverging eudicots, with mixtures of whorled, spiral or irregular patterns in several clades, such as Winteraceae, Ranunculaceae and Nymphaeaceae (Endress, 2001; Ronse De Craene, Soltis and Soltis, 2003). Cyclic flowers are generally trimerous or dimerous, with far fewer parts, and cyclization starts from the perianth onwards (e.g. Magnoliaceae, Monimiaceae: Leins and Erbar, 2007). The perianth may be whorled while androecium and gynoecium have spiral initiation (e.g. *Anemone* in Ranunculaceae; Magnoliaceae). Cyclic flowers, such as Lauraceae, have an inner staminodial whorl, which is involved in nectar production. Trimerous flowers are widespread (as in monocots and some early diverging eudicots) and linked with a whorled initiation. Trimerous and dimerous flowers often coexist in families such as Lauraceae, Hernandiaceae and Papaveraceae, although dimerous patterns may be a continuation of a decussate bract arrangement (e.g. Buxaceae: Von Balthazar and Endress, 2002a, b). Number and arrangement of whorls is generally variable. Differentiation into sepals and petals is rarely clear-cut (e.g. *Cabomba*, Annonaceae), more often gradual or reflected in minor differences (colour, number of vascular traces) with little scope for specializations. A combination of fusion and monosymmetry is found only in Piperales (Aristolochiaceae), where the perianth is reduced to a single trimerous whorl. A much higher numbers of stamens and carpels can arise in different ways, linked with the reduction in size of stamens in some cyclic Annonaceae, by extension of a conical receptacle in Magnoliaceae, or by growth of a hypanthium in Nymphaeaceae.

Austrobaileyales, Magnoliales and Laurales are proterogynous, with separate pistillate and staminate phases developing in succession, or flowers are unisexual. This leads to specializations in floral structure where staminodes play a role in preventing self-pollination (Walker-Larsen and Harder, 2000). Staminodes can take up secondary reward functions, such as providing food-bodies, scent or visual attraction (Endress, 1984, 1994). Syncarpy is rare in basal angiosperms and Ranunculales. There is more generally a pattern to evolve monocarpellate gynoecia, sometimes in the same family with many carpels (e.g. *Idiospermum* in Calycanthaceae). However, single carpels are not the result of pseudomonomery, which involves syncarpy. Endress and Igersheim (2000a) discussed the evolution of early carpel closure, leading to the formation of a compitum and fused ovaries. A hypanthium is rare in basal angiosperms, except in families of Laurales characterized by elaborate cups enclosing floral organs.

A well-differentiated disc nectary is not found in basal angiosperms. Instead, nectarioles (viz. unusual nectar-producing idioblasts or groups of cells on the surface of organs) are present in a few genera (e.g. *Cabomba* in Cabombaceae, *Chimonanthus* in Calycanthaceae: Vogel, 1998a; Endress, 2001, 2008c). This may indicate that nectaries are in an early stage of differentiation.

Monocots

Monocots are easily characterized by a number of characters reflected in their floral diagrams: flowers are trimerous with a more or less stable formula (P3+3 A3+3 G3). Only some families of Alismatales do not follow this strict arrangement, with variability in number of stamen and carpel whorls. However, flowers are never spiral throughout, supporting an early derivation from trimerous basal angiosperms. Although the floral formula is superficially similar to some rosids (see p. 360), origin of trimerous flowers has to be sought in spiral or trimerous basal angiosperms. Monocots are structurally far less variable than dicots. Differences between groups are the result of subtle changes between whorls, rarely with secondary increases of organs.

Trimerous flowers are prevalent among monocots. Pentamery is rare, although occasionally present in *Paris* (Fig. 6.13). It is only more or less stable in the genus *Pentastemona* of Stemonaceae (Pandanales). Tetramery is more common although restricted to Cyclanthaceae, Triuridaceae and Convallariaceae (Dahlgren, Clifford and Yeo, 1985). In monocots, syncarpy is generally the result of a postgenital fusion of carpels with marginal placentation (Remizowa *et al.*,2006). Styles and stigmas are generally carinal, reflecting the lack of fusion or its late occurrence. Contrary to core eudicots where petals can have different homologies (see Ronse De Craene, 2007, 2008), the biseriate perianth of monocots is of a homologous nature. Differences between the inner tepals (petals) and outer tepals (sepals) were interpreted as the result of shifts of gene expression (sliding boundary hypothesis). Nonetheless, gene expression patterns in the perianth can be very variable (see overview in Ronse De Craene, 2007). A biseriate perianth remains constant throughout monocots, occasionally with further reductions in wind-pollinated Poales and Alismatales.

Unique floral or inflorescence characters for monocots include presence of a single bracteole (adaxial or transversal), and septal nectaries. Septal nectaries are an early-diverging feature in monocots and are associated with postgenital carpel fusion (Dahlgren, Clifford and Yeo, 1985; Smets *et al.*, 2000; Rudall, 2002; Remizowa, Sokoloff and Kondo, 2008). Flowers of monocots have undergone the same pattern of hypanthium growth and shifts to inferior ovaries as in dicots. A preliminary condition for an inferior ovary is the fusion of carpels. The shift to an inferior ovary position is strongly correlated with shifts of the septal nectaries and their increase in complexity (e.g. Simpson, 1998a; Rudall, 2002).

In earlier publications I emphasized the different structural nature for the androecium in monocots compared with eudicots and introduced a different terminology (dicycly and monocycly, compared with diplostemony and haplostemony: e.g. Ronse De Craene and Smets, 1993, 1998a; Fig. 1.2). The prevalent trimery in monocots is related to a spiral or polycyclic ancestry found in the basal angiosperms and this is reflected in the evolution of the androecium (see Ronse De Craene and Smets, 1995a, 1998a). Although *Acorus* represents the basalmost monocot, the genus has several unique apomorphies and probably a derived androecium. Multistaminate Alismataceae could have a floral arrangement closer to the ancestral monocots, and could therefore be basal to the common two-whorled androecium. In some Alismataceae, the upper stamens and carpels tend to develop in irregular position or form parastichies (e.g. *Ranalisma*: Charlton, 1991, 2004). Two stamen whorls are most common in all major monocot clades, with occasional loss of inner or outer whorl (Ronse De Craene and Smets, 1995a). Secondary polyandry is rare in monocots and characterizes a few families (Arecaceae, some Poaceae, Velloziaceae). In some Alismatales, it may be superimposed on existing primary polyandry (e.g. Limnocharitaceae). The ovary is generally tricarpellate, except for Alismatales where the gynoecium may consist of several whorls of free carpels. Reductions of carpels are linked to pseudomonomery and are localized in certain families, either as the result of a prevailing monosymmetry, or as a general pattern of a reduction to a lower ovule number. Parietal placentation (occasionally found in some families: Orchidaceae, Melanthiaceae, Hydrocharitaceae) probably has an independent origin, arising from postgenitally fused apocarpous gynoecia with marginal placentae. Contraction of the margins led to parietal placentation, while intrusion and fusion led to axile placentation. In eudicots parietal placentation has a different origin, mostly as a derivation from an axile placentation (Fig. 1.3).

Early diverging eudicots

Early diverging or basal eudicots are a transitional grade between basal angiosperms and core eudicots, as they share morphological characters of both clades. There is a general tendency for flowers to have a weakly differentiated dimerous perianth with a single kind of organs (tepals) or no perianth at all, although there is a high diversification in certain clades (e.g. Sabiales, Ranunculales). Basal eudicots were probably much more diverse in the past, as can be shown in florally diverse Platanaceae (e.g. Crepet, 2008; von Balthazar and Schönenberger, 2009). Especially in the grade following divergence of Ranunculales, there is a general development of small dimerous flowers without petals and with questionable identity of the perianth (e.g. Proteales, Trochodendrales, Buxales, Gunnerales). As the grade immediately precedes the core eudicots, there was much speculation about the origin of core eudicot flowers from such prototypes (e.g. von Balthazar and Endress, 2002b; Soltis *et al.*, 2003; Ronse De Craene, 2004, 2007, 2008). Von Balthazar and Endress (2002b) interpreted the condition in Buxaceae, where flowers generally have bract-like phyllomes, as a precursory stage in the differentiation of a bipartite perianth, while Soltis *et al.* (2003) derived the pentamerous core eudicot flower from a dimerous Gunneraceae-like precursor. Early diverging eudicots represent an evolutionary line with a gradual reduction of flowers and are not precursory to core eudicot floral diversity (see Wanntorp and Ronse De Craene, 2005). Basal eudicots, especially Ranunculales, represent a playground for floral diversification with a far greater variation in perianth forms than core eudicots, and this is reflected in higher diversity of *AP3* and *PI* paralogs (Kramer *et al.*, 2006; Rasmussen, Kramer and Zimmer, 2009). Ranunculales have the highest floral diversity, especially in Ranunculaceae (e.g. Kramer, Di Stilio and Schlüter, 2003). In this family, several elaborations have evolved independently, such as spurred staminodial petals, cyclic flowers, zygomorphy and pentamery (Ronse De Craene, Soltis and Soltis, 2003).

The occurrence of pentamery and dimery appears to be linked with trimerous precursors, as stamens are inserted opposite tepals, not alternating with them (e.g. von Balthazar and Endress, 2002b; Wanntorp and Ronse De Craene, 2006). Fossil pentamerous Platanaceae and Buxaceae have carpels or stamens opposite tepals, as pentamerous Sabiaceae. Pentamery is different in Ranunculaceae where it is derived from more complex spiral flowers.

Floral diagrams of early diverging eudicots illustrate a progressive pauperization of flower structures, with the exception of increased diversification in Ranunculales, Sabiaceae and Nelumbonaceae. Although our understanding of early diverging eudicots has dramatically improved in recent years, it is not possible to throw unequivocal light on the sudden transtion to core eudicot diversity, based on the evidence of extant flower diversity.

13.2   Core eudicots

Flowers of core eudicots generally have a well-defined distinction between bracts and flower, except in basal orders, such as Berberidopsidales. Inclusion of bracts at the base of the flower is a secondary process, leading to an epicalyx and eventually to a secondary calyx. The basic floral formula in core eudicots appears to be K5 C5 A5+5 G5, found in all major clades, except for smaller orders such as Berberidopsidales, Vitales and Dilleniales. The perianth is generally differentiated into sepals and petals (with a transition between undifferentiated perianth to sepals and petals in Berberidopsidales), and diplostemony is found in basal clades of all major orders, except euasterids. The reader is referred to Ronse De Craene (2004, 2007, 2008) for a discussion of the evolution of the bipartite perianth. The ovary shows a strong degree of fusion and is rarely apocarpous, in which case it has frequently evolved secondarily (e.g. Endress, Jenny and Fallen, 1983).

Hypanthial growth strongly shapes the flowers of core eudicots (as in some monocots), with specific tendencies among the major clades. Rosids often have cup-like hypanthia lined with nectary tissue (Malpighiales, Rosales, Myrtales), generally with a superior ovary or none, rarely with an inferior ovary (e.g. Cucurbitales, Myrtaceae). In caryophyllids, hypanthial growth tends to be restricted to the area between androecium and gynoecium and is correlated with an intrastaminal nectary. Asterids have generalized stamen-petal tubes, with variable changes in the position of the ovary.

Nectaries have evolved among core eudicots in relationship with hypanthia. The greatest variation in nectary structures is found in rosids, ranging from a receptacular disc or intrastaminal glands, shifting to extrastaminal nectaries associated with stamens and petals (e.g. Geraniales, Malpighiales). Asterids generally have a broad disc surrounding the ovary base, shifting to an epigynous disc or a gynoecial nectary.

The androecium tends to be basically diplostemonous in core eudicots, as it is found in all major clades, even characterizing major orders (e.g. Geraniales, Malvales, Sapindales, Oxalidales). Polyandry is always secondary (in contrast to basal angiosperms) and is associated with one or two whorls. Stamens tend to develop centrifugally on common primordia or on a ring primordium, more rarely centripetally and this is linked with hypanthial growth (Ronse De Craene and Smets, 1991a, 1992a). One stamen whorl is often reduced to staminodes and consequently lost, leading to either obhaplostemony or haplostemony, with both conditions rarely together. Further reductions lead to single stamens in some groups (e.g. Ochnaceae, Lacistemataceae, Euphorbiaceae). The gynoecium is basically syncarpous, arising congenitally united (Endress, 2006). It is unclear whether five carpels are the basal structure, as trimerous and dimerous gynoecia are firmly settled in the clade. Contrary to basal angiosperms where monocarpellate gynoecia are derived from a reduction of free carpels, a single carpel is nearly always derived through pseudomonomery. Multiplications of carpels tend to occur rarely and lead to irregular carpel closure and deformed symmetry of flowers (e.g. Endress, 2006).

Caryophyllids

Although the clade is relatively small in number of species, its evolutionary patterns are fascinating, especially in perianth and androecium, which hold more scope for diversity than any other major clade of angiosperms. Caryophyllids are basically pentamerous and apetalous with a single sepaloid whorl in a 2/5 arrangement (at least in core Caryophyllales). Major evolutionary patterns are the reinvention of petaloid organs, stamen shifts and reduction in ovary structure (Ronse De Craene, 2008; Brockington *et al.*, 2009). Several Molluginaceae (Endress and Bittrich, 1993) and Aizoaceae (e.g. *Sesuvium*: Brockington, pers. comm., 2008) have their sepals green on the outside and petaloid on the inside and sepals are hooded or have a dorsal appendage. The petaloid perianth shows no expression of *PI* or *AP3* genes, suggesting that petaloidy was independently derived in ancestral wind-pollinated Caryophyllales (Brockington, pers. comm., 2008).

The basic floral formula of Caryophyllales is K5 C0 A5+5+5 G(5). The presence of a third stamen (or staminodial) whorl is not reflected in any other taxon of core eudicots, where petals appear to have the same origin as sepals. As discussed previously, one option is to accept that the outer alternisepalous stamen/staminode/petal was secondarily derived by radial division of a primary stamen. This assumption is supported by the initiation of common primordia and centrifugal development of outer stamen. Two opposite patterns can be recognized in the order: reduction of the initial ten stamens to five or fewer, and centrifugal multiplication of stamens by development of complex primordia or a ring primordium.

While floral evolution within Caryophyllales is guided by this initial Bauplan, a similar evolution occurs in the 'Polygonaceae-clade' of Polygonales, which share a wealth of characters with Caryophyllales *sensu stricto*. Similar sieve-tube plastid forms were found in some Polygonaceae and Amaranthaceae (Behnke, 1999), which may be another convergence. It is reasonable to accept that a basic diplostemonous flower is present in other Polygonales, such as Tamaricaceae or Dioncophyllaceae, but that loss of petals is one major 'cryptic apomorphy' for the clade. A secondary centrifugal stamen increase is widespread and appears to be superimposed on the basic androecial Bauplan.

Near absence of monosymmetry tends to be generalized in caryophyllids (except for some Cactaceae). Occurrence of nectaries is linked with hypanthial growth between androecium and ovary. Nectaries are generally confined to the inside of the filaments or filament tube, occasionally extending centrifugally (see Zandonella, 1977; Smets, 1986, Bernardello, 2007).

Rosids

The larger a clade, the more difficult it becomes to define and circumscribe it on a morphological basis, especially because there is more scope for variation and because the group is morphologically largely understudied. The rosid clade contains around 70 000 species representing 25% of all angiosperm species and 40% of eudicots (Schönenberger and von Balthazar, 2006; Wang *et al.*, 2009). Rosids are a highly diverse grouping of taxa in which many relationships are still poorly resolved, and many non-molecular characters have barely been investigated (Soltis *et al.*, 2005).

A generalized floral diagram does not exist for rosids, although one kind of floral arrangement is more common and tends to be found almost exclusively in rosids: K5 C5 A5+5 G(5). Monosymmetry is uncommon, and when present, often arises in late ontogeny. Flowers are pentamerous to tetramerous, often with a bipartite perianth. A cryptic apomorphy is that petals are free, often small and retarded in growth, or absent, and are generally clawed (Ronse De Craene, 2008). The androecium is (ob)diplostemonous, rarely obhaplostemonous, and when haplostemonous, often with one whorl of staminodes. Staminodes are rarely persistent or take up new functions except in Malvaceae (Walker-Larsen and Harder, 2000). Complex polyandry is common and highly variable. A hypanthium is found on a frequent basis and lifts calyx, corolla and androecium, or only calyx and corolla. The gynoecium is mostly superior, of three to five fused carpels with axile placentation. The nectary is receptacular in most cases and usually develops as a disc or on hypanthial slopes. Free extrastaminal nectar glands appear frequently in different clades (e.g. Geraniales, Malpighiales). Sympetaly and inferior ovaries are rare and restricted to a few families. There are several exceptions to this assemblage of characters as it is difficult to assign synapomorphies for such a large clade. Endress and Matthews (2006a) enumerated characters that are concentrated or exclusively found in certain orders of rosids. In malvids, there is a concentration of (andro)gynophores, petals with ventral elaborations, contorted petals, campylotropous ovules and monosymmetry. In the nitrogen-fixing clade there is a concentration of apetaly, wind pollination and gynoecia with single ovules. Cucurbitales share trimerous inferior ovaries with their sister group Fagales, as well as the presence of wind-pollinated flowers (Matthews and Endress, 2004). Characters shared by the COM-clade among others (see Matthews and Endress, 2005) are: rarity of monosymmetry (Celastraceae: *Apodostigma* with unequal petals: Simmons 2004), absence of petal tubes, fringed petals, two stamen whorls (diplostemony) with a tendency for secondary polyandry (only in *Plagiopteron* of Celastraceae but abundant in Malphighiales), often two lateral ovules per carpel, presence of a broad nectary extending outside the stamens. There are several convergences between Rosaceae and Myrtaceae, such as a deep hypanthium with intrastaminal nectary, centripetal stamen development, and inferior or half-inferior ovary. Convergences between Rosaceae and Malvaceae include the valvate calyx and presence of an epicalyx.

Several rosids share the presence of paired stamens, mostly in antepetalous position (listed for 24 families by Ronse De Craene and Smets, 1996a). Paired stamens are often linked with obdiplostemony and isomerous gynoecia; they appear to be one way to avoid loss of antepetalous stamens by the pressure of the carpels, which are generally in alternisepalous position (see p. 9).

Asterids

Sympetaly seems to be one of the few characters that are predominantly found in asterids and that have led to the high success of the clade. Sympetaly is rare elsewhere and is found only in Crassulaceae, Caricaceae, Plumbaginaceae and Cucurbitaceae.

As discussed above, two steps lead to sympetaly: confluent meristems of initiated petals and stamens, and intercalary elongation below the united petal bases. In Apiaceae, Araliaceae and Pittosporaceae only step one takes place, or it is shortened in some Lamiales and Solanales (e.g. Montiniaceae: Ronse De Craene, Linder and Smets, 2000; *Besseya* in Plantaginaceae: Hufford, 1995). Other micromorphological characters were enumerated by Soltis *et al.* (2005). It is often difficult to unequivocally determine the direction of character evolution (e.g. bitegmy to unitegmy) as the genetic pathways for reduction can be highly complex and reversals are possible. However, for certain structural characters evolution appears to be unidirectional, as for stamen loss affecting major groups of angiosperms, especially in Lamiales (Fig. 11.21).

Lower asterids

Cornales and Ericales appear somewhat anomalous among asterids, having more morphological similarities with rosids: (ob)diplostemony or obhaplostemony are frequent, as well as secondary stamen increases, a relative rarity of petal tubes or stamen-petal tubes, and presence of more than two carpels. Staminodial structures play an important role in the pollination of several Ericales and Cornales by acquiring novel functions (Walker-Larsen and Harder, 2000). There is a syndrome of stamen and carpel increase in asterids concentrated in basal orders Ericales and Cornales (Endress, 2003a, 2006; Jabour, Damerval and Nadot, 2008). While monosymmetry appears to be restricted to few families of Ericales (e.g. Balsaminaceae), it is widespread in euasterids.

However, there are several shared cryptic apomorphies, announcing euasterids, such as (half-)inferior ovaries and elaborate pseudanthia in Cornales, and elaborations of stamen-petal tubes in Ericales.

Cornales appear to be closer to campanulids on floral evidence than any other clade of asterids: there is a similar tendency for small flowers to become associated in complex inflorescences surrounded by involucral bracts; both share a similar initiation of a concave apex early in the development (cf. early sympetaly by formation of a ring primordium) and there is rapid petal growth overtaking the weaker sepals; the placentation is generally apical and axile on an inferior ovary; the nectaries are epigynous.

Euasterids

Euasterids share a number of specific floral morphological characters that makes them easily identifiable as 'sympetalous'. Haplostemony is almost universal (with few exceptions of secondary increase in Araliaceae, Hoplestigmataceae and Rubiaceae). Hypanthia are variously developed but are not cup-shaped and affect the gynoecium and other organs to different degrees. The position of the gynoecium ranges from superior to inferior by progressive invagination in the receptacle. Development of a stamen-petal tube, a petal tube or stamen tube occurs independently of the gynoecium and is rarely absent. The gynoecium consists of two median carpels in most cases (rarely with three to five carpels or a secondary increase) and is surrounded by a disc nectary or gynoecial nectary, or the nectary is epigynous in inferior ovaries. The general floral formula is K5[C(5)A5] G(2). Placentation is generally axile, occasionally parietal, or derived states may be linked with reduction of ovules to a single basal (Asteraceae) or apical structure (Apiaceae).

Lamiids can be identified by superior bicarpellate ovaries (with notable exceptions), surrounded by a well-developed persistent calyx. Nectaries are generally intrastaminal discs or free glands. It is generally assumed that ancestors of asterids were polysymmetric and that monosymmetry evolved several times independently within the clade (e.g. Donoghue, Ree and Baum, 1998). Lamiids contains orders with polysymmetric flowers and contorted petals ('Contortae' – Gentianales) or valvate to plicate petals (Solanales). Monosymmetric flowers are generalized in Lamiales and mostly constructed on a 2:3 (bilabiate) plan with two petals adaxially and sometimes fused into a single unit, and three abaxial petals often fused into a lower lip. It is not possible to differentiate between floral diagrams of several families of Lamiales (such as Scrophulariaceae, Orobanchaceae, Plantaginaceae, Gesneriaceae, Bignoniaceae), except on the basis of the extent of reduction of stamens or ovules (Fig. 11.21). Monosymmetry is probably plesiomorphic in Lamiales, with several reversals (see Endress, 1994; Soltis *et al.* 2005: 235). There appears to be a common tendency for a transition to tetramerous polysymmetric flowers by fusion of the posterior petals and loss of the adaxial stamen (Ronse De Craene and Smets, 1994; Donoghue, Ree and Baum, 1998; Endress, 1999; Bello *et al.*, 2004). However, are polysymmetric flowers always tetramerous? Endress (1998) pointed out that regular flowers with five stamens are not so rare in Lamiales, occurring mainly in basal families, such as Gesneriaceae and Bignoniaceae, where staminodes tend to be well developed. The case for a reversal of the posterior staminode in *Penstemon* was presented by Walker-Larsen and Harder (2000). The possibility of a reversal obviously depends on how far the staminode is reduced (Endress, 1999). While reversals to pentamerous polysymmetry are possible at this stage, they become impossible with further reduction of the staminode; the only option left is the fusion of two posterior petals. Other clades of euasterids (e.g. Campanulids, Solanales, Gentialales) are basically polysymmetric and monosymmetry is only incidental in some groups. However, there is a convergence with Lamiales in the development of monosymmetry and transitions to tetramery (e.g. Caprifoliaceae).

Campanulids share several clearly identifiable characters. Inflorescences are grouped in more or less dense heads or umbels often surrounded by involucral bracts (e.g. Asteraceae, Brunoniaceae, Campanulaceae, Calyceraceae, Caprifoliaceae, Apiaceae, Araliaceae, Goodeniaceae). However, capitula in Asterales have a polytelic origin, while those of Apiales and Dipsacales are monotelic. All campanulids share early sympetaly (Erbar, 1991). As mentioned earlier (p. 34), the term is not well chosen as it reflects an early invagination of the apex forming a ring-like zone on which the petal lobes emerge (Roels and Smets, 1996; Ronse De Craene and Smets, 2000; Ronse De Craene, Linder and Smets, 2000). There is a clear tendency to form one to several hypanthia and fusions in the flower (stamen-petal tube, inferior ovary). Petal tubes occasionally fail to develop as in Apiales. Sepals do not contribute to the protection of the bud, or do marginally so. There is a strong tendency for sepal loss (e.g. Apiales) or their replacement by hairs, scales, or a pappus. Monosymmetry is less common than in lamiids, but appears to develop in a similar manner (0:5 arrangement in Asterales mainly as a slit on one side of the corolla-tube), or a 2:3 (1:4) arrangement in Dipsacales (e.g. Donoghue, Ree and Baum, 1998; Endress, 1999). The androecium is universally haplostemonous (except for increases in Araliaceae and reductions in Caprifoliaceae from five to one).

Carpels are antepetalous when isomerous, with a frequent switch to three or two median carpels, rarely more by a secondary increase (e.g. Araliaceae). Pseudomonomery is common and is linked with rudimentary septa in Asterales or empty locules in Dipsacales. The number of ovules is often low, mostly a single pendent ovule with a rudiment of the other ovule in the same locule (Apiales, Dipsacales), or a basal ovule (Asterales). Placentation is initially axile with septa developing independently from the ovary wall. All taxa have a long empty locular space. Nectaries are generally gynoecial as a disc around or on top of the gynoecium (Asterales, Apiales), or develop as trichomes on the corolla (Dipsacales).

14

Outlook

It is not an easy task to capture floral diversity by floral diagrams. To be fully comprehensive, several volumes would have to be written, comprising several hundreds of drawings. However, characters on floral diagrams are clear enough to reflect where a taxon belongs and can be used for identification at least to family level. Floral diagrams are increasingly important as a principal means in understanding the complexity of flowers and leading to hitherto unexplored floral characters. While the phylogeny of angiosperms is being progressively refined by worldwide collaborative research (APG III will be published before this book will be out), the challenge for studying flowers and their hidden secrets becomes increasingly important. While writing this book, I made several new observations of flower structures by study of fresh flowers, while the information from the literature was either too basic, or restricted to obscure nineteenth-century work. While Eichler had broader access to a wide botanical knowledge, this knowledge has become progressively eroded during the twentieth century because of emphasis on new exciting areas of botany. Nowadays, we know much more about the genetic structure of plant groups than about their floral structure. The scope of morphological research is immense, especially in tropical families for which only a fraction of the diversity is currently known. In the present biodiversity crisis, such studies are a race against time, with the certainty that pertinent scientific knowledge and aesthetic models are lost forever, before even being discovered.

It is high time that morphology receives the place it deserves, as a central focus for systematics, ecology and evolutionary developmental genetics. Hopefully this book on floral diagrams will be inspirational as a synthesis of the floral diversity of this world.

References

Abbe, E. C. (1935). Studies in the phylogeny of the Betulaceae I. Floral and inflorescence anatomy and morphology. *Bot. Gaz.* **95**, 1–67.

Abbe, E. C. (1938). Studies in the phylogeny of the Betulaceae II. Extremes in the range of variation of floral and inflorescence morphology. *Bot. Gaz.* **99**, 431–469.

Abbe, E. C. (1974). Flowers and inflorescences of the 'Amentiferae'. *Bot. Rev.* **40**, 159–261.

Albert, V. A., Gustafsson, M. H. G. and Di Laurenzio, L. (1998). Ontogenetic systematics, molecular developmental genetics, and the Angiosperm petal. In *Molecular Systematics of Plants Ii: DNA Sequencing*, ed. D. E. Soltis, P. S. Soltis and J. J. Doyle, Boston, MA: Kluwer Academic Publishers, pp. 349–374.

Albert, V. A. and Struwe, L. (2002). Gentianaceae in context. In *Gentianaceae: Systematics and Natural History*, ed. L. Struwe and V. A. Albert, Cambridge, UK: Cambridge University Press, pp. 1–20.

Alverson, W. S., Karol, K. G., Baum, D. A., *et al.* (1998). Circumscription of the Malvales and relationships to other Rosidae, evidence from *rbcL* sequence data. *Am. J. Bot.* **85**, 876–887.

Alverson, W. S., Whitlock, B. A., Nyffeler, R., Bayer, C. and Baum, D. A. (1999). Phylogeny of the core Malvales, evidence from *ndhF* sequence data. *Am. J. Bot.* **86**, 1474–1486.

Amaral, M. C. E. (1991). Phylogenetische Systematik der Ochnaceae. *Bot. Jahrb. Syst.* **113**, 105–196.

Ambrose, B. A., Espinosa-Matías, S., Vásquez-Santana, S., *et al.* (2006). Comparative developmental series of the Mexican Triurids support a euanthial interpretation for the unusual reproductive axes of *Lacandonia schismatica* (Triuridaceae). *Am. J. Bot.* **93**, 15–35.

Anderberg, A. A., Rydin, C. and Källersjö, M. (2002). Phylogenetic relationships in the order Ericales s.l., analysis of molecular data from five genes from the plastid and mitochondrial genomes. *Am. J. Bot.* **89**, 677–687.

Anderberg, A. A. and Ståhl, B. (1995). Phylogenetic interrelationships in the order Primulales, with special emphasis on the family circumscriptions. *Can. J. Bot.* **73**, 1699–1730.

Angiosperm Phylogeny Group I (1998). An ordinal classification for the families of flowering plants. *Ann. Missouri Bot. Gard.* **85**, 531–553.

Angiosperm Phylogeny Group II (2003). An update of the Angiosperm Phylogeny Group classification for the orders and families of flowering plants, APG II. *Bot. J. Linn. Soc.* **141**, 399–436.

Appel, O. (1996). Morphology and systematics of the Scytopetalaceae. *Bot. J. Linn. Soc.* **121**, 207–227.

Arber, A. (1925). *Monocotyledons: A Morphological Study.* Cambridge, UK: Cambridge University Press.

Arber, A. (1934). *The Gramineae.* Cambridge, UK: Cambridge University Press.

Arber, E. A. N. and Parkin, J. (1907). The origin of Angiosperms. *Bot. J. Linn. Soc.* **38**, 29–80.

Armstrong, J. E. (1985). The delimitation of Bignoniaceae and Scrophulariaceae based on floral anatomy, and the placement of problem genera. *Am. J. Bot.* **72**, 755–766.

Armstrong, J. E. and Douglas, A. W. (1989). The ontogenetic basis for corolla aestivation in Scrophulariaceae. *Bull. Torrey Bot. Club* **116**, 378–389.

Armstrong, J. E. and Tucker, S. C. (1986). Floral development in *Myristica* (Myristicaceae). *Am. J. Bot.* **73**, 1131–1143.

Armstrong, J. E. and Wilson, T. K. (1978). Floral morphology of *Horsfieldia* (Myristicaceae). *Am. J. Bot.* **65**, 441–449.

Bachelier, J. B. and Endress, P. K. (2007). Development of inflorescences, cupules, and flowers in *Amphipterygium* and comparison with *Pistacia* (Anacardiaceae). *Int. J. Plant Sci.* **168**, 1237–1253.

Backlund, M., Oxelman, B. and Bremer, B. (2000). Phylogenetic relationships within the Gentianales based on *ndhF* and *rbcL* sequences, with particular reference to the Loganiaceae. *Am. J. Bot.* **87**, 1029–1043.

Baillon, H. (1860). Observations organogéniques pour servir à l'histoire des Polygalées. *Adansonia* **1**, 174–180.

Baillon, H. (1862). Organogénie florale des Cordiacées. *Adansonia* **3**, 1–7.

Baillon, H. (1868a). Monographie des Dilléniacées. *Histoire des plantes I, 2*, 133–192. Paris: Hachette.

Baillon, H. (1868b). Monographie des Magnoliacées. *Histoire des Plantes I, 3*, 133–192. Paris: Hachette.

Baillon, H. (1868c). Monographie des Annonacées. *Histoire des Plantes I, 4*. Paris: Hachette, pp. 193–288.

Baillon, H. (1870). Monographie des Elaeagnacées. *Histoire des Plantes II.* Paris: Hachette, pp. 487–495.

Baillon, H. (1871a). Du genre *Garcinia* et de l'origine de la gomme-gutte. *Adansonia* **10**, 283–298.

Baillon, H. (1871b). Observations sur les Rutacées. *Adansonia* **10**, 299–333.

Baillon, H. (1874). Euphorbiacées. *Histoire des Plantes V, 41*. Paris: Hachette, pp. 105–176.

Baillon, H. (1876a). Traité du développement de la fleur et du fruit X. Castanéacées. *Adansonia* **12**, 1–17

Baillon, H. (1876b). Traité du développement de la fleur et du fruit XVI. Stylidiées. *Adansonia* **12**, 354–361.

Barabé, D. and Lacroix, C. (2000). Homeosis in Araceae flowers, the case of *Philodendron melinonii*. *Ann. Bot.* **86**, 479–491.

Bateman, R. M., Hilton, J. and Rudall, P. J. (2006). Morphological and molecular phylogenetic context of the angiosperms: contrasting the 'top-down' and 'bottom-up' approaches used to infer the likely characteristics of the first flowers. *J. Exper. Bot.* **57**, 3471–3503.

Batenburg, L. H. and Moeliono, B. M. (1982). Oligomery and vasculature in the androecium of *Mollugo nudicaulis* Lam. (Molluginaceae). *Acta Bot. Neerl.* **31**, 215–220.

Bauer, R. (1922). Entwicklungsgeschichtliche Untersuchungen an Polygonaceenblüten. *Flora* **115**, 273–292.

Bayer, C. (1999). The bicolor unit – homology and transformation of an inflorescence structure unique to core Malvales. *Plant Syst. Evol.* **214**, 187–198.

Bayer, C., Fay, M. F., De Bruijn, A., Savolainen, V., *et al.* (1999). Support for an expanded family concept of Malvaceae within a recircumscribed order Malvales, a combined analysis of plastid *atp*B and *rbc*L DNA sequences. *Bot. J. Linn. Soc.* **129**, 267–303.

Bayer, C. and Hoppe, J. R. (1990). Die Blütenentwicklung von *Theobroma cacao* L. (Sterculiaceae). *Beitr. Biol. Pflanz.* **65**, 301–312.

Bayer, C. and Kubitzki, K. (2003). Malvaceae. In *The Families and Genera of Vascular Plants, Vol. V*, ed. K. Kubitzki, and C. Bayer. Berlin: Springer Verlag, pp. 225–311.

Bechtel, A. R. (1921). The floral anatomy of the Urticales. *Am. J. Bot.* **8**, 386–410.

Behnke, H. D. (1999). P-type sieve-element plastids present in members of the tribes Triplareae and Coccolobeae (Polygonaceae) renew the links between the Polygonales and the Caryophyllales. *Plant Syst. Evol.* **214**: 15–27.

Bello, M. A., Hawkins, J. A. and Rudall, P. J. (2007). Floral morphology and development in Quillajaceae and Surianaceae (Fabales), the species-poor relatives of Leguminosae and Polygalaceae. *Ann. Bot.* **100**, 1491–1505.

Bello, M. A., Rudall, P. J., González, F. and Fernández-Alonso, J. L. (2004). Floral morphology and development in *Aragoa* (Plantaginaceae) and related members of the order Lamiales. *Int. J. Plant Sci.* **165**, 723–738.

Belsham, S. R. and Orlovich, D. A. (2003). Development of the hypanthium and androecium in *Acmena smithii* and *Syzygium australe* (*Acmena* alliance, Myrtaceae). *Aust. Syst. Bot.* **16**, 621–628.

Bennek, C. (1958). Die morphologische Beurteilung der Staub- und Blumenblätter der Rhamnaceen. *Bot. Jahrb. Syst.* **77**, 423–457.

Bensel, C. R. and Palser, B. F. (1975a). Floral anatomy in the Saxifragaceae sensu lato I. Introduction, Parnassioideae and Brexioideae. *Am. J. Bot.* **62**, 176–185.

Bensel, C. R. and Palser, B. F. (1975b). Floral anatomy in the Saxifragaceae sensu lato II. Saxifragoideae and Iteoideae. *Am. J. Bot.* **62**, 661–675.

Berger, A. (1930). Crassulaceae. In *Die Natürlichen Pflanzenfamilien 18a*, ed. A. Engler and K. Prantl. Leipzig: W. Engelmann, pp. 352–483.

Bernardello, G. (2007). A systematic survey of floral nectaries. In *Nectaries and Nectar*, ed. S. W. Nicolson, M. Nepi and E. Pacini, Dordrecht: Springer Verlag, pp. 19–128.

Bernhard, A. (1999). Flower structure, development, and systematics in Passifloraceae and in *Abatia* (Flacourtiaceae). *Int. J. Plant Sci.* **160**, 135–150.

Bernhard, A. and Endress, P. K. (1999). Androecial development and systematics in Flacourtiaceae s.l. *Plant Syst. Evol.* **215**, 141–155.

Bittrich, V. (1993). Caryophyllaceae. In *The Families and Genera of Vascular Plants, Vol. II*, ed. K. Kubitzki, J. G. Rohwer and V. Bittrich. Berlin: Springer Verlag, pp. 206–236.

Bittrich, V. and Amaral, M. C. E. (1996). Flower morphology and pollination biology of *Clusia* species from the Gran Sabana (Venezuela). *Kew Bull.* **51**, 681–694.

Bogle, A. L. (1989). The floral morphology, vascular anatomy, and ontogeny of the Rhodoleioideae (Hamamelidaceae) and their significance in relation to the 'Lower' hamamelids. In *Evolution, Systematics and Fossil History of the Hamamelidae, Vol. I, Introduction and 'Lower' Hamamelidae*, ed. P. R. Crane and S. Blackmore. Oxford, UK: Clarendon Press, pp. 201–220.

Boke, N. H. (1963). Anatomy and development of the flower and fruit of *Pereskia pititache*. *Am. J. Bot.* **50**, 843–858.

Boke, N. H. (1966). Ontogeny and structure of the flower and fruit of *Pereskia aculeata*. *Am. J. Bot.* **53**, 534–542.

Bowman, J. L. and Smyth, D. R. (1998). Patterns of petal and stamen reduction in Australian species of *Lepidium* L. (Brassicaceae). *Int. J. Plant Sci.* **159**, 65–74.

Box, M. S. and Rudall, P. J. (2006). Floral structure and ontogeny in *Globba* (Zingiberaceae). *Plant Syst. Evol.* **258**, 107–122.

Brandbyge, J. (1993). Polygonaceae. In *The Families and Genera of Vascular Plants, Vol. II*, ed. K. Kubitzki, J. G. Rohwer and V. Bittrich. Berlin: Springer Verlag, pp. 531–544.

Bremer, K., Backlund, A., Sennblad, B., *et al.* (2001). A phylogenetic analysis of 100+ genera and 50+ families of euasterids based on morphological and molecular data with notes on possible higher level morphological synapomorphies. *Plant Syst. Evol.* **229**, 137–169.

Brett, J. F. and Posluszny, U. (1982). Floral development of *Caulophyllum thalictroides* (Berberidaceae). *Can. J. Bot.* **60**, 2133–2141.

Brockington, S. F., Alexandre, R., Ramdial, J., *et al.* (2009). Phylogeny of the Caryophyllales sensu lato: revisiting hypotheses on pollination biology and perianth differentiation in the core Caryophyllales. *Int. J. Plant Sci.*, **170**, 627–643.

Brown, D. K. and Kaul, R. B. (1981). Floral structure and mechanism in Loasaceae. *Am. J. Bot.* **68**, 361–372.

Burtt, B. L. and Dickison, W. C. (1975). The morphology and relationships of *Seemannaralia* (Araliaceae). *Notes Roy. Bot. Gard. Edinburgh* **33**, 449–466.

Busch, A. and Zachgo, S. (2007). Control of corolla monosymmetry in the Brassicaceae *Iberis amara*. *Proc. Natl Acad. Sci.* **104**, 16 714–16 719.

Buxbaum, F. (1961). Vorläufige Untersuchungen über Umfang, systematische Stellung und Gliederung der Caryophyllales (Centrospermae). *Beitr. Biol. Pflanz.* **36**, 1–56.

Buzgo, M. (2001). Flower structure and development of Araceae compared with Alismatids and Acoraceae. *Bot. J. Linn. Soc.* **136**, 393–425.

Buzgo, M., Chanderbali, A. S., Kim, S., *et al.* (2007). Floral developmental morphology of *Persea Americana* (avocado, Lauraceae), the oddities of staminate organ identity. *Int. J. Plant Sci.* **168**, 261–284.

Buzgo, M. and Endress, P. K. (2000). Floral structure and development of Acoraceae and its systematic relationships with basal Angiosperms. *Int. J. Plant Sci.* **161**, 23–41.

Buzgo, M., Soltis, D. E., Soltis, P. S., *et al.* (2006). Perianth development in the basal monocot *Triglochin maritima* (Juncaginaceae). *Aliso* **22**, 107–125.

Buzgo, M., Soltis, P. S. and Soltis, D. E. (2004). Floral developmental morphology of *Amborella trichopoda* (Amborellaceae). *Int. J. Plant Sci.* **165**, 925–947.

Caddick, L. R., Rudall, P. J. and Wilkin, P. (2000). Floral morphology and development in Dioscoreales. *Feddes Repert.* **111**, 189–230.

Cantino, P. D. (1992). Evidence for a polyphyletic origin of the Labiatae. *Ann. Missouri Bot. Gard.* **79**, 361–379.

Caris, P. (1998). *Bloemontogenie en fylogenie van de Myrsinaceae en aanverwante taxa.* Katholieke Universiteit Leuven, Belgium: unpubl. thesis.

Caris, P. L., Geuten K. P., Janssens, S. B. and Smets, E. F. (2006a). Floral development in three species of *Impatiens* (Balsaminaceae). *Am. J. Bot.* **93**, 1–14.

Caris P., Ronse De Craene, L. P., Smets, E. F. and Clinckemaillie, D. (2000). Floral development of three *Maesa* species, with special emphasis on the position of the genus within Primulales. *Ann. Bot.* **86**, 87–97.

Caris, P. and Smets, E. F. (2004). A floral ontogenetic study on the sister group relationship between the genus *Samolus* (Primulaceae) and the Theophrastaceae. *Am. J. Bot.* **91**, 627–643.

Caris, P., Smets, E. F., De Coster, K. and Ronse De Craene, L. P. (2006b). Floral ontogeny of *Cneorum tricoccon* L. (Rutaceae). *Plant Syst. Evol.* **257**, 223–232.

Carlquist, S. (2003). Wood anatomy of Aextoxicaceae and Berberidopsidaceae is compatible with their inclusion in Berberidopsidales. *Syst. Bot.* **28**, 317–325.

Carolin, R. C. (1960). Floral structure and anatomy in the family Stylidiaceae Swartz. *Proc. Linn. Soc. New South Wales* **85**, 189–196.

Carolin, R. C. (1993). Portulacaceae. In *The Families and Genera of Vascular Plants, Vol. II*, ed. K. Kubitzki, J. G. Rohwer and V. Bittrich. Berlin: Springer Verlag, pp. 544–555.

Carrucan, A. E. and Drinnan, A. N. (2000). The ontogenetic basis for floral diversity in the *Baeckea* sub-group. *Kew Bull.* **55**, 593–613.

Chakravarty, M. L. (1958). Morphology of the staminate flowers in the Cucurbitaceae with special reference to the evolution of the stamens. *Lloydia* **21**, 49–87

Chandler, G. T. and Plunkett, G. M. (2004). Evolution in Apiales: nuclear and chloroplast markers together in (almost) perfect harmony. *Bot. J. Linn. Soc.* **144**, 123–147.

Charlton, W. A. (1991). Studies in the Alismataceae IX. Development of the flower in *Ranalisma humile. Can. J. Bot.* **69**, 2790–2796.

Charlton, W. A. (1999a). Studies in the Alismataceae X. Floral organogenesis in *Luronium natans* (L.) Raf. *Can. J. Bot.* **77**, 1560–1568.

Charlton, W. A. (1999b). Studies in the Alismataceae XI. Development of the inflorescence and flowers of *Wiesneria triandra* (Dalzell) Micheli. *Can. J. Bot.* **77**, 1569–1579.

Charlton, W. A. (2004). Studies in the Alismataceae XII. Floral organogenesis in *Damasonium alisma* and *Baldellia ranunculoides*, and comparisons with *Butomus umbellatus. Can. J. Bot.* **82**, 528–539.

Chase, M. W., Fay, M. F., Devey, D. S., *et al.* (2006). Multigene analyses of monocot relationships, a summary. *Aliso* **22**, 63–75.

Chase, M. W., Zmarzty, S., Lledó, M. D., Wurdack, K. J. Swensen, S. M. and Fay, M. F. (2002). When in doubt, put it in Flacourtiaceae: a molecular phylogenetic analysis based on plastid *rbcL* DNA sequences. *Kew Bull.* **57**, 141–181.

Chen, L., Ren, Y., Endress, P. K., Tian, X. H. and Zhang, X. H. (2007). Floral organogenesis in *Tetracentron sinense* (Trochodendraceae) and its systematic significance. *Plant Syst. Evol.* **264**, 183–193.

Church, A. W. (1908). *Types of Floral Mechanism: A Selection of Diagrams and Descriptions of Common Flowers, part I.* Oxford, UK: Clarendon Press.

Citerne, H. L., Moeller, M. and Cronk, Q. C. B. (2000). Diversity of *cycloidea*-like genes in Gesneriaceae in relation to floral symmetry. *Ann. Bot.* **86**, 167–176.

Citerne, H. L., Pennington, R. T. and Cronk, Q. C. B. (2006). An apparent reversal in floral symmetry in the legume *Cadia* is a homeotic transformation. *Proc. Natl Acad. Sci.* **103**, 12 017–12 020.

Classen-Bockhoff, R. (1990). Pattern analysis in pseudanthia. *Plant Syst. Evol.* **171**, 57–88.

Classen-Bockhoff, R. and Heller. A. (2006). Floral synorganization and secondary pollen presentation in four Marantaceae from Costa Rica. *Int. J. Plant Sci.* **169**, 745–760.

Classen-Bockhoff, R., Wester, P. and Tweraser, E. (2003). The staminal lever mechanism in *Salvia* L. (Lamiaceae): a review. *Plant Biol.* **5**, 33–41.

Clinckemaillie, D. and Smets, E. F. (1992). Floral similarities between Plumbaginaceae and Primulaceae, systematic significance. *Belg. J. Bot.* **125**, 151–153.

Cocucci, A. E. and Anton, A. M. (1988). The grass flower, suggestions on its origin and evolution. *Flora* **181**, 353–362.

Coen, E. S. and Meyerowitz, E. M. (1991). The war of the whorls: genetic interactions controlling flower development. *Nature* **353**, 31–37.

Coen, E. S., Nugent, J. M., Luo, D., *et al.* (1995). Evolution of floral symmetry. *Phil. Trans. Roy. Soc. London B* **350**, 35–38.

Costello, A. and Motley, T. J. (2004). The development of the superior ovary in *Tetraplasandra* (Araliaceae). *Am. J. Bot.* **91**, 644–655.

Cox, C. D. K. (1998). Hydrocharitaceae. In *The Families and Genera of Vascular Plants, Vol. IV*, ed. K. Kubitzki. Berlin: Springer Verlag, pp. 234–248.

Crane, P. R., Friis, E. M. and Pedersen, K. R. (1994). Paleobotanical evidence on the early radiation of magnoliid Angiosperms. *Plant Syst. Evol., Suppl.* **8**, 51–72.

Crepet, W. L. (2008). The fossil record of Angiosperms, requiem or renaissance. *Ann. Missouri Bot. Gard.* **95**, 3–33.

Cronquist, A. (1981). *An Integrated System of Classification of Flowering Plants.* New York: Columbia University Press.

Cuénoud, P., Savolainen, V., Chatrou, L. W., Powell, M., Grayer, R. J. and Chase, M. W. (2002). Molecular phylogenetics of Caryophyllales based on nuclear 18S rDNA and plastid *rbc*L, *atp*B, and *mat*K DNA sequences. *Am. J. Bot.* **89**, 132–144.

Dahlgren, R. (1975). A system of classification of the Angiosperms to be used to demonstrate the distribution of characters. *Bot. Not.* **128**, 119–147.

Dahlgren, R. (1983). General aspects of Angiosperm evolution and macrosystematics. *Nord. J. Bot.* **3**, 119–149.

Dahlgren, R., Clifford, T. H. and Yeo, P. (1985). *The Families of the Monocotyledons: Structure, Evolution and Taxonomy.* Berlin: Springer Verlag.

Dahlgren, R. and Thorne, R. F. (1984). The order Myrtales: circumscription, variation and relationships. *Ann. Missouri Bot. Gard.* **71**, 633–699.

De Laet, J., Clinckemaillie, D., Jansen, S. and Smets, E. F. (1995). Floral ontogeny in the Plumbaginaceae. *J. Plant Res.* **108**, 289–304.

De Maggio, A. E. and Wilson, C. L. (1986). Floral structure and organogenesis in *Podophyllum peltatum* (Berberidaceae). *Am. J. Bot.* **73**, 21–32.

De Menezes, N. L. (1980). Evolution in Velloziaceae, with special reference to androecial characters. In *Petaloid Monocotyledons*, ed. C. D. Brickell, D. F. Cutler and M. Gregory. Hortic. and Botan. Research. Linnean Soc. Symposium Series 8. London: Academic Press, pp. 117–138.

Deroin, T. (1985). Contribution à la morphologie comparée du gynécée des Annonaceae – Monodoroideae. *Bull. Mus. Hist. Nat. (Paris)* Sér. IV, **7**, 167–176.

Deroin, T. (1997). Confirmation and origin of the paracarpy in Annonaceae, with comments on some methodological aspects. *Candollea* **52**, 45–58.

Deroin, T. (2000). Floral anatomy of *Toussaintia hallei* Le Thomas, a case of convergence of Annonaceae with Magnoliaceae. In *Proceedings of the International Symposium on the Family Magnoliaceae*, ed. Y.-H. Liu, H.-M. Fan, Z.-Y. Chen, Q.-G. Wu and Q.-W. Zeng. Beijing: Science Press, pp. 168–176.

Deroin, T. (2007). Floral vascular pattern of the endemic Malagasy genus *Fenerivia* Diels (Annonaceae). *Adansonia* Sér. 3, **29**, 7–12.

Deroin, T. and Le Thomas, A. (1989). Sur la systématique et les potentialités évolutives des Annonacées: cas d'*Ambavia gerrardii* (Baill.) Le Thomas, espèce endémique de Madagascar. *C. R. Acad. Sci. Paris* **309**, Sér. III, 647–652.

Derstine, K. S. and Tucker, S. C. (1991). Organ initiation and development of inflorescences and flowers of *Acacia baileyana*. *Am. J. Bot.* **78**, 816–832.

De Wilde, W. J. J. O. (1974). The genera of tribe Passifloreae (Passifloraceae) with special reference to flower morphology. *Blumea* **22**, 37–50.

Dickison, W. C. (1970). Comparative morphological studies in Dilleniaceae VI. Stamens and young stem. *J. Arnold Arbor.* **51**, 403–418.

Dickison, W. C. (1975). Studies on the floral anatomy of the Cunoniaceae. *Am. J. Bot.* **62**, 433–447.

Dickison, W. C. (1978). Comparative anatomy of Eucryphiaceae. *Am. J. Bot.* **65**, 722–735.

Dickison, W. C. (1993). Floral anatomy of the Styracaceae, including observations on intra-ovarian trichomes. *Bot. J. Linn. Soc.* **112**, 223–255.

Dilcher, D. L. (2000). Toward a new synthesis, major evolutionary trends in the Angiosperm fossil record. *Proc. Natl Acad. Sci.* **97**, 7030–7036.

Donoghue, M. J., Bell, C. D. and Winkworth, R. C. (2003). The evolution of reproductive characters in Dipsacales. *Int. J. Plant Sci.* **164** (5 Suppl.), S453–S464.

Donoghue, M. J., Ree, R. H. and Baum, D. A. (1998). Phylogeny and the evolution of flower symmetry in the Asteridae. *Tr. Plant Sci.* **3**, 311–317.

Douglas, A. W. and Tucker, S. C. (1996a). Comparative floral ontogenies among Persoonioideae including *Bellendena* (Proteaceae). *Am. J. Bot.* **83**, 1528–1555.

Douglas, A. W. and Tucker, S. C. (1996b). Inflorescence ontogeny and floral organogenesis in Grevilleoideae (Proteaceae), with emphasis on the nature of the flower pairs. *Int. J. Plant Sci.* **157**, 341–372.

Douglas, A. W. and Tucker, S. C. (1996c). The developmental basis of diverse carpel orientations in Grevilleoideae (Proteaceae). *Int. J. Plant Sci.* **157**, 373–397.

Douglas, A. W. and Tucker, S. C. (1997). The developmental basis of morphological diversification and synorganization in flowers of Conospermeae (*Stirlingia* and Conosperminae, Proteaceae). *Int. J. Plant Sci.* **158**, S13–S48.

Doyle, J. A. (2008). Integrating molecular phylogenetic and paleobotanical evidence on origin of the flower. *Int. J. Plant Sci.* **169**, 816–843.

Doyle, J. A. and Endress, P. K. (2000). Morphological phylogenetic analysis of basal Angiosperms, comparison and combination with molecular data. *Int. J. Plant Sci.* **161** (6 Suppl.), S121–S153.

Drinnan, A. N. and Ladiges, P. Y. (1989). Corolla and androecium development in some *Eudesmia* eucalypts (Myrtaceae). *Plant Syst. Evol.* **165**, 239–254.

Eames, A. J. (1961). *Morphology of the Angiosperms*. New York: McGraw-Hill.

Eckardt, T. (1937). Untersuchungen über Morphologie, Entwicklungsgeschichte und systematische Bedeutung des pseudomonomeren Gynoeciums. *Nova Acta Leopold. N. F.* **5**, 1–112.

Ecklund, H. (2000). Lauraceous flowers from the Late Cretaceous of North Carolina, U.S.A. *Bot. J. Linn. Soc.* **132**, 397–428.

Edgell, T. (2004). *Floral studies of Brexia madagascariensis Thouars (Celastraceae)*. Edinburgh, UK: Royal Botanic Garden Edinburgh: unpubl. MSc thesis.

Eichler, A. W. (1875). *Blüthendiagramme vol. I*. Leipzig: Wilhelm Engelmann.

Eichler, A. W. (1878). *Blüthendiagramme vol. II*. Leipzig: Wilhelm Engelmann.

Eliasson, U. H. 1988. Floral morphology and taxonomic relations among the genera of Amaranthaceae in the New World and the Hawaiian Islands. *Bot. J. Linn. Soc.* **96**, 235–283.

Endress, M. E. and Bittrich, V. (1993). Molluginaceae. In *The Families and Genera of Vascular Plants, Vol. II*, ed. K. Kubitzki, J. G. Rohwer and V. Bittrich. Berlin: Springer Verlag, pp. 419–426.

Endress, M. E., Sennblad, B., Nilsson, S., *et al.* (1996). A phylogenetic analysis of Apocynaceae s.str. and some related taxa in Gentianales, a multidisciplinary approach. *Opera Bot. Belg.* **7**, 59–102.

Endress, P. K. (1967). Systematische Studien über die verwandschaftlichen Beziehungen zwischen den Hamamelidaceen und Betulaceen. *Bot. Jahrb. Syst.* **87**, 431–525.

Endress, P. K. (1976). Die Androeciumanlage bei polyandrischen Hamamelidaceen und ihre systematische Bedeutung. *Bot. Jahrb. Syst.* **97**, 436–457.

Endress, P. K. (1977). Evolutionary trends in the Hamamelidales–Fagales-group. *Plant Syst. Evol.* Suppl. **1**, 321–347.

Endress, P. K. (1978). Blütenontogenese, Blütenabgrenzung und Systematische Stellung der perianthlosen Hamamelidoideae. *Bot. Jahrb. Syst.* **100**, 249–317.

Endress, P. K. (1980a). Floral structure and relationships of *Hortonia* (Monimiaceae). *Plant Syst. Evol.* **133**, 199–221.

Endress, P. K. (1980b). Ontogeny, function and evolution of extreme floral construction in Monimiaceae. *Plant Syst. Evol.* **134**, 79–120.

Endress, P. K. (1980c). The reproductive structures and systematic position of the Austrobaileyaceae. *Bot. Jahrb. Syst.* **101**, 393–433.

Endress, P. K. (1984). The role of inner staminodes in the floral display of some relic Magnoliales. *Plant Syst. Evol.* **146**, 269–282.

Endress, P. K. (1986). Floral structure, systematics, and phylogeny in Trochodendrales. *Ann. Missouri Bot. Gard.* **73**, 297–324.

Endress, P. K. (1987). Floral phyllotaxis and floral evolution. *Bot. Jahrb. Syst.* **108**, 417–438.

Endress, P. K. (1989). Chaotic floral phyllotaxis and reduced perianth in *Achlys* (Berberidaceae). *Bot. Acta* **102**, 159–163.

Endress, P. K. (1990). Patterns of floral construction in ontogeny and phylogeny. *Biol. J. Linn. Soc.* **39**, 153–175.

Endress, P. K. (1992). Evolution and floral diversity, the phylogenetic surroundings of *Arabidopsis* and *Antirrhinum*. *Int. J. Plant Sci.* **153**, S106–S122.

Endress, P. K. (1994). *Diversity and Evolutionary Biology of Tropical Flowers*. Cambridge, UK: Cambridge University Press.

Endress, P. K. (1995a). Major evolutionary traits of monocot flowers. In *Monocotyledons: Systematics and Evolution*, ed. P. J. Rudall, D. F. Cribb and C. J. Humphries. Kew, UK: Royal Botanic Gardens, pp. 43–79.

Endress, P. K. (1995b). Floral structure and evolution in Ranunculanae. *Plant Syst. Evol.* Suppl. 9, 47–61.

Endress, P. K. (1997). Relationships between floral organization, architecture, and pollination mode in *Dillenia* (Dilleniaceae). *Plant Syst. Evol.* **206**, 99–118.

Endress, P. K. (1998). *Antirrhinum* and Asteridae: evolutionary changes of floral symmetry. *Society for Experimental Biology Symposium Series* **51**, 133–140

Endress, P. K. (1999). Symmetry in flowers: diversity and evolution. *Int. J. Plant Sci.* **160** (6 Suppl.), S3–S23.

Endress, P. K. (2001). The flower in extant basal Angiosperms and inferences on ancestral flowers. *Int. J. Plant Sci* **162**, 1111–1140.

Endress, P. K. (2003a). Morphology and Angiosperm systematics in the molecular era. *Bot. Rev.* **68**, 545–570.

Endress, P. K. (2003b). Early floral development and nature of the calyptra in Eupomatiaceae (Magnoliales). *Int. J. Plant Sci.* **164**, 489–503.

Endress, P. K. (2006). Angiosperm floral evolution, morphological developmental framework. *Adv. Bot. Res.* **44**, 1–61.

Endress, P. K. (2008a). My favourite flowering image. *J. Exper. Bot. FNL 2008*, 1–3.

Endress, P. K. (2008b). The whole and the parts, relationships between floral architecture and floral organ shape, and their repercussions on the interpretation of fragmentary floral fossils. *Ann. Missouri Bot. Gard.* **95**, 101–120.

Endress, P. K. (2008c). Perianth biology in the basal grade of extant angiosperms. *Int. J. Plant Sci.* **169**, 844–862.

Endress, P. K. and Doyle, J. A. (2007). Floral phyllotaxis in basal Angiosperms, development and evolution. *Curr. Opin. Plant Biol.* **10**, 52–57.

Endress, P. K. and Doyle, J. A. (2009). Reconstructing the ancestral angiosperm flower and its initial specializations. *Am. J. Bot.* **96**, 22–66.

Endress, P. K. and Hufford, L. D. (1989). The diversity of stamen structures and dehiscence patterns among Magnoliidae. *Bot. J. Linn. Soc.* **100**, 45–85.

Endress, P. K. and Igersheim, A. (1997). Gynoecium diversity and systematics of the Laurales. *Bot. J. Linn. Soc.* **125**, 93–168.

Endress, P. K. and Igersheim, A. (2000a). Gynoecium structure and evolution in basal Angiosperms. *Int. J. Plant Sci.* **161** (6 Suppl.), S211–S223.

Endress, P. K. and Igersheim, A. (2000b). The reproductive structures of the basal Angiosperm *Amborella trichopoda* (Amborellaceae). *Int. J. Plant Sci.* **161** (6 Suppl.), S237–S248.

Endress, P. K., Jenny, M. and Fallen, M. (1983). Convergent elaboration of apocarpous gynoecia in higher advanced dicotyledons. *Nord. J. Bot.* **3**, 293–300.

Endress, P. K. and Lorence, D. H. (2004). Heterodichogamy of a novel type in *Hernandia* (Hernandiaceae) and its structural basis. *Int. J. Plant Sci.* **165**, 753–763.

Endress, P. K. and Matthews, M. L. (2006a). First steps towards a floral structural characterization of the major rosid subclades. *Plant Syst. Evol.* **260**, 223–251.

Endress, P. K. and Matthews, M. L. (2006b). Elaborate petals and staminodes in eudicots, diversity, function, and evolution. *Org. Div. Evol.* **6**, 257–293.

Engler, A. and Krause, K. (1935). Loranthaceae. In *Die Natürlichen Pflanzenfamilien 16b*, ed. A. Engler, and K. Prantl, 2nd edn. Leipzig: W. Engelmann, pp. 98–203.

Engler, A. and Prantl, K., eds. (1887–1909). *Die Natürlichen Pflanzenfamilien I–IV*. Leipzig: W. Engelmann.

Erbar, C. (1986). Untersuchungen zur Entwicklung der spiraligen Blüte von *Stewartia pseudocamellia* (Theaceae). *Bot. Jahrb. Syst.* **106**, 391–407.

Erbar, C. (1991). Sympetaly: a systematic character? *Bot. Jahrb. Syst.* **112**, 417–451.

Erbar, C. (1992). Floral development of two species of *Stylidium* (Stylidiaceae) and some remarks on the systematic position of the family Stylidiaceae. *Can. J. Bot.* **70**, 258–271.

Erbar, C. (1993). Studies on the floral development and pollen presentation in *Acicarpha tribuloides* with a discussion of the systematic position of the family Calyceraceae. *Bot. Jahrb. Syst.* **115**, 325–350.

Erbar, C. (1994). Contributions to the affinities of *Adoxa* from the viewpoint of floral development. *Bot. Jahrb. Syst.* **116**, 259–282.

Erbar, C. (1998). Coenokarpie ohne und mit Compitum, ein Vergleich der Gynoeceen von *Nigella* (Ranunculaceae) and *Geranium* (Geraniaceae). *Beitr. Biol. Pflanz.* **71**, 13–39.

Erbar, C., Kusma, S. and Leins, P. (1998). Development and interpretation of nectary organs in Ranunculaceae. *Flora* **194**, 317–332.

Erbar, C. and Leins, P. (1981). Zur Spirale in Magnolienblüten. *Beitr. Biol. Pflanz.* **56**, 225–241.

Erbar, C. and Leins, P. (1983). Zur Sequenz von Blütenorganen bei einigen Magnoliiden. *Bot. Jahrb. Syst.* **103**, 433–449.

Erbar, C. and Leins, P. (1985). Studien zur Organsequenz in Apiaceen-Blüten. *Bot. Jahrb. Syst.* **105**, 379–400.

Erbar, C. and Leins, P. (1988a). Blütenentwicklungsgeschichtliche Studien an *Aralia* und *Hedera* (Araliaceae). *Flora* **180**, 391–406.

Erbar, C. and Leins, P. (1988b). Studien zur Blütenentwicklung und Pollenpräsentation bei *Brunonia australis* Smith (Brunoniaceae). *Bot. Jahrb. Syst.* **110**, 263–282

Erbar, C. and Leins, P. (1989). On the early floral development and the mechanisms of secondary pollen presentation in *Campanula, Jasione and Lobelia*. *Bot. Jarhb. Syst.* **111**, 29–55.

Erbar, C. and Leins, P. (1994). Flowers in Magnoliidae and the origin of flowers in other subclasses of the Angiosperms I. The relationships between flowers of Magnoliidae and Alismatidae. *Plant Syst. Evol. Suppl.* 8, 193–208.

Erbar, C. and Leins, P. (1995a). An analysis of the early floral development of *Pittosporum tobira* (Thunb.) Aiton and some remarks on the systematic position of the family Pittosporaceae. *Feddes Repert.* **106**, 463–473.

Erbar, C. and Leins, P. (1995b). Portioned pollen release and the syndromes of secondary pollen presentation in the Campanulales–Asterales complex. *Flora* **190**, 323–338.

Erbar, C. and Leins, P. (1997). Different patterns of floral development in whorled flowers, exemplified by Apiaceae and Brassicaceae. *Int. J. Plant Sci.* **158** (6 Suppl.), S49–S64.

Ernst, W. R. (1967). Floral morphology and systematics of *Platystemon* and its allies *Hesperomecon* and *Meconella* (Papaveraceae, Platystemonoideae). *Univ. Kansas Sci. Bull.* **47**, 25–70.

Evans, R. C. and Dickinson, T. A. (1996). North American black-fruited hawthorns II. Floral development of 10- and 20-stamen morphotypes in *Crateaegus* section *douglasii* (Rosaceae, Maloideae). *Am. J. Bot.* **83**, 961–978.

Evans, R. C. and Dickinson, T. A. (2005). Floral ontogeny and morphology in *Gillenia* ('Spiraeoideae') and subfamily Maloideae C. Weber (Rosaceae). *Int. J. Plant Sci.* **166**, 427–447.

Eyde, R. H. (1977). Reproductive structures and evolution in *Ludwigia* (Onagraceae) I. androecium, placentation, merism. *Ann. Missouri Bot. Gard.* **64**, 644–655.

Eyde, R. H. and Morgan, J. T. (1973). Floral structure and evolution in Lopezieae (Onagraceae). *Am. J. Bot.* **60**, 771–787.

Faden, R. B. (2000). Floral biology of Commelinaceae. In *Monocots: Systematics and Evolution*, ed. K. L. Wilson and D. A. Morrison. Melbourne: CSIRO, pp. 309–317.

Fay, M. M., Swensen, S. M. and Chase, M. W. (1997). Taxonomic affinities of *Medusagyne oppositifolia* (Medusagynaceae). *Kew Bull.* **52**, 111–120.

Fey, B. S. and Endress, P. K. (1983). Development and morphological interpretation of the cupule in Fagaceae. *Flora* **173**, 451–468.

Fiedler, H. (1910). Beiträge zur Kenntnis der Nyctaginaceen. *Engl. Bot. Jahrb.* **44**, 572–605.

Franz, E. (1908). Beiträge zur Kenntnis der Portulacaceen und Basellaceen. *Bot. Jahrb. Syst.* **42**, Beibl. 97, 1–28

Freitas L., Bernardello, G., Galetto, L. and Paoli, A. A. S. (2001). Nectaries and reproductive biology of *Croton sarcopetalus* (Euphorbiaceae). *Bot. J. Linn. Soc.* **136**, 267–277.

Friedman, J. and Barrett, S. C. (2008). A phylogenetic analysis of the evolution of wind pollination in the Angiosperms. *Int. J. Plant Sci.* **169**, 49–58.

Friedrich, H.-C. (1956). Studien über die natürliche Verwandtschaft der Plumbaginales und Centrospermae. *Phyton (Austria)* **6**, 220–263.

Friis, E. M. (1984). Preliminary Report of Upper Cretaceous Angiosperm reproductive organs from Sweden and their level of organisation. *Ann. Missouri Bot. Gard.* **71**, 403–418.

Friis, E. M., Pedersen, K. L. and Crane, P. R. (2006). Cretaceous Angiosperm flowers, innovation and evolution in plant reproduction. *Paleogeogr. Palaeoclimat. Palaeoecol.* **232**, 251–293.

Friis, E. M., Pedersen, K. L. and Schönenberger, J. (2006). *Normapolles* plants, a prominent component of the Cretaceous rosid diversification. *Plant Syst. Evol.* **260**, 107–140.

Froebe, H.-A. and Magin, N. (1993). Pattern analysis in the inflorescence of *Dalechampsia* L. (Euphorbiaceae). *Bot. Jahrb. Syst.* **115**, 27–44.

Fukuoka, N., Ito, M. and Iwatsuki, K. (1986). Floral anatomy of the mangrove genus *Lumnizera* (Combretaceae). *Acta Phytotax. Geobot.* **37**, 69–81.

Gallant, J. B., Kemp, J. R. and Lacroix, C. R. (1998). Floral development of dioecious staghorn sumac, *Rhus hirta* (Anacardiaceae). *Int. J. Plant Sci.* **159**, 539–549.

Galle, P. (1977). Untersuchungen zur Blütenentwicklung der Polygonaceen. *Bot. Jahrb. Syst.* **98**, 449–489.

Gandhi, K. N. and Dale Thomas, R. (1983). A note on the androecium of the genus *Croton* and flowers in general of the family Euphorbiaceae. *Phytologia* **54**, 6–8.

Gandolfo, M. A., Nixon, K. C. and Crepet, W. L. (1998). *Tylerianthus crossmanensis* gen. et sp. nov. (Aff. Hydrageaceae) from the Upper Cretaceous of New Jersey. *Am. J. Bot.* **85**, 376–386.

Gauthier, R. and Arros, J. (1963). L'anatomie de la fleur staminée de l'*Hillebrandia sandwicensis* Oliver et la vascularisation de l'étamine. *Phytomorphology* **13**, 115–127.

Ge, L.-P., Lu, A.-M. and Gong, C.-R. (2007). Ontogeny of the fertile flower in *Platycrater arguta* (Hydrangeaceae). *Int. J. Plant Sci.* **168**, 835–844.

Geitler, L. (1929). Zur Morphologie der Blüten von *Polygonum. Österr. Bot. Zeit.* **78**, 229–241.

Gelius, L. (1967). Studien zur Entwicklungsgeschichte an Blüten der Saxifragales sensu lato mit besonderer Berücksichtigung des Androeceum. *Bot. Jahrb. Syst.* **87**, 253–303.

Gemmeke, V. (1982). Entwicklungsgeschichtliche Untersuchungen an Mimosaceen-Blüten. *Bot. Jahrb. Syst.* **103**, 185–210.

Geuten, K., Becker, A., Kaufmann, K., *et al.* (2006). Petaloidy and petal identity MADS-box genes in the balsaminoid genera *Impatiens* and *Marcgravia*. *The Plant Journal* **47**, 501–518.

Glover, B. J. (2007). *Understanding Flowers and Flowering: An Integrated Approach*. Oxford, UK: Oxford University Press.

González, F. and Bello, M. A. (2009). Intraindividual variation of flowers in *Gunnera* subgenus *Panke* (Gunneraceae) and proposed apomorphies for Gunnerales. *Bot. J. Linn. Soc.*, **160**, 262–283.

González, F. and Stevenson, D. W. (2000a). Perianth development and systematics of *Aristolochia. Flora* **195**, 370–391.

González, F. and Stevenson, D. W. (2000b). Gynostemium development in *Aristolochia* (Aristolochiaceae). *Bot. Jahrb. Syst.* **122**, 249–291

Gottschling, M. (2004). Floral ontogeny in *Bourreria* (Ehretiaceae, Boraginales). *Flora* **199**, 409–423.

Graf, J. (1975). *Tafelwerk zur Pflanzensystematik mit Euartiger Bildmethode*. Munich: J. F. Lehmanns Verlag.

Grey-Wilson, C. (1980). Some observations on the floral vascular antomy of *Impatiens* (Studies in Balsaminaceae VI). *Kew Bull.* **35**, 221–227.

Groeninckx, I., Vrijdaghs, A., Huysmans, S., Smets, E. and Dessein, S. (2007). Floral ontogeny of the Afro-Madagascan genus *Mitrasacmopsis* with comments on the development of superior ovaries in Rubiaceae. *Ann. Bot.* **100**, 41–49.

Guédès, M. (1979). *Morphology of Seed-Plants*. Vaduz, Liechtenstein: J. Cramer.

Gustafsson, M. H. G. (1995). Petal venation in Asterales and related orders. *Bot. J. Linn. Soc.* **118**, 1–18.

Gustafsson, M. H. G. (2000). Floral morphology and relationships of *Clusia gundlachii* with a discussion of floral organ identity and diversity in the genus *Clusia. Int. J. Plant Sci.* **161**, 43–53.

Gustafsson, M. H. G. and Albert, V. A. (1999). Inferior ovaries and Angiosperm diversification. In *Molecular Systematics and Plant Evolution*, ed. P. M. Hollingsworth, R. M. Bateman and R. J. Gornall. London: Taylor and Francis, pp. 403–431.

Gustafsson, M. H. G. and Bittrich, V. (2002). Evolution of morphological diversity and resin secretion in flowers of *Clusia* (Clusiaceae), insights from ITS sequence variation. *Nord. J. Bot.* **22**, 183–203.

Gustafsson, M. H. G., Bittrich, V. and Stevens, P. F. (2002). Phylogeny of Clusiaceae based on *rbc*L sequences. *Int. J. Plant Sci.* **163**, 1045–1054.

Haas, P. (1976). Morphologische, anatomische und entwicklungsgeschichtliche Untersuchungen an Blüten und Früchten hochsukkulenter Mesembryanthemaceen-Gattungen – ein Beitrag zu ihrer Systematik. *Diss. Bot.* **33**, 1–256.

Haber, J. M. (1966). The comparative anatomy and morphology of the flowers and inflorescences of the Proteaceae III. Some African taxa. *Phytomorphology* **16**, 490–527.

Hall, J. C., Sytsma, K. J. and Iltis, H. H. (2002). Phylogeny of Capparaceae and Brassicaceae based on chloroplast sequence data. *Am. J. Bot.* **89**, 1826–1842.

Hardy, C. R., Davis, J. R. and Stevenson, D. W. (2004). Floral organogenesis in *Plowmanianthus* (Commelinaceae). *Int. J. Plant Sci.* **165**, 511–519.

Hardy, C. R. and Stevenson, D. W. (2000a). Development of the gametophytes, flower, and floral vasculature in *Cochliostema odoratissimum* (Commelinaceae). *Bot. J. Linn. Soc.* **134**, 131–157.

Hardy, C. R. and Stevenson, D. W. (2000b). Floral organogenesis in some species of *Tradescantia* and *Callisia* (Commelinaceae). *Int. J. Plant Sci.* **161**, 551–562.

Harris, E. M. (1995). Inflorescence and floral ontogeny in Asteraceae, a synthesis of historical and current concepts. *Bot. Rev.* **61**, 94–278.

Hayes, V., Schneider, E. L. and Carlquist, S. (2000). Floral development of *Nelumbo nucifera* Nelumbonaceae). *Int. J. Plant Sci.* **161** (6 Suppl.), S183–S191.

Hiepko, P. (1964). Das zentrifugale Androeceum der Paeoniaceae. *Ber. Dtsch. Bot. Ges.* **77**, 427–435.

Hiepko, P. (1965). Vergleichend-morphologische und entwicklungsgeschichtliche Untersuchungen über das Perianth bei den Polycarpicae. *Bot. Jahrb. Syst.* **84**, 359–508

Hiepko, P. (1966). Zur Morphologie, Anatomie und Funktion des Diskus der Paeoniaceae. *Ber. Dtsch. Bot. Ges.* **79**, 233–245.

Hileman, L. C., Kramer, E. M. and Baum, D. A. (2003). Differential regulation of symmetry genes and the evolution of floral morphologies. *Proc. Natl Acad. Sci.* **100**, 12 814–12 819.

Hilger, H. H. (1984). Wachstum und Ausbildungsformen des Gynoeceums von *Rochelia* (Boraginaceae). *Plant Syst. Evol.* **146**, 123–139.

Hofmann, U. (1973). Centrospermen-Studien 6, Morphologische Untersuchungen zur Umgrenzung und Gliederung der Aizoaceen. *Bot. Jahrb. Syst.* **93**, 247–324.

Hofmann, U. (1993). Flower morphology and ontogeny. In *Caryophyllales: Evolution and Systematics*, ed. H.-D. Behnke and T. J. Mabry. Berlin: Springer Verlag, pp. 123–166.

Hofmann, U. and Göttmann, J. (1990). *Morina* L. und *Triplostegia* Wall. ex DC. im Vergleich mit Valerianaceae und Dipsacaceae. *Bot. Jahrb. Syst.* **111**, 499–553.

Horn J. W. (2007). Dilleniaceae. In *The Families and Genera of Vascular Plants, Vol. IX*, ed. K. Kubitzki. Berlin: Springer Verlag, pp. 132–154.

Howarth, D. G. and Donoghue, M. J. (2005). Duplications in *cyc*-like genes from Dipsacales correlate with floral form. *Int. J. Plant Sci.* **166**, 357–370.

Hufford, L. D. (1989a). Structure of the inflorescence and flower of *Petalonix linearis* (Loasaceae). *Plant Syst. Evol.* **163**, 211–226.

Hufford, L. D. (1989b). The structure and potential loasaceous affinities of *Schismocarpus*. *Nord. J. Bot.* **9**, 217–227.

Hufford, L. D. (1990). Androecial development and the problem of monophyly of Loasaceae. *Can. J. Bot.* **68**, 402–419.

Hufford, L. D. (1992a). Rosidae and their relationships to other non-magnoliid Dicotyledons: a phylogenetic analysis using morphological and chemical data. *Ann. Missouri Bot. Gard.* **79**, 218–248.

Hufford, L. D. (1992b). Floral structure of *Besseya* and *Synthyris* (Scrophulariaceae). *Int. J. Plant Sci.* **153**, 217–229.

Hufford, L. D. (1995). Patterns of ontogenetic evolution in perianth diversification of *Besseya* (Scrophulariaceae). *Am. J. Bot.* **82**, 655–680.

Hufford, L. D. (1998). Early development of androecia in polystemonous Hydrangeaceae. *Am. J. Bot.* **85**, 1057–1067.

Hufford, L. D. (2001). Ontogeny and morphology of the fertile flowers of *Hydrangea* and allied genera of tribe Hydrangeeae (Hydrangeaceae). *Bot. J. Linn. Soc.* **137**, 139–187.

Hufford, L. D. (2003). Homology and developmental transformation, models for the origins of the staminodes of Loasaceae subfamily Loasoideae. *Int. J. Plant Sci.* **164** (5 Suppl.), S409–S439.

Igersheim, A., Puff, C., Leins, P. and Erbar, C. (1994). Gynoecial development of *Gaertnera* Lam. and of presumably allied taxa of the Psychotrieae (Rubiaceae), secondarily 'superior' vs. inferior ovaries. *Bot. Jahrb. Syst.* **116**, 401–414.

Ihlenfeldt, H. D. (1960). Entwicklungsgeschichtliche, morphologische und systematische Untersuchungen an Mesembryanthemen. *Feddes Repert.* **63**, 1–104.

Innes, R. L., Remphrey, W. R. and Lenz, L. M. (1989). An analysis of the development of single and double flowers in *Potentilla fruticosa*. *Can. J. Bot.* **67**, 1071–1079.

Ito, M. (1986a). Studies in the floral morphology and anatomy of Nymphaeales III. Floral anatomy of *Brasenia schreberi* Gmel. and *Cabomba caroliniana* A. Gray. *Bot. Mag. Tokyo* **99**, 169–184.

Ito, M. (1986b). Studies in the floral morphology and antomy of Nymphaeales IV. Floral anatomy of *Nelumbo nucifera*. *Acta Phytotax.. Geobot.* **37**, 82–96.

Jabour, F., Damerval, C. and Nadot, S. (2008). Evolutionary trends in the flowers of Asteridae: is polyandry an alternative to zygomorphy? *Ann. Bot.* **102**, 153–165.

Jäger-Zürn, I. (1966). Infloreszenz- und blütenmorphologische, sowie embryologische Untersuchungen an *Myothamnus* Welw. *Beitr. Biol. Pflanz.* **42**, 241–271.

Janka, H., von Balthazar, M., Alverson, W. S., Baum, D. A., Semir, J. and Bayer, C. (2008). Structure, development and evolution of the androecium in Adansonieae (core Bombacoideae, Malvaceae s.l.). *Plant Syst. Evol.* **275**, 69–91.

Jansen, R. K., Cai, Z., Raubeson, L. A., *et al.* (2007). Analysis of 81 genes from 64 plastid genomes resolves relationships in angiosperms and identifies genome-scale evolutionary patterns. *Proc. Natl Acad. Sci.* **104**, 19 369–19 374.

Jaramillo, M. A. and Kramer, E. M. (2004). *APETALA3* and *PISTILLATA* homologs exhibit novel expression patterns in the unique perianth of *Aristolochia* (Aristolochiaceae). *Evol. Devel.* **6**, 449–458.

Jaramillo, M. A. and Manos, P. S. (2001). Phylogeny and patterns of floral diversity in the genus *Piper* (Piperaceae). *Am. J. Bot.* **88**, 706–716.

Jaramillo, M. A., Manos, P. and Zimmer, E. A. (2004). Phylogenetic relationships of the perianthless Piperales, reconstructing the evolution of floral development. *Int. J. Plant Sci.* **165**, 403–416.

Jenny, M. (1988). Different gynoecium types in Sterculiaceae: ontogeny and functional aspects. In *Aspects of Floral Development*, ed. P. Leins, S. C. Tucker and P. K. Endress. Berlin: J. Cramer, pp. 225–236.

Judd, W. S., Campbell, C. S., Kellogg, E. A., Stevens, P. F. and Donoghue, M. J. (2002). *Plant Systematics: A Phylogenetic Approach*, 2nd edn. Sunderland, MA: Sinauer.

Judd, W. S. and Olmstead, R. G. (2004). A survey of tricolpate (eudicot) phylogenetic relationships. *Am. J. Bot.* **91**, 1627–1644.

Judd, W. S., Sanders, R. W. and Donoghue, M. J. (1994). Angiosperm family pairs, preliminary phylogenetic analyses. *Harvard Pap. Bot.* **5**, 1–51.

Källersjö, M., Bergqvist, G. and Anderberg, A. A. (2000). Generic realignment in primuloid families of the Ericales s.l.: a phylogenetic analysis based on DNA sequences from three chloroplast genes and morphology. *Am. J. Bot.* **87**, 1325–1341.

Kania, W. (1973). Entwicklungsgeschichtliche Untersuchungen an Rosaceenblüten. *Bot. Jahrb. Syst.* **93**, 175–246.

Karrer, A. B. (1991). *Blütenentwicklung und systematische Stellung der Papaveraceae und Capparaceae.* University of Zürich, Switzerland: unpubl. thesis.

Kaul, R. B. (1968). Floral morphology and phylogeny in the Hydrocharitaceae. *Phytomorphology* **18**, 13–35.

Kelly, L. M. (2001). Taxonomy of *Asarum* section *Asarum* (Aristolochiaceae). *Syst. Bot.* **26**, 17–53.

Kelly, L. M. and González, F. (2003). Phylogenetic relationships in Aristolochiaceae. *Syst. Bot.* **28**, 236–249.

Kessler, P. J. A. (1993). Menispermaceae. In *The Families and Genera of Vascular Plants, Vol. II*, ed. K. Kubitzki, J. G. Rohwer and V. Bittrich. Berlin: Springer Verlag, pp. 402–418.

Kim, S., Yoo, M.-J., Kong, H., *et al.* (2005). Expression of floral MADS-box genes in basal Angiosperms: implications for the evolution of floral regulators. *The Plant Journal* **43**, 724–744.

Kirchoff, B. K. (1983). Floral organogenesis in five genera of the Marataceae and in *Canna* (Cannaceae). *Am. J. Bot.* **70**, 508–523.

Kirchoff, B. K. (1988). Inflorescence and flower development in *Costus scaber* (Costaceae). *Can. J. Bot.* **62**, 339–345.

Kirchoff, B. K. (1992). Ovary structure and anatomy in the Heliconiaceae and Musaceae (Zingiberales). *Can. J. Bot.* **70**, 2490–2508.

Kirchoff, B. K. (1997). Inflorescence and flower development in the Hedychieae (Zingiberaceae), *Hedychium. Can. J. Bot.* **75**, 581–594.

Kirchoff, B. K. (2000). Hofmeister's rule and primordium shape, influences on organ position in *Hedychium coronarium* (Zingiberaceae). In *Monocots: Systematics and Evolution*, ed. K. L. Wilson and D. A. Morrison. Melbourne: CSIRO, pp. 75–83.

Klopfer, K. (1973). Florale Morphogenese und Taxonomie der Saxifragaceae sensu lato. *Feddes Repert.* **84**, 475–516.

Knapp, S. (2002). Floral diversity and evolution in the Solanaceae. In *Developmental Genetics and Plant Evolution*, ed. Q. C. B. Cronk, R. M. Bateman and J. A. Hawkins. London: Taylor and Francis, pp. 267–297.

Kocyan, A. and Endress, P. K. (2001a). Floral structure and development and systematic aspects of some 'lower' Asparagales. *Plant Syst. Evol.* **229**, 187–216.

Kocyan, A. and Endress, P. K. (2001b). Floral structure and development of *Apostasia* and *Neuwiedia* (Apostasioideae) and their relationships to other Orchidaceae. *Int. J. Plant Sci.* **162**, 847–867.

Köhler, E. (2003). Simmondsiaceae. In *The Families and Genera of Vascular Plants, Vol. II*, eds. K. Kubitzki and C. Bayer. Berlin: Springer Verlag, pp. 355–358.

Kosuge, K. (1994). Petal evolution in Ranunculaceae. *Plant Syst. Evol.* **Suppl. 8**, 185–191.

Kramer, E. M., Di Stilio, V. S. and Schlüter, P. M. (2003). Complex patterns of gene duplication in the *APETALA3* and *PISTILLATA* lineages of the Ranunculaceae. *Int. J. Plant Sci.* **164**, 1–11.

Kramer, E. M., Su, H.-J. Wu, C.-C. and Hu, J.-M. (2006). A simplified explanation for the frameshift mutation that created a novel C-terminal motif in the *APETALA3* gene lineage. *BMC Evol. Biol.* **6**, 30.

Kress, W. J. (1990). Phylogeny and classification of Zingiberales. *Ann. Missouri Bot. Gard.* **77**, 698–721.

Kron, K. A., Judd, W. S., Stevens, P. F., *et al.* (2002). A phylogenetic classification of Ericaceae: molecular and morphological evidence. *Bot. Rev.* **68**, 335–423.

Krosnick, S. E., Harris, E. M. and Freudenstein, J. V. (2006). Patterns of anomalous floral development in the Asian *Passiflora* (subgenus *Decaloba*, supersection *Disemma*). *Am. J. Bot.* **93**, 620–636.

Krüger, H. and Robbertse, P. J. (1988). Floral ontogeny of *Securidaca longepedunculata* Fresen (Polygalaceae) including inflorescence morphology. In *Aspects of Floral Development*, ed. P. Leins, S. C. Tucker and P. K. Endress. Berlin: J. Cramer., pp. 159–167.

Kubitzki, K. (1969). Monographie der Hernandiaceen. *Bot. Jahrb. Syst.* **89**, 78–148.

Kubitzki, K. (1987). Origin and significance of trimerous flowers. *Taxon* **36**, 21–28

Kubitzki, K. (1993). Hernandiaceae. In *The Families and Genera of Vascular Plants, Vol. II*, ed. K. Kubitzki, J. G. Rohwer and V. Bittrich. Berlin: Springer Verlag, pp. 334–338.

Kubitzki, K. (2004). Introduction. In *The Families and Genera of Vascular Plants, Vol. VI*, ed. K. Kubitzki. Berlin: Springer Verlag, pp. 1–11.

Kubitzki, K. (2007a). Berberidopsidaceae. In *The Families and Genera of Vascular Plants, Vol. IX*, ed. K. Kubitzki. Berlin: Springer Verlag, pp. 33–35.

Kubitzki, K. (2007b). Haloragaceae. In *The Families and Genera of Vascular Plants, Vol. IX*, ed. K. Kubitzki. Berlin: Springer Verlag, pp. 184–190.

Kuzoff, R. K., Hufford, L. and Soltis, D. E. (2001). Structural homology and developmental transformations associated with ovary diversification in *Lithophragma* (Saxifragaceae). *Am. J. Bot.* **88**, 196–205.

Lamb-Frye, A. S. and Kron, K. A. (2003). Phylogeny and character evolution in Polygonaceae. *Syst. Bot.* **21**, 17–29.

Laubengayer, R. A. (1937). Studies in the anatomy and morphology of the Polygonaceous flower. *Am. J. Bot.* **24**, 329–343.

Lehmann, N. L. and Sattler, R. (1992). Irregular floral development in *Calla palustris* (Araceae) and the concept of homeosis. *Am. J. Bot.* **79**, 1145–1157.

Lehmann, N. L. and Sattler, R. (1993). Homeosis in floral development of *Sanguinaria canadensis* and *S. canadensis* 'Multiplex' (Papaveraceae). *Am. J. Bot.* **80**, 1323–1335.

Lehmann, N. L. and Sattler, R. (1994). Floral development and homeosis in *Actaea rubra* (Ranunculaceae). *Int. J. Plant Sci.* **155**, 658–671.

Lei, L.-G. and Liang, H.-X. (1998. Floral development of dioecious species and trends of floral evolution in *Piper* sensu lato. *Bot. J. Linn. Soc.* **127**, 225–237

Leins, P. (1964a). Entwicklungsgeschichtliche Studien an Ericales-Blüten. *Bot. Jahrb. Syst.* **83**, 57–88.

Leins, P. (1964b). Die frühe Blütenentwicklung von *Hypericum hookerianum* Wight et Arn. und *H. aegypticum* L. *Ber. Dtsch. Bot. Ges.* **77**, 112–123.

Leins, P. (1965). Die Infloreszenz und Frühe Blütenentwicklung von *Melaleuca nesophila* F. Muell. (Myrtaceae). *Planta* **65**, 195–204.

Leins, P. (1967). Die frühe Blütenentwicklung von *Aegle marmelos* (Rutaceae). *Ber. Dtsch. Bot. Ges.* **80**, 320–325

Leins, P. (1988). Das zentripetale Androeceum von *Punica*. *Bot. Jahrb. Syst.* **109**, 555–561

Leins, P. and Erbar, C. (1985). Ein Beitrag zur Blütenentwicklung der Aristolochiaceen, einer Vermittlergruppe zu den Monokotylen. *Bot. Jahrb. Syst.* **107**, 343–368.

Leins, P. and Erbar, C. (1987). Studien zur Blütenentwicklung an Compositen. *Bot. Jahrb. Syst.* **108**, 381–401.

Leins, P and Erbar, C. (1988). Einige Bemerkungen zur Blütenentwicklung und systematische Stellung der Wasserpflanzen *Callitriche*, *Hippuris* und *Hydrostachys*. *Beitr. Biol. Pflanz.* **63**, 157–178.

Leins, P. and Erbar, C. (1989). Zur Blütenentwicklung und sekundären Pollenpräsentation bei *Selliera radicans*. Cav. (Goodeniaceae). *Flora* **182**, 43–56.

Leins, P. and Erbar, C. (1991). Fascicled androecia in Dilleniidae and some remarks on the *Garcinia* androecium. *Bot. Acta* **104**, 336–344.

Leins, P. and Erbar, C. (1995). Das frühe Differenzierungsmuster in den Blüten von *Saruma henryi* Oliv. (Aristolochiaceae). *Bot. Jahrb. Syst.* **117**, 365–376.

Leins, P. and Erbar, C. (1996). Early floral developmental studies in Annonaceae. In *Reproductive Morphology in Annonaceae*, ed. W. Morawetz and H. Winkler. Vienna: Akademie der Wissenschaften, Biosystematics and Ecology Series 10, pp. 1–27.

Leins, P. and Erbar, C. (2000). Die frühesten Entwicklungsstadien der Blüten bei den Asteraceae. *Bot. Jahrb. Syst.* **122**, 503–515.

Leins, P. and Erbar, C. (2007). *Blüte und Frucht*, 2nd edn. Stuttgart: E. Schweizerbart'sche Verlagsbuchhandlung.

Leins, P., Erbar, C. and van Heel, W. A. (1988). Note on the floral development of *Thottea* (Aristolochiaceae). *Blumea* **33**, 357–370.

Leins, P. and Galle, P. (1971). Entwicklungsgeschichtliche Untersuchungen an Cucurbitaceen-Blüten. *Österr. Bot. Zeit.* **119**, 531–548.

Leins, P. and Schwitalla, S. (1985). Studien an Cactaceen-Blüten I. Einige Bermerkungen zur Blütenentwicklung von *Pereskia*. *Beitr. Biol. Pflanz.* **60**, 313–323.

Leins, P. and Stadler, P. (1973). Entwicklungsgeschichtliche Untersuchungen am Androecium der Alismatales. *Österr. Bot. Zeit.* **122**, 145–165.

Leins, P. and Winhard, W. (1973). Entwicklungsgeschichtliche Studien an Loasaceenblüten. *Österr. Bot. Zeit.* **122**, 145–165.

Leredde, C. (1955). Sur la position des étamines chez quelques *Echium*. *Bull. Soc. Hist. Nat. Toulouse* **90**, 369–372.

Le Roux, L. G. and Kellogg, E. A. (1999). Floral development and the formation of unisexual spikelets in the Andropogoneae (Poaceae). *Am. J. Bot.* **86**, 354–366.

Levyns, M. R. (1949). The floral morphology of some South African members of Polygalaceae. *J. S. Afr. Bot.* **15**, 79–92.

Leyser, O. and Day, S. (2003). *Mechanisms in Plant Development*. Oxford, UK: Blackwell.

Li, P. and Johnston, M. O. (2000). Heterochrony in plant evolutionary studies through the twentieth century. *Bot. Rev.* **66**, 57–88.

Liang, H.-X. and Tucker, S. C. (1989). Floral development in *Gymnotheca chinensis*. *Am. J. Bot.* **76**, 806–819.

Lindenhofer, A. and Weber, A. (1999a). Polyandry in Rosaceae: evidence for a spiral origin of the androecium in Spiraeoideae. *Bot. Jahrb. Syst.* **121**, 553–582.

Lindenhofer, A. and Weber, A. (1999b). The spiraeoid androecium of Pyroideae and Amygdaloideae (Rosaceae). *Bot. Jahrb. Syst.* **121**, 583–605.

Lindenhofer, A. and Weber, A. (2000). Structural and developmental diversity of the androecium of Rosoideae (Rosaceae). *Bot. Jahrb. Syst.* **122**, 63–91.

Linder, H. P. (1991). A review of the southern African Restionaceae. *Contr. Bolus Herb.* **13**, 209–264.

Linder, H. P. (1992a). The gynoecia of Australian Restionaceae, morphology, anatomy and systematic implications. *Aust. Syst. Bot.* **5**, 227–245.

Linder, H. P. (1992b). The structure and evolution of the pistillate flower of the African Restionaceae. *Bot. J. Linn. Soc.* **109**, 401–425.

Linder, H. P. (1998). Morphology and the evolution of wind pollination. In *Reproductive Biology in Systematics, Conservation and Economic Botany*, ed. S. J. Owens and P. J. Rudall. Kew, UK: Royal Botanic Gardens, pp. 123–135.

Linder, H. P. and Rudall, P. J. (2005). Evolutionary history of Poales. *Annu. Rev. Evol. Syst.* **36**, 107–124.

Lindsey, A. A. (1940). Floral anatomy in the Gentianaceae. *Am. J. Bot.* **27**, 640–652.

Litt, A. and Stevenson, D. W. (2003a). Floral development and morphology of Vochysiaceae I. The structure of the gynoecium. *Am. J. Bot.* **90**, 1533–1547.

Litt, A. and Stevenson, D. W. (2003b). Floral development and morphology of Vochysiaceae II. The position of the single fertile stamen. *Am. J. Bot.* **90**, 1548–1559.

Lorence, D. H. (1985). A monograph of the Monimiaceae (Laurales) in the Malagasy region (Southwest Indian Ocean). *Ann. Missouri Bot. Gard.* **72**, 1–165.

Lüders, H. (1907). Systematische Untersuchungen über die Caryophyllaceen mit einfachem Diagramm. *Bot. Jahrb. Syst.* **40**, Beibl. 91, 1–37.

Luo, D., Carpenter, R., Vincent, C., Copsey, L. and Coen, E. (1996). Origin of floral asymmetry in *Antirrhinum*. *Nature* **383**, 794–799.

Ma, O. S. W. and Saunders, R. M. K. (2003). Comparative floral ontogeny of *Maesa* (Maesaceae), *Aegiceras* (Myrsinaceae) and *Embelia* (Myrsinaceae), taxonomic and phylogenetic implications. *Plant Syst. Evol.* **243**, 39–58.

Maas, P. J. M. and Rübsamen, T. (1986). *Triuridaceae, Flora neotropica* no. 40. New York: Hafner.

Mabberley, D. (2000). *Arthur Harry Church: The Anatomy of Flowers*. London: The Natural History Museum.

Mabry, T. J. (1977). The order Centrospermae. *Ann. Missouri Bot. Gard.* **64**, 210–220.

Macfarlane, J. M. (1908). Nepenthaceae. In *Das Pflanzenreich IV, 3*, ed. A. Engler. Leipzig: W. Engelmann, pp. 1–92.

Magallón, S. (2007). From fossils to molecules, phylogeny and the core eudicot floral groundplan in Hamamelidoideae (Hamamelidaceae, Saxifragales). *Syst. Bot.* **32**, 317–347.

Malécot, V. and Nickrent, D. L. (2008). Molecular phylogenetic relationships of Olacaceae and related Santalales. *Syst. Bot.* **33**, 97–106.

Marazzi, B. and Endress, P. K. (2008). Patterns and development of floral asymmetry in *Senna* (Leguminosae, Cassiinae). *Am. J. Bot.* **95**, 22–40.

Marazzi, B., Endress, P. K., Paganucci de Quieroz, L. and Conti, E. (2006). Phylogenetic relationships within *Senna* (Leguminosae, Cassiinae) based on three chloroplast DNA regions: patterns in the evolution of floral symmetry and extrafloral nectaries. *Am. J. Bot.* **93**, 288–303.

Matthews, M. L. and Endress, P. K. (2002). Comparative floral structure and systematics in Oxalidales (Oxalidaceae, Connaraceae, Brunelliaceae, Cephalotaceae, Cunoniaceae, Elaeocarpaceae, Tremandraceae). *Bot. J. Linn. Soc.* **140**, 321–381.

Matthews, M. L. and Endress, P. K. (2004). Comparative floral structure and systematics in Cucurbitales (Corynocarpaceae, Coriariaceae, Tetramelaceae,

Datiscaceae, Begoniaceae, Cucurbitaceae, Anisophylleaceae). *Bot. J. Linn. Soc.* **145**, 129–185.

Matthews, M. L. and Endress, P. K. (2005). Comparative floral structure and systematics in Celastrales (Celastraceae, Parnassiaceae, Lepidobotryaceae). *Bot. J. Linn. Soc.* **149**, 129–194.

Matthews, M. L. and Endress, P. K. (2008). Comparative floral struture and systematics in Chrysobalanaceae s.l. (Chrysobalanaceae, Dichapetalaceae, Euphroniaceae, Trigoniaceae; Malpighiales). *Bot. J. Linn. Soc.* **157**, 249–309.

Mayr, B. (1969). Ontogenetische Studien an Myrtales-Blüten. *Bot. Jahrb. Syst.* **89**, 210–271.

Mayr, E. M. and Weber, A. (2006). Calceolariaceae: floral development and systematic implications. *Am. J. Bot.* **93**, 327–343.

Medan, D. and Hilger, H. H. 1992. Comparative flower and fruit morphogenesis in *Colubrina* (Rhamnaceae) with special reference to *C. asiatica*. *Am. J. Bot.* **79**, 809–819.

Melchior, H. (1925). Violaceae. In *Die natürlichen Pflanzenfamilien XXI*, 2nd edn, ed. A. Engler and K. Prantl. Leipzig: Wilhelm Engelmann, pp. 329–377.

Melchior, H. (1964). *Engler's Syllabus der Pflanzenfamilien*. Berlin: Gebr. Borntraeger.

Melville, R. (1984). The affinity of *Paeonia* and a second genus of Paeoniaceae. *Kew Bull.* **38**, 87–105.

Merckx, V., Schols, V., Maas-van de Kamer, H., Maas, P., Huysmans, S. and Smets, E. (2006). Phylogeny and evolution of Burmanniaceae (Dioscoreales) based on nuclear and mitochondrial data. *Am. J. Bot.* **93**, 1684–1698.

Merino Sutter, D., Foster, P. I. and Endress, P. K. (2006). Female flowers and systematic position of Picodendraceae (Euphorbiaceae s.l., Malpighiales). *Plant Syst. Evol.* **261**, 187–215.

Michaelis, P. (1924). Blütenmorphologische Untersuchungen an den Euphorbiaceen, unter besonderer Berücksichtigung der Phylogenie der Angiospermenblüte. *Goebel Bot. Abhandl.* **3**, 1–150.

Milby, T. H. (1980). Studies in the floral anatomy of *Claytonia*. *Am. J. Bot.* **67**, 1046–1050.

Mione, T. and Bogle, A. L. (1990). Comparative ontogeny of the inflorescence and flower of *Hamamelis virginiana* and *Loropetalum chinense* (Hamamelidaceae). *Am. J. Bot.* **77**, 77–91.

Mitchell, C. H. and Diggle, P. K. (2005). The evolution of unisexual flowers: morphological and functional convergence results from diverse developmental transitions. *Am. J. Bot.* **92**, 1068–1076.

Moncur, M. W. (1988). *Floral Development of Tropical and Subtropical Fruit and Nut Species: An Atlas of Scanning Electron Micrographs*. Melbourne: CSIRO.

Mondragón-Palomino, M. and Theissen, G. (2008). MADS about the evolution of orchid flowers. *Tr. Plant Sci.* **13**, 51–59.

Moody, M. and Hufford, L. (2000a). Floral development and structure of *Davidsonia* (Cunoniaceae). *Can. J. Bot.* **78**, 1034–1043.

Moody, M. and Hufford, L. (2000b). Floral ontogeny and morphology of *Cevallia*, *Fuertesia*, and *Gronovia* (Loasaceae subfamily Gronovioideae). *Int. J. Plant Sci.* **161**, 869–883.

Moore, H. E. (1973). The major groups of palms and their distribution. *Gentes Herb.* **11**, 27–141.

Moore, H. E. and Uhl, N. W. (1982). Major trends of evolution in palms. *Bot. Rev.* **48**, 1–69.

Moore, M. J., Bell, C. D., Soltis, P. S. and Soltis, D. E. (1997). Using plastid genome-scale data to resolve enigmatic relationships among basal angiosperms. *Proc. Natl Acad. Sci.* **104**, 19 363–19 368.

Morgan, D. R. and Soltis, D. E. (1993). Phylogenetic relationships among members of Saxifragaceae sensu lato based on *rbcL* sequence data. *Ann. Missouri Bot. Gard.* **80**, 631–660.

Mort, M. E., Soltis, D. E., Soltis, P. S., Francisco-ortega, J. and Santos-Guerra, A. (2001). Phylogenetic relationships and evolution of Crassulaceae inferred from *matK* sequence data. *Am. J. Bot.* **88**, 76–91.

Moylan, E. C., Rudall, P. J. and Scotland, R. W. (2004). Comparative floral anatomy of Strobilanthinae (Acanthaceae), with particular reference to internal partitioning of the flower. *Plant Syst. Evol.* **249**, 77–98.

Murbeck, S. (1912). Üntersuchungen über den Blütenbau der Papaveraceen. *Kungl. Sv. Vet. Akad. Handl.* **50**, 1–168.

Nair, N. C. and Abraham, V. (1962). Floral morphology of a few species of Euphorbiaceae. *Proc. Indian Acad. Sci.* **B,56**, 1–12.

Nandi, O. I. (1998). Floral development and systematics of Cistaceae. *Plant Syst. Evol.* **212**, 107–134.

Narayana, L. L. and Rao, D. (1966). Floral morphology of Linaceae. *J. Jap. Bot.* **41**, 1–10.

Narayana, L. L. and Rao, D. (1969). Contributions to the floral anatomy of Linaceae 1. *J. Jap. Bot.* **44**, 289–294.

Narayana, L. L. and Rao, D. (1973). Contributions to the floral anatomy of Linaceae 5. *J. Jap. Bot.* **48**, 205–208.

Narayana, L. L. and Rao, D. (1976). Contributions to the floral anatomy of Linaceae 6. *J. Jap. Bot.* **51**, 92–96.

Narayana, L. L. and Rao, D. (1978). Contributions to the floral anatomy of Linaceae 11. *J. Jap. Bot.* **53**, 12–14.

Narayana, R. (1958a). Morphological and embryological studies in the family Loranthaceae II. *Lysiana exocarpi* (Behr.) van Tieghem. *Phytomorphology* **8**, 147–168.

Narayana, R. (1958b). Morphological and embryological studies in the family Loranthaceae III. *Nuytsia floribunda* (Labill.) R. Br. *Phytomorphology* **8**, 306–323.

Narita, M. and Takahashi, H. (2008). A comparative study of shoot and floral development in *Paris tetraphylla* and *P. verticillata* (Trilliaceae). *Plant Syst. Evol.* **272**, 67–78.

Ng, F. (1991). The relationship of the Sapotaceae within the Ebenales. In *The Genera of Sapotaceae*, ed. T. D. Pennington. Kew, UK: Royal Botanic Garden and Bronx, NY: New York Botanic Garden, pp. 1–14.

Nicholas, A. and Baijnath, H. (1994). A consensus classification for the order Gentianales with additional details on the suborder Apocynineae. *Bot. Rev.* **60**, 440–482.

Nickrent, D. L. and Malécot. V. (2001). A molecular phylogeny of Santalales. In *Proceedings of the 7th International Parasitic Weed Symposium*, ed. A. Fer, P. Thalouarn, D. M. Joel, L. J. Musselman, C. Parker and J. A. C. Verkleij. Nantes: Faculté des Sciences, Université de Nantes, pp. 69–74.

Niedenzu, F. (1897). Malpighiaceae. In *Die natürlichen Pflanzenfamilien III, 4*, 1st edn., ed. A. Engler and K. Prantl. Leipzig: W. Engelmann, pp. 41–74.

Niedenzu, F. (1925). Frankeniaceae. In *Die Natürlichen Pflanzenfamilien II, 1*, 2nd edn., ed. A. Engler and K. Prantl. Leipzig: W. Engelmann, pp. 276–281.

Nishino, E. (1983a). Corolla tube formation in the Tubiflorae and Gentianales. *Bot. Mag. Tokyo* **96**: 223–243.

Nishino, E. (1983b). Corolla tube formation in the Primulales and Ericales. *Bot. Mag. Tokyo* **96**: 319–342.

Nishino, E. (1988). Early floral organogenesis in *Tripetaleia* (Ericaceae). In *Aspects of Floral Development*, ed. P. Leins, S. C. Tucker and P. K. Endress. Berlin: J. Cramer, pp. 181–190.

Nooteboom, H. P. (1993). Magnoliaceae. In *The Families and Genera of Vascular Plants, Vol. II*, ed. K. Kubitzki, J. G. Rohwer and V. Bittrich. Berlin: Springer Verlag, pp. 391–401.

Oh, S.-H. and Manos, P. S. (2008). Molecular phylogenetics and cupule evolution in Fagaceae as inferred from nuclear *CRABS CLAW* sequences. *Taxon* **57**, 434–451.

Okamoto, M. (1983). Floral development of *Castanopsis cuspidata* var. *sieboldii. Acta Phytotax. Geobot.* **34**, 10–17.

Okamoto, M., Kosuge, K. and Fukuoka, N. (1992). Pistil development and parietal placentation in the pseudomomerous ovary of *Zelkova serrata* (Ulmaceae). *Am. J. Bot.* **79**, 921–927.

Olmstead, R. G., DePamphilis, C. W., Wolfe, A. D., Young, N. D., Elisons, W. J. and Reeves, P. A. (2001). Disintegration of the Scrophulariaceae. *Am. J. Bot.* **88**, 348–361.

Olson, M. E. (2003). Ontogenetic origins of floral bilateral symmetry in Moringaceae (Brassicales). *Am. J. Bot.* **90**, 49–71.

Orlovich, D. A., Drinnan, A. N. and Ladiges, P. Y. (1996). Floral development in the *Metrosideros* group (Myrtaceae) with special emphasis on the androecium. *Telopea* **6**, 689–719.

Orlovich, D. A., Drinnan, A. N. and Ladiges, P. Y. (1999). Floral development in *Melaleuca* and *Callistemon* (Myrtaceae). *Aust. Syst. Bot.* **11**, 689–710.

Pabón-Mora, N. and González, F. (2008). Floral ontogeny of *Telipogon* spp. (Orchidaceae) and insights on the perianth symmetry in the family. *Int. J. Plant Sci.* **169**, 1159–1173.

Pacini, E., Nepi, M. and Vesprini, J. L. (2003). Nectar biodiversity: a short review. *Plant Syst. Evol.* **238**, 7–21.

Pai, R. M. (1965). Morphology of the flower in the Cannaceae. *J. Biol. Sciences* **8**, 4–8.

Pai, R. M. and Tilak, V. D. (1965). Septal nectaries in the Scitamineae. *J. Biol. Sciences* **8**, 1–3.

Pauwels, L. (1993). *Nzayilu N'ti: Guide des Arbres et Arbustes de la Région de Kinshasa-Brazzaville*. Meise: Jardin Botanique National de Belgique.

Pauzé, F. and Sattler, R. (1978). l'Androcée centripète d'*Ochna atropurpurea*. *Can. J. Bot.* **56**, 2500–2511.

Payer, J. B. (1857). *Traité d'Organogénie Comparée de la Fleur*. Paris: Victor Masson.

Pennington, T. D. (2004). Sapotaceae. In *The Families and Genera of Vascular Plants, Vol. VI*, ed. K. Kubitzki. Berlin: Springer Verlag, pp. 390–421.

Philipson, W. R. (1970). Constant and variable features of the Araliaceae. *Bot. J. Linn. Soc.* **63** (Suppl. 1), 87–100.

Philipson, W. R. (1985). Is the grass gynoecium monocarpellary? *Am. J. Bot.* **72**, 1954–1961.

Philipson, W. R. (1993). Monimiaceae. In *The Families and Genera of Vascular Plants, Vol. II*, ed. K. Kubitzki, J. G. Rohwer and V. Bittrich. Berlin: Springer Verlag, pp. 426–437.

Pilger, R. (1935). Santalaceae. In *Die Natürlichen Pflanzenfamilien 16b*, ed. A. Engler. and K. Prantl. Leipzig: W. Engelmann, pp. 52–91.

Pluys, T. (2002). *Bloemontogenetische studie van de Rosaceae, Dipsacaceae en Malvaceae met bijzondere aandacht voor de bijkelk*. Katholieke Universiteit Leuven, Belgium: unpubl. thesis.

Plunkett, G. M. (2001). Relationship of the order Apiales to subclass Asteridae: a re-evaluation of morphological characters based on insights from molecular data. *Edinburgh J. Bot.* **58**, 183–200.

Plunkett, G. M., Soltis, D. E. and Soltis, P. S. (1996). Higher level relationships of Apiales (Apiaceae and Araliaceae) based on phylogenetic analysis of *rbcL* sequences. *Am. J. Bot.* **83**, 499–515.

Prance, G. T. and Mori, S. A. (2004). Lecythidaceae. In *The Families and Genera of Vascular Plants, Vol. VI*, ed. K. Kubitzki. Berlin: Springer Verlag, pp. 221–232.

Prenner, G. (2004a). New aspects in floral development of Papilionoideae: initiated but suppressed bracteoles and variable initiation of sepals. *Ann. Bot.* **93**, 537–545.

Prenner, G. (2004b). Floral development in *Polygala myrtifolia* (Polygalaceae) and its similarities with Leguminosae. *Plant Syst. Evol.* **249**, 67–76.

Prenner, G. (2004c). Floral ontogeny in *Calliandra angustifolia* (Leguminosae, Mimosoideae, Ingeae) and its systematic implications. *Int. J. Plant Sci.* **165**, 417–426.

Prenner, G. and Klitgaard, B. B. (2008). Towards unlocking the deep nodes of Leguminosae: floral development and morphology of the enigmatic *Duparquetia orchidacea* (Leguminosae, Caesalpinioideae). *Am. J. Bot.* **95**: 1349–1365.

Prenner, G. and Rudall, P. (2007). Comparative ontogeny of the cyathium in *Euphorbia* (Euphorbiaceae) and its allies, exploring the organ-flower-inflorescence boundary. *Am. J. Bot.* **94**, 1612–1629.

Prenner, G. and Rudall, P. (2008). The branching stamens of *Ricinus* and the homologies of the Angiosperm stamen fascicle. *Int. J. Plant Sci.* **169**, 735–744.

Puff, C. and Igersheim, A. (1991). The flowers of *Paederia* L. (Rubiaceae-Paederieae). *Opera Bot. Belg.* **3**, 55–75.

Puglisi, C. (2007). *Multiplications of floral parts in the genus* Conostegia *(Melastomataceae)*. Royal Botanic Garden Edinburgh: unpubl. MSc thesis.

Qiu, Y.-L., Lee, J., Bernasconi-Quadroni, F., *et al.* (1999). The earliest Angiosperms: evidence from mitochondrial, plastid and nuclear genomes. *Nature* **402**, 404–407.

Rama Devi, D. (1991a). Floral anatomy of *Hypseocharis* (Oxalidaceae) with a discussion on its systematic position. *Plant Syst. Evol.* **177**, 161–164.

Rama Devi, D. (1991b). Floral anatomy of six species of *Impatiens*. *Feddes Repert.* **102**, 395–398.

Ramirez-Domenech, J. I. and Tucker, S. C. (1990). Comparative ontogeny of the perianth in Mimosoid legumes. *Am. J. Bot.* **77**, 624–635.

Rao, V. S. (1953). The floral anatomy of some bicarpellatae 1. Acanthaceae. *J. Univ. Bombay* **21**, 1–34.

Rao, V. S. (1963). The epigynous glands of Zingiberaceae. *New Phytol.* **62**, 342–349.

Rao, V. S. (1974). The nature of the perianth in *Eleagnus* on the basis of floral anatomy with some comments on the systematic position of Eleagnaceae. *J. Indian Bot. Soc.* **53**, 156–161.

Rao, V. S., Karnik, H. and Gupte, K. (1954). The floral anatomy of some Scitamineae I. *J. Indian Bot. Soc.* **33**, 118–147.

Rasmussen, D. A., Kramer, E. M. and Zimmer, E. A. (2009). One size fits all? Molecular evidence for a commonly inherited petal identity program in Ranunculales. *Am. J. Bot.* **96**, 96–109.

Reeves, P. A. and Olmstead, R. G. (1998). Evolution of novel morphological and reproductive traits in a clade containing *Antirrhinum majus* (Scrophulariaceae). *Am. J. Bot.* **85**, 1047–1056.

Reinheimer, R., Pozner, R. and Vegetti, A. C. (2005). Inflorescence, spikelet, and floral development in *Panicum maximum* and *Urochloa plantaginea* (Poaceae). *Am. J. Bot.* **92**, 565–575.

Remizowa, M. and Sokoloff, D. (2003). Inflorescence and floral morphology in *Tofieldia* (Tofieldiaceae) compared with Araceae, Acoraceae and Alismatales s.str. *Bot. Jahrb. Syst.* **124**, 255–271.

Remizowa, M., Sokoloff, D. and Kondo, K. (2008). Floral evolution in the monocot family Nartheciaceae (Dioscoreales), evidence from anatomy and development in *Metanarthecium luteo-viride* Maxim. *Bot. J. Linn. Soc.* **158**, 1–18.

Remizowa, M., Sokoloff, D. and Rudall, P. J. (2006). Evolution of the monocot gynoecium, evidence from comparative morphology and development in *Tofieldia, Japanolirion, Petrosavia* and *Narthecium*. *Plant Syst. Evol.* **258**, 183–209.

Renner, S. S. (1999). Circumscription and phylogeny of the Laurales, evidence from molecular and morphological data. *Am. J. Bot.* **86**, 1301–1315.

Richardson, F. C. (1969). Morphological studies of the Nymphaeae IV. Structure and development of the flower of *Brasenia schreberi* Gmel. *Univ. Calif. Publ. Bot.* **47**, 1–101.

Richardson, J. E., Fay, M. F., Cronk, Q. C. B., Bowman, D. and Chase, M. W. (2000). A phylogenetic analysis of Rhamnaceae using *RbcL* and *trnL-F* plastid DNA sequences. *Am. J. Bot.* **87**, 1309–1324.

Robbrecht, E. (1988). Tropical woody Rubiaceae. *Opera Bot. Belg.* **1**, 1–271.

Rodman, J. E., Soltis, P. S., Soltis, D. E., Sytsma, K. J. and Karol, K. G. (1998). Parallel evolution of glucosinolate biosynthesis inferred from congruent nuclear and plastid gene phylogenies. *Am. J. Bot.* **85**, 997–1006.

Roels, P., Ronse De Craene, L. P. and Smets, E. F. (1997). A floral ontogenetic investigation of the Hydrangeaceae. *Nord. J. Bot.* **17**, 235–254.

Roels, P. and Smets, E. F. (1994). A comparative floral ontogenetical study between *Adoxa moschatellina* and *Sambucus ebulus*. *Belg. J. Bot.* **127**, 157–170.

Roels, P. and Smets, E. F. (1996). A floral ontogenetic study in the Dipsacales. *Int. J. Plant Sci.* **157**, 203–218.

Rohrer, J. R., Robertson, K. R. and Phipps, J. B. (1994). Floral morphology of Maloideae (Rosaceae) and its systematic relevance. *Am. J. Bot.* **81**, 574–581.

Rohweder, O. (1965). Centrospermen-Studien 2. Entwicklung und morphologische Deutung des Gynöciums bei *Phytolacca*. *Bot. Jahrb. Syst.* **84**, 509–526.

Rohweder, O. and Huber, K. (1974). Centrospermen-Studien 7. Beobachtungen und Anmerkungen zur Morphologie und Entwicklungsgeschichte einiger Nyctaginaceen. *Bot. Jahrb. Syst.* **94**, 327–359.

Rohwer, J. (1993a). Lauraceae. In *The Families and Genera of Vascular Plants, Vol. II*, ed. K. Kubitzki, J. G. Rohwer and V. Bittrich. Berlin: Springer Verlag, pp. 366–391.

Rohwer, J. (1993b). Phytolaccaceae. In *The Families and Genera of Vascular Plants, Vol. II*, ed. K. Kubitzki, J. G. Rohwer and V. Bittrich. Berlin: Springer Verlag, pp. 506–515.

Ronse De Craene, L. P. (1988). Two types of ringwall formation in the development of complex polyandry. *Bull. Soc. Roy. Bot. Belg.* **121**, 122–124.

Ronse De Craene, L. P. (1989a). The flower of *Koenigia islandica* L. (Polygonaceae), an interpretation. *Watsonia* **17**, 419–423.

Ronse De Craene, L. P. (1989b). The floral development of *Cochlospermum tinctorium* and *Bixa orellana* with special emphasis on the androecium. *Am. J. Bot.* **76**, 1344–1359.

Ronse De Craene, L. P. (1990). Morphological studies in Tamaricales I. Floral ontogeny and anatomy of *Reaumuria vermiculata* L. *Beitr. Biol. Pflanz.* **65**, 181–203.

Ronse De Craene, L. P. (2002). Floral development and anatomy of *Pentadiplandra* (Pentadiplandraceae): a key genus in the identification of floral morphological trends in the core Brassicales. *Can. J. Bot.* **80**, 443–459.

Ronse De Craene, L. P. (2003). The evolutionary significance of homeosis in flowers, a morphological perspective. *Int. J. Plant Sci.* **164** (5 Suppl.), S225–S235.

Ronse De Craene, L. P. (2004). Floral development of *Berberidopsis corallina*: a crucial link in the evolution of flowers in the core eudicots. *Ann. Bot.* **94**, 1–11.

Ronse De Craene, L. P. (2005). Floral developmental evidence for the systematic position of *Batis* (Bataceae). *Am. J. Bot.* **92**, 752–760.

Ronse De Craene, L. P. (2007). Are petals sterile stamens or bracts? The origin and evolution of petals in the core eudicots. *Ann. Bot.* **100**, 621–630.

Ronse De Craene, L. P. (2008). Homology and evolution of petals in the core eudicots. *Syst. Bot.* **33**, 301–325.

Ronse De Craene, L. P. and Akeroyd, J. R. (1988). Generic limits in *Polygonum* and related genera (Polygonaceae) on the basis of floral characters. *Bot. J. Linn. Soc.* **98**, 321–371.

Ronse De Craene, L. P., Clinckemaillie, D. and Smets, E. F. (1993). Stamen-petal complexes in Magnoliatae. *Bull. Jard. Bot. Nat. Belg.* **62**, 97–112.

Ronse De Craene, L. P., De Laet, J. and Smets, E. F. (1996). Morphological studies in Zygophyllaceae II. The floral development and vascular anatomy of *Peganum harmala*. *Am. J. Bot.* **83**, 201–215.

Ronse De Craene, L. P., De Laet, J. and Smets, E. F. (1998). Floral development and anatomy of *Moringa oleifera* (Moringaceae): what is the evidence for a capparalean or sapindalean affinity? *Ann. Bot.* **82**, 273–284.

Ronse De Craene, L. P. and Haston, E. (2006). The systematic relationships of glucosinolate-producing plants and related families: a cladistic investigation based on morphological and molecular characters. *Bot. J. Linn. Soc.* **151**, 453–494.

Ronse De Craene, L. P., Hong, S.-P. and Smets, E. F. (2004). What is the taxonomic status of *Polygonella*? Evidence from floral morphology. *Ann. Missouri Bot. Gard.* **91**, 320–345.

Ronse De Craene, L. P., Linder, H. P., Dlamini, T. and Smets, E. F. (2001). Evolution and development of floral diversity of Melianthaceae, an enigmatic Southern African family. *Int. J. Plant Sci.* **162**: 59–82.

Ronse De Craene, L. P., Linder, H. P. and Smets, E. F. (2000). The questionable relationship of *Montinia* (Montiniaceae): evidence from a floral ontogenetic and anatomical study. *Am. J. Bot.* **87**, 1408–1424.

Ronse De Craene, L. P., Linder, H. P. and Smets, E. F. (2001). Floral ontogenetic evidence in support of the *Willdenowia* clade of South African Restionaceae. *J. Plant Res.* **114**, 329–342.

Ronse De Craene, L. P. Linder, H. P. and Smets, E. F. (2002). Ontogeny and evolution of the flower of South African Restionaceae with special emphasis on the gynoecium. *Plant Syst. Evol.* **231**, 225–258.

Ronse De Craene, L. P. and Miller, A. G. (2004). Floral development and anatomy of *Dirachma socotrana* (Dirachmaceae), a controversial member of the Rosales. *Plant Syst. Evol.* **249**, 111–127.

Ronse De Craene, L. P. and Smets, E. F. (1987). The distribution and the systematic relevance of the androecial characters Oligomery and Polymery in the Magnoliophytina. *Nord. J. Bot.* **7**, 239–253.

Ronse De Craene, L. P., and Smets, E. F. (1990a). The floral development of *Popowia whitei* (Annonaceae). *Nord. J. Bot.* **10**, 411–420. [Correction in *Nord. J. Bot.* **11** (1991), 420.]

Ronse De Craene, L. P. and Smets, E. F. (1990b). The systematic relationship between Begoniaceae and Papaveraceae, a comparative study of their floral development. *Bull. Jard. Bot. Nat. Belg.* **60**, 229–273.

Ronse De Craene, L. P. and Smets, E. F. (1991a). The impact of receptacular growth on polyandry in the Myrtales. *Bot. J. Linn. Soc.* **105**, 257–269.

Ronse De Craene, L. P. and Smets, E. F. (1991b). The floral nectaries of *Polygonum* s.l. and related genera (Persicarieae and Polygoneae), position, morphological nature and semophylesis. *Flora* **185**, 165–185.

Ronse De Craene, L. P. and Smets, E. F. (1991c). The floral ontogeny of some members of the Phytolaccaceae (subfamily Rivinoideae) with a discussion of the evolution of the androecium in the Rivinoideae. *Biol. Jahrb. Dodonaea* **59**, 77–99.

Ronse Decraene, L. P. and Smets, E. F. (1991d). Androecium and floral nectaries of *Harungana madagascariensis* (Clusiaceae). *Plant Syst. Evol.* **178**, 179–194.

Ronse De Craene, L. P. and Smets, E. F. (1991e). Morphological studies in Zygophyllaceae I. The floral development and vascular anatomy of *Nitraria retusa*. *Am. J. Bot.* **78**, 1438–1448.

Ronse De Craene, L. P. and Smets, E. F. (1992a). Complex polyandry in the Magnoliatae, definition, distribution and systematic value. *Nord. J. Bot.* **12**, 621–649.

Ronse De Craene, L. P. and Smets, E. F. (1992b). An updated interpretation of the androecium of the Fumariaceae. *Can. J. Bot.* **70**, 1765–1776.

Ronse De Craene, L. P. and Smets, E. F. (1993). The distribution and systematic relevance of the androecial character polymery. *Bot. J. Linn. Soc.* **113**, 285–350.

Ronse De Craene, L. P. and Smets, E. F. (1994). Merosity, definition, origin and taxonomic significance. *Plant Syst. Evol.* **191**, 83–104.

Ronse De Craene, L. P. and Smets, E. F. (1995a). The androecium of monocotyledons. In *Monocotyledons: Systematics and Evolution*, ed. P. J. Rudall, P. Cribb, D. F. Cutler, and C. J. Hymphries. Kew, UK: Royal Botanic Gardens, pp. 243–254.

Ronse De Craene, L. P. and Smets, E. F. (1995b). The distribution and systematic relevance of the androecial character oligomery. *Bot. J. Linn. Soc.* **118**, 193–247.

Ronse De Craene, L. P. and Smets, E. F. (1995c). Evolution of the androecium in the Ranunculiflorae. *Plant Syst. Evol.* Suppl. **9**, 63–70.

Ronse Decraene, L. P. and Smets, E. F. (1996a). The morphological variation and systematic value of stamen pairs in the Magnoliatae. *Feddes Repert.* **107**, 1–17.

Ronse Decraene, L. P. and Smets, E. F. (1996b). The floral development of *Neurada procumbens* L. (Neuradaceae). *Acta Bot. Neerl.* **45**, 229–241.

Ronse De Craene, L. P. and Smets, E. F. (1997a). A floral ontogenetic study of some species of *Capparis* and *Boscia*, with special emphasis on the androecium. *Bot. Jahrb. Syst.* **119**, 231–255.

Ronse De Craene, L. P. and Smets, E. F. (1997b). Evidence for carpel multiplications in the Capparaceae. *Belg. J. Bot.* **130**, 59–67.

Ronse De Craene, L. P. and Smets, E. F. (1998a). Notes on the evolution of androecial organisation in the Magnoliophytina (Angiosperms). *Bot. Acta* **111**, 77–86.

Ronse De Craene, L. P. and Smets, E. F. (1998b). Meristic changes in gynoecium morphology, exemplified by floral ontogeny and anatomy. In *Reproductive Biology in Systematics, Conservation and Economic Botany*, ed. S. J. Owens and P. J. Rudall. Kew, UK: Royal Botanic Gardens, pp. 85–112.

Ronse De Craene, L. P. and Smets, E. F. (1999a). The floral development and anatomy of *Carica papaya* (Caricaceae). *Can. J. Bot.* **77**, 582–598.

Ronse De Craene, L. P. and Smets E. F. (1999b). Similarities in floral ontogeny and anatomy between the genera *Francoa* (Francoaceae) and *Greyia* (Greyiaceae). *Int. J. Plant Sci.* **160**, 377–393.

Ronse De Craene, L. P. and Smets E. F. (2000). Floral development of *Galopina tomentosa* with a discussion of sympetaly and placentation in the Rubiaceae. *Syst. Geogr. Plants* **70**, 155–170.

Ronse De Craene, L. P. and Smets E. F. (2001a). Staminodes: their morphological and evolutionary significance. *Bot. Rev.* **67**, 351–402.

Ronse De Craene, L. P., and Smets, E. F. (2001b). Floral developmental evidence for the systematic relationships of *Tropaeolum* (Tropaeolaceae). *Ann. Bot.* **88**, 879–892.

Ronse De Craene, L. P., Smets, E. F. and Clinckemaillie, D. (1995). The floral development and floral anatomy of *Coris monspeliensis*. *Can. J. Bot.* **73**, 1687–1698.

Ronse De Craene, L. P., Smets, E. F. and Clinckemaillie, D. (2000). Floral ontogeny and anatomy in *Koelreuteria* with special emphasis on monosymmetry and septal cavities. *Plant Syst. Evol.* **223**, 91–107.

Ronse De Craene, L. P., Smets, E. F. and Vanvinckenroye, P. (1998). Pseudodiplostemony, and its implications for the evolution of the androecium in the Caryophyllaceae. *J. Plant Res.* **111**, 25–43.

Ronse De Craene, L. P., Soltis, P. S. and Soltis, D. E. (2003). Evolution of floral structures in basal Angiosperms. *Int. J. Plant Sci.* **164** (5 Suppl.), S329–S363.

Ronse De Craene, L. P., Vanvinckenroye, P. and Smets, E. F. (1997). A study of the floral morphological diversity in *Phytolacca* (Phytolaccaceae) based on early floral ontogeny. *Int. J. Plant Sci.* **158**, 56–72.

Ronse De Craene, L. P. Volgin, S. A. and Smets, E. F. (1999). The floral development of *Pleuropetalum darwinii*, an anomalous member of the Amaranthaceae. *Flora* **194**, 189–199.

Ronse De Craene, L. P., Yang, T. Y., Schols, P. and Smets, E. F. (2002). Floral anatomy and systematics of *Bretschneidera* (Bretschneideraceae). *Bot. J. Linn. Soc.* **139**, 29–45.

Ronse De Craene, L. P. and Wanntorp, L. (2006). Evolution of floral characters in *Gunnera* (Gunneraceae). *Syst. Bot.* **31**, 671–688.

Ronse De Craene, L. P. and Wanntorp, L. (2008). Morphology and anatomy of the flower of *Meliosma* (Sabiaceae), implications for pollination biology. *Plant Syst. Evol.* **271**, 79–91.

Ronse De Craene, L. P. and Wanntorp, L. (2009). Floral development and anatomy of Salvadoraceae. *Ann. Bot.* **104**, 913–923.

Ross, R. (1982). Initiation of stamens, carpels and receptacle in the Cactaceae. *Am. J. Bot.* **69**, 369–379

Rudall, P. J. (2002). Homologies of inferior ovaries and septal nectaries in monocotyledons. *Int. J. Plant Sci.* **163**, 261–276.

Rudall, P. J. (2003). Monocot pseudanthia revisited, floral structure of the mycoheterotrophic family Triuridaceae. *Int. J. Plant Sci.* **164** (5 Suppl.), S307–S320.

Rudall, P. J. and Bateman, R. M. (2003). Evolutionary change in flowers and inflorescences, evidence from naturally occurring terata. *Trends Plant Sci.* **8**, 76–82.

Rudall, P. J. and Bateman, R. M. (2004). Evolution of zygomorphy in monocot flowers, iterative patterns and developmental constraints. *New Phytol.* **162**, 25–44.

Rudall, P. J., Bateman, R. M., Fay, M. F. and Eastman, A. (2002). Floral anatomy and systematics of Alliaceae with particular reference to *Gilliesia*, a presumed insect mimic with strongly zygomorphic flowers. *Am. J. Bot.* **89**, 1867–1883.

Rudall, P. J., Sokoloff, D. D., Remizowa, M. V., *et al.* (2007). Morphology of Hydatellaceae, an anomalous aquatic family recently recognized as an early-divergent Angiosperm lineage. *Am. J. Bot.* **94**, 1073–1092.

Rudall, P. J., Stuppy, W., Cunniff, J., Kellogg, E. A. and Briggs, B. G. (2005). Evolution of reproductive structures in grasses (Poaceae) inferred by sister-group comparison with their putative closest living relatives, Ecdeiocoleaceae. *Am. J. Bot.* **92**, 1432–1443.

Rutishauser, R., Ronse De Craene, L. P., Smets, E. F. and Mendoza-Heuer, I. (1998). *Theligonum cynocrambe*, developmental morphology of a peculiar rubiaceous herb. *Plant Syst. Evol.* **210**, 1–24.

Sajo, M. G., Longhi-Wagner, H. and Rudall, P. J. (2007). Floral development and embryology in the early-divergent grass *Pharus*. *Int. J. Plant Sci.* **168**, 181–191.

Sajo, M. G., Longhi-Wagner, H. M. and Rudall, P. J. (2008). Reproductive morphology of the early-divergent grass *Streptochaeta* and its bearing on the homologies of the grass spikelet. *Plant Syst. Evol.* **275**, 245–255.

Sampson, F. B. (1969). Studies on the Monimiaceae II. Floral morphology of *Laurelia novae-zelandiae* A.Cunn. (Subfamily Atherospermoideae). *New Zeal. J. Bot.* **7**, 214–240.

Sanchez, A. and Kron, K. A. (2008). Phylogenetics of Polygonaceae with an emphasis on the evolution of Eriogonoideae. *Syst. Bot.* **33**, 87–96.

Sattler, R. (1962). Zur frühen Infloreszenz und Blütenentwicklung der Primulales sensu lato mit besonderer Berücksichtigung der Stamen-Petalum-Entwicklung. *Bot. Jahrb. Syst.* **81**, 385–396

Sattler, R. (1973). *Organogenesis of Flowers: A Photographic Text-Atlas*. Toronto and Buffalo: University of Toronto Press.

Sattler, R. (1977). Kronröhrenentstehung bei *Solanum dulcamara* L. und 'kongenitale Verwachsung'. *Ber. Dtsch. Bot. Ges.* **90**, 29–38.

Sattler, R. (1978). 'Fusion' and 'Continuity' in floral morphology. *Notes Roy. Bot. Gard. Edinburgh* **36**, 397–405.

Sattler, R. and Perlin, L. (1982). Floral development of *Bougainvillea spectabilis* Willd., *Boerhaavia diffusa* L. and *Mirabilis jalapa* L. (Nyctaginaceae). *Bot. J. Linn. Soc.* **84**, 161–182.

Sattler, R. and Singh, V. (1973). Floral development of *Hydrocleis nymphoides*. *Can. J. Bot.* **51**, 2455–2458.

Sattler, R. and Singh, V. (1977). Floral organogenesis of *Limnocharis flava*. *Can. J. Bot.* **55**, 1076–1086.

Sattler, R. and Singh, V. (1978). Floral organogenesis of *Echinodorus amazonicus* Rataj and floral construction of the Alismatales. *Bot. J. Linn. Soc.* **77**, 141–156.

Saunders, E. R. (1937, 1939). *Floral Morphology. A New Outlook with Special Reference to the Interpretation of the Gynoecium, Vols I and II*. Cambridge, UK: W. Heffer and Sons.

Sauquet, H. (2003). Androecium diversity and evolution in Myristicaceae (Magnoliales), with a description of a new Malagasy genus, *Doyleanthus* gen. nov. *Am. J. Bot.* **90**, 1293–1305.

Schaeppi, H. (1976). Über die männlichen Blüten einiger Menispermaceen. *Beitr. Biol. Pflanz.* **52**, 207–215.

Schindler, A. K. (1905). Halorrhagaceae. In *Das Pflanzenreich IV, 225*, ed. A. Engler. Leipzig: W. Engelmann, pp. 1–133.

Schmid, R. (1980). Comparative anatomy and morphology of *Psiloxylon* and *Heteropyxis*, and the subfamilial and tribal classification of Myrtaceae. *Taxon* **29**, 559–595.

Schmidt, E. (1928). Untersuchungen über Berberidaceen. *Beih. Bot. Centralbl.* **45**, 329–396.

Schneider, E. L. (1976). The floral anatomy of *Victoria* Schomb. (Nymphaeaceae). *Bot. J. Linn. Soc.* **72**, 115–148.

Schneider, E. L., Tucker, S. C. and Williamson, P. S. (2003). Floral development in the Nymphaeales. *Int. J. Plant Sci.* **164** (5 Suppl.), S279–S292.

Schöffel, K. (1932). Untersuchungen über den Blütenbau der Ranunculaceen. *Planta* **17**, 315–371.

Schönenberger, J., Anderberg, A. A. and Sytsma, K. J. (2005). Molecular phylogenetics and patterns of floral evolution in the Ericales. *Int. J. Plant Sci.* **166**, 265–288.

Schönenberger, J. and Conti, E. (2003). Molecular phylogeny and floral evolution of Penaeaceae, Oliniaceae, Rhychocalycaceae, and Alzateaceae (Myrtales). *Am. J. Bot.* **90**, 293–309

Schönenberger, J. and Endress, P. K. (1998). Structure and development of the flowers in *Mendoncia, Pseudocalyx*, and *Thunbergia* (Acanthaceae) and their systematic implications. *Int. J. Plant Sci.* **159**, 446–465.

Schönenberger, J. and Friis, E. M. (2001). Fossil flowers of ericalean s.l. affinity from the Late Cretaceous of southern Sweden. *Am. J. Bot.* **88**, 467–480.

Schönenberger, J., Friis, E. M., Matthews, M. L. and Endress, P. K. (2001). Cunoniaceae in the Cretaceous of Europe: evidence from fossil flowers. *Ann. Bot.* **88**, 423–437.

Schönenberger, J. and Grenhagen, A. (2005). Early floral development and androecium organization in Fouquieriaceae (Ericales). *Plant Syst. Evol.* **254**, 233–249.

Schönenberger, J. and von Balthazar, M. (2006). Reproductive structures and phylogenetic framework of the rosids: progress and prospects. *Plant Syst. Evol.* **260**, 87–106.

Scotland, R. W., Endress, P. K. and Lawrence, T. J. (1994). Corolla ontogeny and aestivation in the Acanthaceae. *Bot. J. Linn. Soc.* **114**, 49–65.

Scribailo, R. W. and Posluszny, U. (1985). Floral development of *Hydrocharis morsus-ranae* L. (Hydrocharitaceae). *Am. J. Bot.* **72**, 1578–1589.

Sérsic, A. N. and Cocucci, A. A. (1999). An unusual kind of nectary in the oil flowers of *Monttea*, its structure and function. *Flora* **194**, 393–404.

Simmons, M. P. (2004). Celastraceae. In *The Families and Genera of Vascular Plants, Vol. VI*, ed. K. Kubitzki. Berlin: Springer Verlag, pp. 29–64.

Simpson, M. G. (1990). Phylogeny and classification of Haemodoraceae. *Ann. Missouri Bot. Gard.* **77**, 722–784.

Simpson, M. G. (1998a). Reversal in ovary position from inferior to superior in the Haemodoraceae: evidence from floral ontogeny. *Int. J. Plant. Sci.* **159**, 466–479.

Simpson, M. G. (1998b). Haemodoraceae. In *The Families and Genera of Vascular Plants,
Vol. IV*, ed. K. Kubitzki. Berlin: Springer Verlag, pp. 212–222.

Simpson, M. G. (2006). *Plant Systematics*. Amsterdam: Elsevier.

Singh, V. and Sattler, R. (1973). Nonspiral androecium and gynoecium of *Sagittaria
latifolia*. *Can. J. Bot.* **51**, 1093–1095.

Singh, V. and Sattler, R. (1977a). Development of the inflorescence of *Sagittaria
cuneata*. *Can. J. Bot.* **55**, 1087–1105.

Singh, V. and Sattler, R. (1977b). Floral development of *Aponogeton natans* and *A.
undulatus*. *Can. J. Bot.* **55**, 1106–1120.

Sleumer, H. (1935). Olacaceae. In *Die Natürlichen Pflanzenfamilien 16b*, ed. A. Engler and
K. Prantl. Leipzig: W. Engelmann, pp. 5–32.

Smets, E. (1986). Localization and systematic importance of the floral nectaries in
the Magnoliatae (Dicotyledons). *Bull. Jard. Bot. Nat. Belg.* **56**, 51–76.

Smets, E. (1988). La présence des 'nectaria persistentia' chez les Magnoliophytina
(Angiospermes). *Candollea* **43**, 709–716.

Smets, E. F., Ronse De Craene, L. P., Caris, P. and Rudall, P. J. (2000). Floral
nectaries in monocotyledons, distribution and evolution. In *Monocots:
Systematics and Evolution*, ed. K. L. Wilson and D. A. Morrison. Melbourne:
CSIRO, pp. 230–240.

Sokoloff, D., Oskolski, A. A., Remizowa, M. V. and Nuraliev, M. S. (2007). Flower
structure and development in *Tupidanthus calyptratus* (Araliaceae), an extreme
case of polymery among asterids. *Plant Syst. Evol.* **268**, 209–234.

Sokoloff, D., Rudall, P. J. and Remizowa, M. (2006). Flower-like terminal structures
in racemose inflorescences, a tool in morphogenetic and evolutionary research.
J. Exper. Bot. **57**, 3517–3530.

Soltis, D. E., Senters, A. E., Zanis, M. J., *et al.* (2003). Gunnerales are sister to other core
eudicots: implications for the evolution of pentamery. *Am. J. Bot.* **90**, 461–470.

Soltis, P. S. and Soltis, D. E. (2004). The origin and diversification of Angiosperms. *Am.
J. Bot.* **91**, 1614–1626.

Soltis, D. E., Soltis, P. S., Endress, P. K. and Chase, M. W. (2005). *Phylogeny and Evolution of
Angiosperms*. Sunderland, MA: Sinauer.

Spichiger, R.-E., Savolainen, V. V., Figeat, M. and Jeanmonod, D. (2002). *Botanique
Systématique des Plantes à Fleurs*, 2nd edn. Lausanne: Presses polytechniques et
universitaires Romandes.

Stace, C. A. (2007). Combretaceae. In *The Families and Genera of Vascular Plants, Vol. IX*,
ed. K. Kubitzki. Berlin: Springer Verlag, pp. 67–82.

Staedler, Y. M. and Endress, P. K. (2009). Diversity and lability of floral phyllotaxis
in the pluricarpellate families of core Laurales (Gomortegaceae,
Atherospermataceae, Siparunaceae, Monimiaceae). *Int. J. Plant Sci.* **170**,
522–550.

Staedler, Y. M., Weston, P. H. and Endress, P. K. (2007). Floral phyllotaxis and floral
architecture in Calycanthaceae (Laurales). *Int. J. Plant Sci.* **168**, 285–306.

Stauffer, F. W. and Endress, P. K. (2003). Comparative morphology of female flowers
and systematics in Geonomeae (Arecaceae). *Plant Syst. Evol.* **242**, 171–203.

Stauffer, F. W., Rutishauser, R. and Endress, P. K. (2002). Morphology and development of the female flowers in *Geonoma interrupta* (Arecaceae). *Am. J. Bot.* **89**, 220–229.

Steeves, T. A., Steeves, M. W. and Randall Olson, A. (1991). Flower development in *Amelanchier alnifolia* (Maloideae). *Can. J. Bot.* **69**, 844–857.

Steinecke, H. (1993). Embryologische, morphologische und systematische Untersuchungen ausgewählter Annonaceae. *Diss. Bot.* **205**, 1–237.

Stern, K. (1917). Beiträge zur Kenntnis der Nepenthaceae. *Flora* **109**, 213–282.

Stevens, P. F. (2001 onwards). Angiosperm Phylogeny Website. Version 9, June 2008 [and more or less continuously updated since]. www.mobot.org/MOBOT/research/APweb/.

Strange, A., Rudall, P. J. and Prychid, C. J. (2004). Comparative floral anatomy of Pontederiaceae. *Bot. J. Linn. Soc.* **144**, 395–408.

Struwe, L., Kadereit, J. W., Klackenberg, J., *et al.* (2002). Systematics, character evolution, and biogeography of Gentianaceae, including a new tribal and subtribal classification. In *Gentianaceae: Systematics and Natural History*, ed. L. Struwe and V. A. Albert. Cambridge, UK: Cambridge University Press, pp. 21–309.

Stützel, T. (2006). *Botanische Bestimmungsübungen*, 2nd edn. Stuttgart: Ulmer.

Suessenguth, K. (1938). *Neue Ziele der Botanik: Über das Vorkommen getrennter Kronblätter bei den Sympetalen*. München-Berlin, pp. 32–36.

Sugiyama, M. (1991). Scanning electron microscopy observation on early ontogeny of the flower of *Camellia japonica* L. *J. Jap. Bot.* **66**, 295–299.

Sutter, D. and Endress, P. K. (1995). Aspects of gynoecium structure and macrosystematics in Euphorbiaceae. *Bot. Jahrb. Syst.* **116**, 517–536.

Svoma, E. (1991). The development of the bicarpellate gynoecium of *Paederia* L. species (Rubiaceae–Paederieae). *Opera Bot. Belg.* **3**, 77–86.

Sweeney, P. W. (2008). Phylogeny and floral diversity in the genus *Garcinia* (Clusiaceae) and relatives. *Int. J. Plant Sci.* **169**, 1288–1303.

Takahashi, H. (1994). A comparative study of floral development in *Trillium apetalon* and *T. camtschaticum* (Liliaceae). *J. Plant Res.* **107**, 237–245.

Takhtajan, A. (1997). *Diversity and Classification of Flowering Plants*. New York: Columbia University Press.

Terabayashi, S. (1983). Studies in the morphology and systematics of Berberidaceae VI. Floral anatomy of *Diphylleia* Michx., *Podophyllum* L. and *Dyosma* Woodson. *Acta Phytotax. Geobot.* **34**, 27–47.

Theissen, G., Becker, A., Winter, K.-U., Münster, T., Kirchner, C. and Saedler, H. (2002). How the land plants learned their floral ABCs, the role of MADS box genes in the evolutionary origin of flowers. In *Developmental Genetics and Plant Evolution*, ed. A. C. B. Cronk, R. M. Bateman and J. A. Hawkins. London: Taylor and Francis, pp. 173–205.

Thorne, R. F. (1992). An updated phylogenetic classification of the flowering plants. *Aliso* **13**, 365–389.

Tiagi, Y. D. (1969). Vascular anatomy of the flower of certain species of the Combretaceae. *Bot. Gaz.* **130**, 150–157.

Tillson, A. H. (1940). The floral anatomy of the Kalanchoideae. *Am. J. Bot.* **27**, 595–600.

Tobe, H., Graham, S. A. and Raven, P. H. (1998). Floral morphology and evolution in Lythraceae sensu lato. In *Reproductive Biology in Systematics, Conservation and Economic Botany*, ed. S. J. Owens and P. J. Rudall. Kew, UK: Royal Botanic Gardens, pp. 329–344.

Todzia, C. A. (1993). Ulmaceae. In *The Families and Genera of Vascular Plants, Vol. II*, ed. K. Kubitzki, J. G. Rohwer and V. Bittrich. Berlin: Springer Verlag, pp. 603–611.

Tokuoka, T. (2008). Molecular phylogenetic analysis of Violaceae (Malpighiales) based on plastid and nuclear DNA sequences. *J. Plant Res.* **121**, 253–260.

Tokuoka, T. and Tobe, H. (2006). Phylogenetic analyses of Malpighiales using plastid and nuclear DNA sequences, with particular reference to the embryology of Euphorbiaceae *sensu stricto*. *J. Plant Res.* **119**, 599–616.

Trimbacher, C. (1989). Der Aussenkelch der Rosaceen. In *Symposium Morphologie, Anatomie und Systematik, Zusammenfassungen der Vorträge, Vol. 9*, ed. A. Weber, E. Vitek and M. Kiehn. Vienna: Institute of Botany, University of Vienna, p. 66.

Troll, W. (1956). Die Urbildlichkeit der organischen Gestaltung und Goethes prinzip der 'Variablen Proportionen' *Neu Hefte zur Morphologie* **2**, 64–76.

Tsou, C.-H. (1998). Early floral development of Camellioideae (Theaceae). *Am. J. Bot.* **85**, 1531–1547.

Tsou, C.-H. and Mori, S. A. (2007). Floral organogenesis and floral evolution of the Lecythidoideae. *Am. J. Bot.* **94**, 716–736.

Tucker, S. C. (1984). Unidirectional organ initiation in leguminous flowers. *Am. J. Bot.* **71**, 1139–1148

Tucker, S. C. (1985). Initiation and development of inflorescence and flower in *Anemopsis californica* (Saururaceae). *Am. J. Bot.* **72**, 20–31.

Tucker, S. C. (1988a). Dioecy in *Bauhinia* resulting from organ suppression. *Am. J. Bot.* **75**, 1584–1597.

Tucker, S. C. (1988b). Loss versus suppression of floral organs. In *Aspects of Floral Development*, ed. P. Leins, S. C. Tucker, and P. K. Endress. Berlin: J. Cramer, pp. 69–82.

Tucker, S. C. (1992). The developmental basis for sexual expression in *Ceratonia siliqua* (Leguminosae, Caesalpinioideae, Cassieae). *Am. J. Bot.* **79**, 318–327.

Tucker, S. C. (1996). Trends in evolution of floral ontogeny in *Cassia* sensu stricto, *Senna*, and *Chamaecrista* (Leguminosae, Caesalpinioideae, Cassieae, Cassiinae): a study in convergence. *Am. J. Bot.* **83**, 687–711

Tucker, S. C. (1997). Floral evolution, development, and convergence, the hierarchical-significance hypothesis. *Int. J. Plant Sci.* **158** (6 Suppl.), S143–S161.

Tucker, S. C. (1998). Floral ontogeny in Legume genera *Petalostylis, Labichea*, and *Dialium* (Caesalpinioideae, Cassieae), a series in floral reduction. *Am. J. Bot.* **85**, 184–208.

Tucker, S. C. (1999a). The inflorescence: introduction. *Bot. Rev.* **65**, 303–316.

Tucker, S. C. (1999b). Evolutionary lability of symmetry in early floral development. *Int. J. Plant Sci.* **160** (6 Suppl.), S25–S39.

Tucker, S. C. (2000a). Evolutionary loss of sepals and/or petals in Detarioid legume taxa (*Aphanocalyx, Brachystegia*, and *Monopetalanthus* (Leguminosae, Caesalpinioideae). *Am. J. Bot.* **87**, 608–624.

Tucker, S. C. (2000b). Floral development in tribe Detarieae (Leguminosae, Caesalpinioideae), *Amherstia, Brownea*, and *Tamarindus*. *Am. J. Bot.* **87**, 1385–1407.

Tucker, S. C. (2000c). Floral development and homeosis in *Saraca* (Leguminosae, Caesalpinioideae, Detarieae). *Int. J. Plant Sci.* **161**, 537–549.

Tucker, S. C. (2001a). The ontogenetic basis for missing petals in *Crudia* (Leguminosae, Caesalpinioideae, Detarieae). *Int. J. Plant Sci.* **162**, 83–89.

Tucker, S. C. (2001b). Floral development in *Schotia* and *Cynometra* (Leguminosae, Caesalpinioideae, Detarieae). *Am. J. Bot.* **88**, 1164–1180.

Tucker, S. C. (2002a). Floral ontogeny in Sophoreae (Leguminosae, Papilionoideae) III. Radial symmetry and random petal aestivation in *Cadia purpurea*. *Am. J. Bot.* **89**, 748–757.

Tucker, S. C. (2002b). Comparative floral ontogeny in Detarieae (Leguminosae, Caesalpinioideae) II. zygomorphic taxa with petal and stamen suppression. *Am. J. Bot.* **89**, 888–907.

Tucker, S. C. (2003a). Floral development in Legumes. *Plant Physiology* **131**: 911–926

Tucker, S. C. (2003b). Floral ontogeny in *Swartzia* (Leguminosae, Papilionoideae, Swartzieae): distribution and role of the ring meristem. *Am. J. Bot.* **90**, 1274–1292.

Tucker, S. C. and Bernhardt, P. (2000). Floral ontogeny, pattern formation, and evolution in *Hibbertia* and *Adrastea* (Dilleniaceae). *Am. J. Bot.* **87**, 1915–1936.

Tucker, S. C. and Douglas, A. W. (1996). Floral structure, development, and relationships of paleoherbs, *Saruma, Cabomba, Lactoris* and selected Piperales. In *Flowering Plant Origin, Evolution and Phylogeny*, ed. D. W. Taylor and L. J. Hickey. New York: Chapman and Hall, pp. 141–175.

Tucker, S. C., Douglas, A. W. and Liang, H.-X. (1993). Utility of ontogenetic and conventional characters in determining phylogenetic relationships of Saururaceae and Piperaceae (Piperales). *Syst. Bot.* **18**, 614–641.

Uhl, N. W. (1976a). Developmental studies in *Ptychosperma* (Palmae) I. The inflorescence and flower cluster. *Am. J. Bot.* **63**, 82–96.

Uhl, N. W. (1976b). Developmental studies in *Ptychosperma* (Palmae) II. The staminate and pistillate flowers. *Am. J. Bot.* **63**, 97–109.

Uhl, N. W. and Moore, H. E. (1980). Androecial development in six polyandrous genera representing five major groups of palms. *Ann. Bot.* **45**, 57–75.

Urban, I. (1892). Blüthen- und Fruchtbau der Loasaceen. *Ber. Dtsch. Bot. Ges.* **10**, 259–265.

Vaes, E., Vrijdaghs, A., Smets, E. F. V. and Dessein, S. (2006). Elaborate petals in Australian *Spermacoce* (Rubiaceae) species, morphology, ontogeny and function. *Ann. Bot.* **98**, 1167–1178.

Van Heel, W. A. (1966). Morphology of the androecium in Malvales. *Blumea* **13**, 177–394.

Van Heel, W. A. (1978). Morphology of the pistil in Malvaceae – Ureneae. *Blumea* **24**, 123–137.

Van Heel, W. A. (1987). Androecium development in *Actinidia chinensis* and *A. melanandra* (Actinidiaceae). *Bot. Jahrb. Syst.* **109**, 17–23

Van Heel, W. A. (1993). Floral ontogeny of *Archidendron luceyi* (Mimosaceae), with remarks on *Amherstia nobilis* (Caesalpiniaceae). *Bot. Jahrb. Syst.* **114**, 551–560.

Van Heel, W. A. (1995). Morphology of the gynoecium of *Kitaibelia vitifolia* Willd. and *Malope trifida* L. (Malvaceae-Malopeae). *Bot. Jahrb. Syst.* **117**, 485–493.

Vanvinckenroye, P., Cresens, E., Ronse De Craene, L. P. and Smets, E. F. (1993). A comparative floral developmental study in *Pisonia, Bougainvillea* and *Mirabilis* (Nyctaginaceae) with special emphasis on the gynoecium and floral nectaries. *Bull. Jard. Bot. Nat. Belg.* **62**, 69–96.

Vanvinckenroye, P., Ronse De Craene, L. P. and Smets, E. F. (1997). The floral development of *Monococcus echinophorus* (Phytolaccaceae). *Can. J. Bot.* **75**, 1941–1950.

Vanvinckenroye, P. and Smets, E. F. (1996). Floral ontogeny of five species of *Talinum* and of related taxa (Portulacaceae). *J. Plant Res.* **109**, 387–402.

Vanvinckenroye, P. and Smets, E. F. (1999). floral ontogeny of *Anacampseros* subg. *Anacampseros* sect. *Anacampseros* (Portulacaceae). *Syst. Geogr. Pl.* **68**, 173–194.

Venkata Rao, C. (1952). Floral anatomy of some Malvales and its bearing on the affinities of families included in the order. *J. Indian Bot. Soc.* **31**, 171–203.

Venkata Rao, C. (1963). On the morphology of the calyculus. *J. Indian Bot. Soc.* **42**, 618–628.

Vergara-Silva, F. Espinosa-Matías, S., Ambrose, B. A., *et al.* (2003). Inside-out flowers characteristic of *Lacandonia schismatica* evolved at least before its divergence from a closely related taxon, *Triuris brevistylis*. *Int. J. Plant Sci.* **164**, 345–357.

Vink, W. (1995). Revision of *Magodendron* (Sapotaceae) with observations on floral development and morphology. *Blumea* **40**, 91–107.

Vogel, S. (1977). Nektarien und ihre ökologische Bedeutung. *Apidology* **8**, 321–335.

Vogel, S. (1997). Remarkable nectaries, structure, ecology, organophyletic perspectives I. Substitutive nectaries. *Flora* **192**, 305–333.

Vogel, S. (1998a). Remarkable nectaries, structure, ecology, organophyletic perspectives II. Nectarioles. *Flora* **193**, 1–29.

Vogel, S. (1998b). Remarkable nectaries, structure, ecology, organophyletic perspectives III. Nectar ducts. *Flora* **193**, 113–131.

Vogel, S. (1998c). Remarkable nectaries, structure, ecology, organophyletic perspectives IV. Miscellaneous cases. *Flora* **193**, 225–248.

Vogel, S. (2000). The floral nectaries of Malvaceae sensu lato: a conspectus. *Kurtziana* **28**, 155–171.

Von Balthazar, M., Alverson, W. S., Schönenberger, J. and Baum, D. A. (2004). Comparative floral development and androecium structure in Malvoideae (Malvaceae s.l.). *Int. J. Plant Sci.* **165**, 445–473.

Von Balthazar, M. and Endress, P. K. (2002a). Reproductive structures and systematics of Buxaceae. *Bot. J. Linn. Soc.* **140**, 193–228.

Von Balthazar, M. and Endress, P. K. (2002b). Development of inflorescences and flowers in Buxaceae and the problem of perianth interpretation. *Int. J. Plant Sci.* **163**, 847–876.

Von Balthazar, M. and Schönenberger, J. (2009). Floral structure and organization in Platanaceae. *Int. J. Plant Sci.* **170**, 210–225.

Von Balthazar, M., Schönenberger, J., Alverson, W. S., Janka, H., Bayer, C. and Baum, D. A. (2006). Structure and evolution of the androecium in the *Malvatheca* clade (Malvaceae s.l.) and implications for Malvaceae and Malvales. *Plant Syst. Evol.* **260**, 171–197.

Wagenitz, G. and Laing, B. (1984). The nectaries of the Dipsacales and their systematic significance. *Bot. Jahrb. Syst.* **104**, 483–507.

Walker-Larsen, J. and Harder, L. D. (2000). The evolution of staminodes in Angiosperms: patterns of stamen reduction, loss, and functional re-invention. *Am. J. Bot.* **87**, 1367–1384.

Wallnöfer, B. (2004). Ebenaceae. In *The Families and Genera of Vascular Plants, Vol. VI*, ed. K. Kubitzki and C. Bayer. Berlin: Springer Verlag, pp. 125–130.

Walter, H. (1906). Die Diagramme der Phytolaccaceen. *Bot. Jahrb. Syst.* **37**, Beibl. 85, 1–57.

Wang, H., Meng, A., Li, J., Feng, M., Chen, Z. and Wang, W. (2006). Floral organogenesis of *Cocculus orbiculatus* and *Stephania dielsiana* (Menispermaceae). *Int. J. Plant Sci.* **167**, 951–960.

Wang, H., Moore, M. J., Soltis, P. S., *et al.* (2009). Rosid radiation and the rapid rise of angiosperm-dominated forests. *Proc. Natl Acad. Sci.* **106**, 3853–3858.

Wannan, B. C. and Quinn, C. J. (1991). Floral structure and evolution in the Anacardiaceae. *Bot. J. Linn. Soc.* **107**, 349–385.

Wanntorp, L. and Ronse De Craene, L. P. (2005). The *Gunnera* flower: key to eudicot diversification or response to pollination mode? *Int. J. Plant Sci.* **166**, 945–953.

Wanntorp, L. and Ronse De Craene, L. P. (2007). Floral development of *Meliosma* (Sabiaceae): evidence for multiple origins of pentamery in the eudicots. *Am. J. Bot.* **94**, 1828–1836.

Wanntorp, L. and Ronse De Craene, L. P. (2009). Perianth evolution in the Sandalwood order Santalales. *Am. J. Bot.* **96**, 1361–1371.

Weberling, F. (1989). *Morphology of Flowers and Inflorescences.* Cambridge, UK: Cambridge University Press.

Weigend, M. (2007). Grossulariaceae. In *The Families and Genera of Vascular Plants, Vol. IX*, ed. K. Kubitzki and C. Bayer. Berlin: Springer Verlag, pp. 168–176.

Williams, S. E., Albert, V. A. and Chase, M. W. (1994). Relationships of Droseraceae: a cladistic analysis of *rbc*L sequence and morphological data. *Am. J. Bot.* **81**, 1027–1037.

Williamson, P. S. and Moseley, M. F. (1989). Morphological studies of the Nymphaeaceae sensu lato XVII. Floral anatomy of *Ondinea purpurea* ssp. *purpurea* (Nymphaeaceae). *Am. J. Bot.* **76**, 1779–1794.

Worberg, A., Alford, M. H., Quandt, D. and Borsch, T. (2008). The Huertales clade identified to contain *Dipentodon, Perrottetia, Huertea, Tapiscia* and *Gerrardina*, and to be sister to Brassicales plus Malvales. In *Systematics 2008: Programme and Abstracts*, Göttingen 7–11 April 2008, ed. S. R. Gradstein *et al.* Göttingen: Universitätsverlag, p. 350.

Worberg, A., Quandt, D., Barniske, A.-M., Löhne, C., Hilu, K. W. and Borsch, T. (2007). Phylogeny of basal eudicots: insights from non-coding and rapidly evolving DNA. *Org. Div. Evol.* **7**, 55–77.

Wu, H.-C., Su, H.-J. and Hu, J.-M. (2007). The identification of A-, B-, C-, and E-class Mads-Box genes and implications for perianth evolution in the basal eudicot *Trochodendron aralioides* (Trochodendraceae). *Int. J. Plant Sci.* **168**, 775–799.

Xu, F.-X. (2006). Floral ontogeny of two species of *Magnolia* L. *J. Integr. Plant Biol.* **48**, 1197–1203.

Xu, F.-X. and Rudall, P. J. (2006). Comparative floral anatomy and ontogeny in Magnoliaceae. *Plant Syst. Evol.* **258**, 1–15.

Zandonella, P. (1977). Apports de l'étude comparée des nectaires floraux à la conception phylogénétique de l'ordre des Centrospermes. *Ber. Dtsch. Bot. Ges.* **90**, 105–125.

Zanis, M. J., Soltis, P. S., Qiu, Y. L., Zimmer, E. and Soltis, D. E. (2003). Phylogenetic analyses and perianth evolution in basal Angiosperms. *Ann. Missouri Bot. Gard.* **90**, 129–150.

Zhang, X.-H. and Ren, Y. (2008). Floral morphology and development in *Sargentodoxa* (Lardizabalaceae). *Int. J. Plant Sci.* **169**, 1148–1158.

Zhou, Q., Wang, Y. and Xiaobai, J. (2002). Ontogeny of floral organs and morphology of floral apex in *Phellodendron amurense* (Rutaceae). *Aust. J. Bot.* **50**, 633–644.

Zohary, M. and Baum, B. (1965). On the androecium of *Tamarix* flower and its evolutionary trends. *Israel J. Bot.* **14**, 101–111.

Glossary

Abaxial: the side of an organ away from the axis to which it is attached. Corresponds to the dorsal side or lower surface (for leaf).

Abortion: the process of arrested growth.

Accrescent: enlarging with age (e.g. calyx of Olacaceae).

Acropetalous: in the direction of the apex, opposite of basipetalous.

Actinomorphic: see polysymmetry.

Acuminate: gradually tapering to an acute apex.

Acyclic: spiral, not in whorls.

Adaxial: the side of an organ towards the axis to which it is attached. Corresponds to the ventral side or upper surface (of leaf).

Adhesion (adherent): attached but not fused; refers to different organs.

Adnate: attached to; refers to fusion of different organs, e.g. stamens and petals. Compare with connate.

Aestivation: the mutual positional relationship of perianth parts in bud.

Alternipetalous: alternating with the petals.

Alternisepalous: alternating with the sepals.

Anatropous: refers to a curved ovule with the micropyle close to the attachment region of the funiculus.

Androecium: the totality of the stamens in a flower.

Androgynophore: an extension of the receptacle bearing stamens and carpels.

Androphore: an extension of the receptacle bearing stamens in a staminate flower without gynoecium.

Ante(i)petalous: opposite a petal.

Ante(i)sepalous: opposite a sepal.

Anther: the pollen-bearing part of a stamen, consists generally of two thecae and four pollen sacs.

Anthesis: the flowering phase, when all organs are exposed, pollen is presented and the stigma is receptive.

Anthocarp: basal persistent part of the perianth in Nyctaginaceae homologous to a hypanthium and enclosing the fruit.

Anthophore: extension of the receptacle between the calyx and the rest of the organs in a flower.

Apert: aestivation of perianth parts not touching each other.

Apetalous (-y): without petals.

Apex: the tip or distal structure of a leaf or axis.

Apical placentation: type of placentation with ovules attached to the top of the locule.

Apocarpous: refers to a gynoecium with distinct carpels.

Apomorphy (-ic): refers to a character state that is derived from an ancestral state.

Ascending: kind of cochleate aestivation, with innermost organs adaxial.

Ascidiate: refers to young carpel that is congenitally urn-shaped, opposed to plicate, a carpel that is postgenitally urn-shaped.

Asymmetrical: refers to a structure that cannot be divided into two equal halves (e.g. flower of *Centranthus*, Caprifoliaceae).

Axile placentation: type of placentation in a syncarpous ovary with septa in which the ovules are attached in the middle section of the locule.

Basal placentation: type of placentation with ovules attached at the base of the locule.

Basipetalous: in the direction away from the apex, opposite of acropetalous.

Blade: the expanded section of a leaf-like organ. Synonymous with lamina.

Bract: a differentiated or reduced leaf associated with inflorescences or subtending a flower; synonym of pherophyll.

Bracteole: a small reduced leaf associated with the flower and occurring singly (monocots) or in pairs (dicots); synonym of prophyll.

Bauplan: the form or structural construction of a flower that tends to be constant for larger taxa.

Buzz pollination: kind of pollination in which the vibration of bee wings triggers the release of pollen. Associated with a syndrome of characters in the flower such as poricidal anthers and hanging flowers.

Caducous: falling off early.

Calyculus: rim-like structure at the base of flowers of Loranthaceae and some Olacaceae.

Calyptra(-te): refers to the fusion of perianth parts by their distal parts, falling off as a whole.

Calyx: the outer perianth whorl in flowers; collective term for sepals.

Campylotropous: refers to a curved ovule with the nucellus included in the curvature in a more or less right angle to the funiculus.

Capitulum: an inflorescence with flowers densely packed in a globular structure or head (e.g. Asteraceae).

Capsule: a dry dehiscent fruit that develops from a syncarpous ovary. The capsule opens in the middle of the locule (loculicidal), at the level of the septa (septicidal), or both.

Carinal: refers to the position of style or stigma in line with the carpel or locule; opposite of commissural.

Carpel: the basic unit of the gynoecium, corresponding to a transformed megasporophyll enclosing the ovules.

Carpophore: an extension of the receptacle to which the carpels are attached (e.g. Apiaceae).

Centrifugal: refers to the sequence of development of floral organs from the centre towards the periphery.

Centripetal: refers to the sequence of development of floral organs from the periphery towards the centre.

Chasmogamous (-y): refers to a flower that opens and is generally cross-fertilized, as opposed to cleistogamous.

Cincinnus: cymose monochasial inflorescence with lateral flowers developing sequentially left or right in one plane, resulting in a zigzag pattern.

Clade: a group of taxa that share features derived from a common ancestor; a monophyletic evolutionary line.

Claw(-ed): a constricted base of a petal.

Cleistogamous (-y): refers to flowers that do not open and are self-fertilizing, as opposed to chasmogamous.

Cochleate (cochlear): refers to the aestivation with one perianth member completely outside and one member completely inside, and the rest in between. The orientation of symmetry is either adaxial (ascending) or abaxial (descending).

Colleter: glandular structure on vegetative or floral buds that secretes a fluid protecting the developing organs from dehydration.

Commissural: refers to the position of style or stigma in alternation with the carpel or locule; opposite of carinal.

Common (complex) primordium: a single primordium dividing in two or more secondary primordia. Each secondary primordium develops into a single organ. Division of a common stamen primordium leads to an increase in stamens. A stamen-petal primordium results from the congenital fusion of petals and opposite stamen.

Compitum: a common zone shared by all carpels in a syncarpous ovary, functioning as pollen-transmitting area.

Congenital fusion: fusion of structures from the onset of initiation (zonal growth).

Connate: refers to the same organs that are marginally coherent.

Contorted (convolute): refers to the perianth aestivation with neighbouring organs overlapping each other on one side only, as tiles on a roof. The flower has a twisted appearance in some cases (e.g. Malvaceae, Apocynaceae).

Convergent (evolution): the independent acquisition of similar characteristics in two of more unrelated groups.

Convolute: see contorted.

Corolla: the second whorl of the perianth; collective term for petals.

Corona: a showy structure located between perianth and androecium, with variable origin (e.g. Passifloraceae, Amaryllidaceae).

Cucullate: hood-shaped.

Cupule: a cup-like structure that usually subtends a fruit and consists of sterile bracts (e.g. Fagaceae); see also involucre.

Cyathium: an inflorescence consisting of several naked staminate flowers and a single pistillate flower, enclosed in a cup-shaped involucre (e.g. Euphorbiaceae).

Cyme: a determinate inflorescence (monotelic), with each major branch ending in a flower.

Deciduous: falling off at a given stage in the development of a plant or flower.

Decussate: refers to arrangement of opposite leaves on the stem, with successive pairs arising at right angles of the pair below.

Dédoublement: the process of splitting of an organ in two equal parts that are identical to the original organ.

Descending: kind of cochleate aestivation, with innermost organs abaxial.

Dichasium (-al): cymose inflorescence with lateral branches developing equally.

Dichogamy: refers to a flower in which pollen is released and stigmas are receptive at different times. See proterandry and proterogyny.

Dicliny: refers to separate staminate and pistillate flowers. Similar to unisexual flowers. See also dioecious and monoecious.

Didynamous: with two pairs of stamens of different length (e.g. Acanthaceae).

Dimerous (-y): with parts in two.

Dimorphic: see heteromorphic.

Dioecious (-y): with staminate and pistillate flowers on different plants. Compare with monoecious.

Diplostemony (-ous): arrangement of stamens in two whorls, with the outer opposite the sepals and the inner opposite the petals. Compare with obdiplostemony.

Disc nectary: the development of a fleshy nectariferous circular protrusion of the receptacle. The disc can be intrastaminal or extrastaminal.

Distal: at the top of an organ or structure or away from the place of attachment, opposite of proximal.

Distichous: two-ranked or in two rows, referring to the arrangement of leaves or bracts on the stem.

Disymmetry (-ic): refers to a flower with two symmetry planes (median and transversal).

Divergence angle: the angle between two organs arising in succession on a spiral.

Elaiophore: outgrowth in the flower that produces oils collected by specialized pollinators.

Enantiostyly: refers to flowers whose styles are oriented alternate to the right and left in mirror-image flowers.

Epicalyx: a whorl of bracts or structures of unknown origin inserted at the base of the calyx.

Epigynous: refers to a flower or organ with its parts inserted on top of the ovary.

Extant: still in existence; having living representatives, as opposed to fossil.

Extrastaminal: outside the stamen whorl.

Extrorse: directed outwards; refers to the dehiscence of an anther. Compare with introrse and latrorse.

False septum (plur. septa): partition in the ovary that does not include the margin of a carpel.

Fascicle: refers to a tight cluster arising from the same point; in flowers often linked to stamens (e.g. Clusiaceae).

Filament: the stalk of a stamen that bears the anther.

Fimbriate: fringes of multicellular trichomes on the margin of an organ.

Floret: in Asteraceae, a small flower of a capitulum; in Poaceae, a small flower of a spikelet.

Free-central placentation: type of placentation in a syncarpous ovary without septa in which the ovules are attached on a central column.

Funiculus: the stalk of the ovule.

Grade: a group of organisms without phylogenetic implication.

Gynobasic: refers to the position of the style at the base of the ovary (e.g. Chrysobalanaceae).

Gynoecium: the collective term for the female parts of the flower.

Gynophore: an extension of the receptacle lifting the gynoecium (e.g. Capparaceae).

Gynostegium: a common structure formed by postgenital fusion of stamens and stigma in Apocynaceae.

Gynostemium: a common structure formed by congenital fusion of stamens and stigma in Orchidaceae and Aristolochiaceae.

Halophytic: adapted to saline conditions.

Haplostemony: arrangement of stamens in a single whorl opposite the sepals. Compare with obhaplostemony.

Helicoid cyme: cymose monochasial inflorescence with lateral flowers developing on the same side in one plane, resulting in a coiled appearance (also called scorpioid cyme).

Herkogamy: spatial separation of male and female organs in a flower, avoiding self-fertilization.

Heteranth(er)y: refers to two distinct types of stamens in a flower; one type bears fertile pollen and the other sterile (fodder) pollen.

Heterochrony: evolutionary change caused by variation in the relative time of appearance and rate of development of a structure.

Heteromorphic: refers to structures of the same organism or species that differ in form and size; includes dimorphic with two forms. Synonym of polymorphic.

Heterostyly: condition in which style and stamens differ in length among individuals of a species, favouring cross-pollination.

Heterotopy: symptom in which an organ develops in a position other than its normal one.

Homeosis: the replacement of an organ by another organ that is not necessarily homologous to it.

Hydathode: water pore; modification in the epidermis allowing release of water through an opening.

Hypanthium: receptacular tube, often cup-shaped and bearing perianth and androecium.

Hypogynous: refers to a flower with the floral parts inserted below the ovary.

Imbricate: refers to the aestivation pattern with overlapping margins of neighbouring organs.

Integument: the outer wall of the ovule which later develops into the seed wall; it may be composed of two layers (bitegmic ovules) or a single layer (unitegmic ovule).

Interstaminal: positioned between the stamens, i.e. within the stamen whorl.

Intrastaminal: positioned inside the stamens; often refers to a nectary.

Introrse: directed inwards; refers to the dehiscence of an anther. Compare with extrorse and latrorse.

Involucre, involucral bract: a series of fused or overlapping bracts that subtend inflorescences, sometimes confused with cupules, which have the same origin (e.g. Apiaceae, Asteraceae).

Isomerous: refers to the number of parts of a whorl being the same as the other whorls in the flower.

Labellum: a lip-like petaloid structure, such as the inner petal of orchids or staminodial structure of Zingiberaceae.

Laminar-diffuse placentation: type of placentation without clearly localized placentae, with ovules often located on the lamina of the carpels; also called laminar placentation.

Latrorse: directed towards the sides; refers to the dehiscence of an anther. Compare with introrse and extrorse.

Lemma: the lower of two bracts enclosing the floret in Poaceae (with palea).

Ligule: a strap-like organ; refers to ligulate flowers with a strap-like corolla in Asteraceae or an outgrowth of the corolla or leaves.

Locule: the cavity (-ies) within the ovary, containing the ovules.

Loculicidal capsule: see capsule.

Lodicule: a small scale in the floret of Poaceae, corresponding to a reduced perianth.

Marginal placentation: type of placentation of apocarpous ovary, in which the ovules arise on the margins of the carpel. Compare with laminar or diffuse placentation.

Median monosymmetry: monosymmetry with the two equal halves running along the median line.

Megasporangium: structure bearing megaspores; in angiosperms, the carpels.

Merism (merosity): refers to the number of parts within a whorl of organs in flowers, generally used as '-merous'.

Meristem: an undifferentiated area of tissue with the potential of cell division, differentiation and growth.

Micropyle: an opening between the integuments of the ovule through which the pollen tube reaches the nucellus.

Microsporangium: structure bearing microspores; in angiosperms, the stamens.

Midrib: the primary vein of a leaf, bract, perianth part or carpel.

Mimic (-ry): the close resemblance of an organism or a part of an organism to another organism (e.g. pistillate flowers of Caricaceae and Begoniaceae resembling their staminate counterparts).

Monadelphous: refers to stamens united by their filaments in a tube.

Moniliform: refers to a hair or filament with constrictions at regular intervals, like beads on a string (e.g. Commelinaceae).

Monocarpellate gynoecium: with a single carpel (unicarpellate).

Monochasium (-ial): cymose inflorescence with lateral branches developing unequally, one branch being generally reduced.

Monoecious (-y): with staminate and pistillate flowers in the same plant. Compare with dioecious.

Monophyletic group: a group comprising a common ancestor and all its descendants.

Monosymmetry (-ic): refers to a bilateral symmetry, dividing the flower in two equal halves; synonym of zygomorphic.

Monothecal: refers to anthers with a single theca, as derived from dithecal anthers.

Monotypic: refers to a family/genus with a single genus/species.

Multilocular ovary: refers to a gynoecium with many locules (septate ovary).

Nectar glands: the development of separate nectariferous appendages. In flowers, these can be receptacular or of staminodial origin.

Nectar guide: markings on the petals that guide pollinators to hidden nectar.

Nectary: a tissue or structure that develops nectar. A distinction is made between extrafloral and floral nectaries.

Nucellus: Tissue of the ovule in which the embryo develops.

Obdiplostemony (-ous): arrangement of stamens in two whorls, with the outer opposite the petals and the inner opposite the sepals. Compare with diplostemony.

Obhaplostemony: arrangement of stamens in a single whorl opposite the petals. Compare with haplostemony.

Oblique monosymmetry: monosymmetry with the two equal halves being divided obliquely in the flower.

Obturator: outgrowth of the placenta, style or ovule facilitating pollen transfer from the style to the ovule (e.g. Euphorbiaceae, Sapindaceae).

Ontogeny (etic): the course of initiation and development of an organism and organ until maturity.

Orthostichies: set of straight lines connecting organs in whorled flowers.

Orthotropous: refers to the straight orientation of an ovule with the funiculus and micropyle along the same straight line.

Osmophore: tissue in the flower or inflorescence that emanates scent (e.g. spadix of Araceae).

Ovary: the section of the gynoecium that contains the ovules.

Ovule: the megaspore of seed plants that becomes the seed after fertilization.

Palea: the upper of two bracts enclosing the floret in Poaceae (with lemma) or bract associated with florets in Asteraceae.

Panicle: a compound raceme, with lateral branches repeating the main axes but terminated by a flower.

Pappus: a modified calyx, consisting of hairs, bristles or scales typical of Asteraceae and some Caprifoliaceae.

Paraphyletic group: a group including a common ancestor, but not all its descendants.

Parastichies: set of contact-spirals resulting from a helical initiation of organs in whorled and spiral flowers.

Parietal placentation: type of placentation found in syncarpous unilocular ovaries where the placentas are located on the fusion of carpel margins.

Pedicel: the stalk supporting a flower.

Peduncle: the primary axis of an inflorescence.

Pentamerous (-y): with parts in five.

Perianth: the collective term for sepals, tepals and petals.

Perigon(e): defines the perianth when no distinction can be made between sepals and petals; collective term for tepals.

Perigynous: refers to a flower with the floral parts inserted somewhere in the middle of the ovary.

Perisperm: nutrient-rich tissue derived from the nucellus in seeds of some angiosperms (e.g. Caryophyllales).

Petaliferous: with petals.

Petaloid: resembling a petal (with pigmentation).

Petiole: the stalk of a leaf.

Phyllotaxis: the arrangement pattern of leaves or floral organs on an axis.

Phylogenetic tree: diagram showing the evolutionary relationships among groups of organisms. Also called evolutionary tree.

Pistil: the female part of the flower, referring to ovary, style and stigma. See also gynoecium.

Pistillate: refers to female part, specifically to unisexual flowers with functional carpels and without functional stamens.

Pistillode: a reduced or aborted gynoecium.

Placenta (plur. -ae): the area in the ovary to which the ovules are attached.

Placentation: the insertion of ovules in the ovary.

Plastochron: the time interval between the initiation of two successive organs on a spiral.

Plesiomorphy (-ic character): ancestral condition.

Plicate: folded into longitudinal plaits (e.g. corolla of Solanaceae).

Pluricarpellate: with numerous carpels in a flower, sometimes as the result of a secondary increase (e.g. Araliaceae).

Pollen flower: flowers offering pollen as reward.

Pollinium: pollen mass from a pollen sac moved as a single unit during pollination (e.g. Orchidaceae, Apocynaceae).

Polyandry: the presence of numerous stamens (more than double the number of petals); primary polyandry refers to many stamens arising singly in a spiral or in whorls; secondary or complex polyandry refers to the division of primary (common) primordia into several stamens.

Polymerization: increase in number of parts.

Polyphyletic group: a phylogenetically artificial group, not including the common ancestor of the group.

Polysymmetry (-ic): refers to the radially symmetrical flowers that can be divided in two equal halves from any angle; synonym of actinomorphic.

Poricidal: opening by pores; refers to anther dehiscence.

Postgenital fusion: fusion of structures that were free from each other at the onset of initiation.

Prot(er)androus (-y): refers to a flower in which pollen is released before the stigmas are receptive (e.g. Campanulaceae).

Prot(er)ogynous (-y): refers to a flower in which stigmas are receptive before the stamens shed pollen (e.g. Calycanthaceae).

Protruding-diffuse placentation: axile or parietal placentation with placentae expanding in the locule and covered with many ovules.

Proximal: at the base of an organ or structure, or at the place of attachment; opposite of distal.

Pseudanthium: an inflorescence of several small or reduced flowers resembling a flower

Pseudomonocarpous gynoecium: gynoecium with single carpel, derived by reduction of more carpels in a syncarpous gynoecium (pseudomonomery).

Pseudomonomery: see pseudomonocarpous gynoecium.

Pseudonectaries: a glistening structure resembling a nectary but lacking any sugary secretion.

Pseudostaminodes: receptacular emergences that resemble staminodial structures.

Quincuncial: refers to an imbricate aestivation in pentamerous flowers, with two organs completely outside, two completely inside, and one intermediate member (2/5 arrangement).

Raceme (-ose): an indeterminate inflorescence (polytelic), with the uppermost flowers the youngest.

Receptacle: the more or less expanded axis bearing the floral parts.

Replum: persistent placental ridge between the two compartments of the fruit in Brassicaceae.

Resupinate (-ion): turned 45°, 90° to 180° during development by the twisting of the pedicel (e.g. Orchidaceae, Balsaminaceae).

Ring primordium (ring wall): the initiation of a circular primordium giving rise to many stamens or carpels; this is phylogenetically derived from the concrescence of separate common primordia.

Sepaloid: resembling a sepal (without pigmentation).

Septum (plur. septa): a partition derived from the carpel margins dividing the ovary in locules.

Septal nectary: a nectary arising within the septa of the gynoecium, characteristic of monocots.

Spike: racemose inflorescence with sessile flowers.

Spikelet: partial inflorescence of grasses, cyperoids and restios; in grasses consisting of a variable number of florets subtended by bracts (palea and lemma).

Spiral: successively arising organs with a same divergence angle.

Sporophyll: spore-bearing leaf; it is understood that flowers have been derived from ancestral forms bearing spores on specialized leaves attached on an axis. In seed plants, there is a differentiation between microsporophylls (stamens), bearing microsporangia (anthers) with microspores (pollen grains) and megasporophylls (carpels), bearing megasporangia (ovules) with megaspores (embryosac).

Spur: tubular, generally nectariferous outgrowth in the flower.

Stamen: the male organ in the flower, bearing microspores (pollen), usually comprised of anther and filament.

Stamen-petal tube: a kind of hypanthium lifting congenitally fused petals and stamens into a tube, often referred to as 'sympetaly with epipetalous stamens'.

Staminate: refers to male part, specifically to unisexual flowers with functional stamens and without functional gynoecium.

Staminode: sterile stamen, sometimes strongly modified.

Stigma: the part of the gynoecium, which is receptive to pollen.

Stipule: a reduced leaf-like appendage, single or in pairs at the base of the petiole of a leaf.

Stomate (plur. stomata): specialized epidermal cells which allow air passage or nectar through an opening (stoma).

Style: the extended part of the gynoecium between ovary and stigma.

Stylode: a generally free style branch.

Stylopodium: enlargement of the base of the style in Apiaceae and Araliaceae.

Sympetalous (-y): having petals fused for at least part of their length.

Symplicate zone: describes the upper zone of a syncarpous gynoecium above the attachment of ovules.

Synandrium: a columnar structure consisting of fused stamens in the absence of a gynoecium in staminate flowers (e.g. Myristicaceae).

Synascidiate zone: describes the lower zone of a syncarpous gynoecium where the ovules are attached.

Syncarpy: refers to a gynoecium with fused carpels.

Syndrome: a combination of characters or features.

Synorganization: spatial and functional connections between organs of the same or different kind leading to a homogeneous functional structure.

Taxon (plur. taxa): taxonomic unit of any rank; name designating an organism or group of organisms.

Tepal: an undifferentiated perianth part; can be sepaloid or petaloid.

Tetramerous (-y): with parts in four.

Theca: the chamber enclosing the pollen grains; two thecae generally make up an anther and one theca consists of two pollen sacs.

Thyrse: a compound inflorescence with several cymose lateral branches.

Thyrsoid: thyrse-like.

Transversal monosymmetry: monosymmetry with the two equal halves being divided in the transversal plane of the flower.

Trichome (-atic): a hair-like outgrowth of the epidermis.

Trimerous (-y): with parts in three.

Torus: a receptacle swollen in a distinct cushion (e.g. Capparaceae, Ochnaceae).

Umbel: a flat inflorescence with all pedicels arising from the same point; can be derived from racemes or cymes.

Unidirectional development: developmental sequence of organs from one side of the flower (generally the abaxial side) to the other.

Unilocular ovary: refers to an ovary with a single locule (absence of septa).

Urceolate: urn-shaped, as a swollen tube contracted at the top and with a narrow rim.

Valvate: aestivation of perianth with margins touching but not overlapping. Also refers to opening of thecae by curved flaps of tissue.

Whorl: group of organs arranged more or less at the same level and arising simultaneously or in a rapid spiral.

Zygomorphic: see monosymmetric

Taxonomic index

Subject index